Springer Proceedings in Mathematics and Statistics

Volume 22

For further volumes:
http://www.springer.com/series/10533

Springer Proceedings in Mathematics and Statistics

This book series features volumes composed of select contributions from workshops and conferences in all areas of current research in mathematics and statistics, including OR and optimization. In addition to an overall evaluation of the interest, scientific quality, and timeliness of each proposal at the hands of the publisher, individual contributions are all refereed to the high quality standards of leading journals in the field. Thus, this series provides the research community with well-edited, authoritative reports on developments in the most exciting areas of mathematical and statistical research today.

Laurent Decreusefond • Jamal Najim
Editors

Stochastic Analysis and Related Topics

In Honour of Ali Süleyman Üstünel
Paris, June 2010

Springer

Editors
Laurent Decreusefond
Jamal Najim
Telecom ParisTech
Paris, France

ISSN 2194-1009 ISSN 2194-1017 (electronic)
ISBN 978-3-642-44803-4 ISBN 978-3-642-29982-7 (eBook)
DOI 10.1007/978-3-642-29982-7
Springer Heidelberg New York Dordrecht London

Mathematical Subject Classification (2010): 60H07, 60H15, 60G22, 60G55

Printed on acid-free paper

Springer is part of Springer Science+Business Media (www.springer.com)

Foreword

The present contributed volume resulted from the "9th Workshop on Stochastic Analysis and Related Topics," part of a series of biannual workshops initiated by H. Körezlioglu and A.S. Üstünel in 1986 and then continued with the help of B. Øksendal until 2003. This event, held on the 14th and 15th of June, 2010, was an ideal occasion to celebrate the 60th birthday of A.S. Üstünel and his contributions to mathematics.

We would like to thank the Institut Telecom and Gérard Memmi, head of the "Computer Science and Networking" department, who fully sponsored this event.

Paris, France

L. Decreusefond
J. Najim

Preface

After brilliant studies in the most renowned turkish institutions, Ali Suleyman Üstünel was longing to become a physicist when Hayri Körezlioglu convinced him to switch to mathematics. This was the beginning of a long and deep collaboration and friendship.

A.S. Üstünel finally defended his Ph.D. in probability in Paris in 1981 with Laurent Schwarz as an examiner. He first began to work at Centre National d'Études en Télécommunications (now Orange Labs) and then at Ecole Nationale Supérieure des Télécommunications (now Télécom Paristech). His first works were related to nuclear-valued processes. The strong topological properties of nuclear spaces induce that many properties only have to be verified "cylindrically" to hold in full generality: For instance, a process $(X(t),\ t \geq 0)$ with values in the set of tempered distributions is continuous if and only if for any φ rapidly decreasing, the real-valued process $(<X(t), \varphi>,\ t \geq 0)$ is continuous. The work of A.S. Üstünel culminated in the "three operators lemma" which states that when three Hilbert–Schmidt operators are applied in a row to a cylindrical semi-martingale, it becomes a true semi-martingale.

In the mid-1980s, he was one of the pioneering researchers to investigate thoroughly the newly born Malliavin Calculus, a field where he quickly became (and still is!) a world renowned expert. From 1986, H. Korezlioglu and A.S. Üstünel organized the "Stochastic analysis and related topics" worskhop whose first occurrences took place in Silivri (Turkey) every 2 years. The "Silivri band" (mainly M. Chaleyat-Maurel, A. Grorud, A. Millet, D. Nualart, E. Pardoux, M. Pontier, M. Sanz) played a major role in the development of Malliavin calculus and its applications. At the same time, A.S. Üstünel and Moshe Zakai started a collaboration which was to last for the next 20 years. Their main subject of investigation has been the absolute continuity of shift transformations in the Wiener space. It is well known that the law of Brownian motion with an adapted, square integrable drift is absolutely continuous with respect to the law of the Brownian motion. They devoted their whole energy to extend the family of admissible drifts, that is to say drifts such that the absolute continuity property still holds. The main question is to get rid of the adaptability. They showed that this can be replaced by, for instance, either monotony or some regularity on the Malliavin derivative of the drift. Most of their results are contained in the beautiful book they coauthored.

The reciprocal problem can be informally stated as: Given a measure equivalent to the Wiener measure, does there exist a shift transformation which realizes this measure? It turned out that the optimal transportation theory which was regaining interest after the work of Brenier in the end of the previous millenium yielded an answer to this problem. Using his previously defined notion of H-convexity, A.S. Üstünel, in collaboration with Denis Feyel, solved the so-called Monge–Kantorovitch problem in the Wiener space for the Cameron–Martin cost. Once again, Malliavin calculus provided the convenient concepts to generalize almost word for word, the results known in finite dimension. Surprisingly, the proofs of some results such as Talagrand or Poincaré inequalities appeared to be even simpler in infinite dimension due to the availability of the Itô calculus. In several papers, they showed different properties of the solution of the Monge–Kantorovitch problem, which yielded in turn several functional inequalities.

Combining all his earlier results, A.S. Üstünel found a criterion which ensures the invertibility of a shift transformation on the Wiener space: If the kinetic energy of the drift u is equal to the entropy of the measure induced by the corresponding shift transformation, then the map $\omega \mapsto \omega + u(\omega)$ is invertible. Such a result can be interpreted as a construction of a strong solution of the stochastic differential equation $dX(t) = -\dot{u}(X,t)\,dt + dB(t)$ for very general u.

This quick glance at A.S. Üstünel's work does not give justice to his other numerous contributions to control, filtering, functional inequalities, fractional Brownian motion, etc. but it shows a strong line of thought and a constant will to focus on deep problems. To borrow one of his favorite metaphor: Instead of looking to the hole which corresponds to the key, he rather prefers to seek for the key which fits into the hole.

Besides his own research activities, A.S. Üstünel has been the professor of several generations of students at Telecom ParisTech and the Ph.D. advisor of many students. We have all been impressed by not only his passion to mathematics, his wide knowledge, but also his generosity, his kindness and the relevance of his advice.

On a more artistic side, Süleyman and his wife, Jacqueline, became world-renowned specialists of the Turkish painter Fikret Moualla, many paintings of whom can be seen at their gallery in Paris. But that is another story. We take this opportunity to deeply thank Jacqueline, whose help was crucial to organize this workshop, or more precisely, to convince Süleyman to participate in this workshop organized on the occasion of his 60th birthday.

Happy birthday Süleyman!

Paris, France

L. Decreusefond
J. Najim

Contents

Contents

Contributors

Boubacar Bah LERSTAD, UFR S.A.T, Université Gaston Berger, BP 234, Saint-Louis, Senegal, bbah12@yahoo.fr

Alain Bensoussan University of Texas at Dallas, Box 830688, Richardson, TX 75083-0688, USA

The Hong Kong Polytechnic Institute, Graduate School of Business, M921, Li Ka Shing Tower, Hung Hom, Kowloon, Hong Kong, Alain.Bensoussan@utdallas.edu

Zdzisław Brzeźniak Department of Mathematics, University of York, Heslington, York YO10 5DD, UK, zb500@york.ac.uk

Caroline Hillairet Laboratoire du CMAP Ecole Polytechnique, Route de Saclay, 91128 Palaiseau Cedex, France, hillaire@cmapx.polytechnique.fr

Annie Millet SAMM, EA 4543, Université Paris 1 Panthéon Sorbonne, 90 Rue de Tolbiac, 75634 Paris Cedex 13

Laboratoire PMA (CNRS UMR 7599), Universités Paris 6-Paris 7, Boîte Courrier 188, 4 place Jussieu, 75252 Paris Cedex 05, France, amillet@univ-paris1.fr, annie.millet@upmc.fr

Etienne Pardoux CMI, LATP-UMR 6632, Université de Provence, 39 rue F. Joliot Curie, Marseille cedex 13, France, pardoux@cmi.univ-mrs.fr

Monique Pontier Institut Mathématique de Toulouse, Laboratoire de Statistique et Probabilités, Université Paul Sabatier, 31062 Toulouse Cedex 9, France, pontier@math.univ-toulouse.fr

Nicolas Privault Division of Mathematical Sciences School of Physical and Mathematical Sciences Nanyang Technological University SPMS-MAS-05-43, 21 Nanyang Link Singapore 637371, Singapore, nprivault@ntu.edu.sg

Ahmadou Bamba Sow LERSTAD, UFR S.A.T, Université Gaston Berger, BP 234, Saint-Louis, Senegal, ahmadou-bamba.sow@ugb.edu.sn

Denis Talay INRIA, 2004 Route des Lucioles, B.P. 93, 06902 Sophia-Antipolis, France, denis.talay@inria.fr

Samy Tindel Institut Élie Cartan Nancy, B.P. 239, 54506 Vandœuvre-lès-Nancy Cedex, France, tindel@iecn.u-nancy.fr

Iván Torrecilla Facultat de Matemàtiques, Universitat de Barcelona, Gran Via 585, 08007 Barcelona, Spain

Facultat de Ciències Econòmiques i Empresarials, Universitat Pompeu Fabra, C/Ramon Trias Fargas, 25-27, 08005 Barcelona, Spain, itorrecillatarantino@gmail.com

Ali Süleyman Üstünel Télécom-Paristech, 46 rue Barrault, 75013 Paris, France, ustunel@telecom-paristech.fr

Chapter 1
A Look-Down Model with Selection

Boubacar Bah, Etienne Pardoux, and Ahmadou Bamba Sow

Abstract The goal of this paper is to study a new version of the look-down construction with selection. We show (see Theorem 1.2) convergence in probability, locally uniformly in t, as the population size N tends to infinity, towards the Wright–Fisher diffusion with selection.

1.1 Introduction and Preliminaries

In this paper we consider the simplest look-down (also called by some authors the "modified look-down") model with selection. We consider the case of two alleles b and B, where B has a selective advantage over b. This selective advantage is modeled by a death rate α for the type b individuals, while the type B individuals are not subject to that specific death mechanism. The look-down construction is due to Donnelly and Kurtz, see [3, 5] in the neutral case. Those authors extended their construction to the selective case in [4].

Our selective look-down construction is slightly different from theirs. We will consider the proportion of b individuals. Hence type b individuals are coded by 1 and B by 0. We assume that individuals are placed at time 0 on levels $1, 2, \ldots$, each one being, independently from the others, 1 with probability x, 0 with probability $1 - x$, for some $0 < x < 1$. For any $t \geq 0$, $i \geq 1$, let $\eta_t(i)$ denote the type of the individual sitting on site i at time t. Clearly $\eta_t(i) \in \{0, 1\}$. The evolution of the population is governed by the two following mechanisms:

B. Bah · A.B. Sow
LERSTAD, UFR S.A.T, Université Gaston Berger, BP 234, Saint-Louis, Senegal
e-mail: bbah12@yahoo.fr; ahmadou-bamba.sow@ugb.edu.sn

E. Pardoux (✉)
CMI, LATP-UMR 6632, Université de Provence, 39 rue F. Joliot Curie,
Marseille cedex 13, France
e-mail: pardoux@cmi.univ-mrs.fr

L. Decreusefond and J. Najim (eds.), *Stochastic Analysis and Related Topics*, Springer Proceedings in Mathematics & Statistics 22, DOI 10.1007/978-3-642-29982-7_1,

1. *Births.* For any $1 \leq i < j$, arrows are placed from i to j according to a rate one Poisson process, independently of the other pairs $i' < j'$. Suppose there is an arrow from i to j at time t. Then a descendent (of the same type) of the individual sitting on level i at time t^- occupies the level j at time t, while for any $k \geq j$, the individual occupying the level k at time t^- is shifted to level $k+1$ at time t. In other words, $\eta_t(k) = \eta_{t^-}(k)$ for $k < j$, $\eta_t(j) = \eta_{t^-}(i)$, $\eta_t(k) = \eta_{t^-}(k-1)$ for $k > j$.
2. *Deaths.* Any type 1 individual dies at rate α, his vacant level being occupied by his right neighbor, who himself is replaced by his right neighbor, etc. In other words, independently of the above arrows, crosses are placed on each level according to a rate α Poisson process, independently of the other levels. Suppose there is a cross at level i at time t. If $\eta_{t^-}(i) = 0$, nothing happens. If $\eta_{t^-}(i) = 1$, then $\eta_t(k) = \eta_{t^-}(k)$ for $k < i$ and $\eta_t(k) = \eta_{t^-}(k+1)$ for $k \geq i$.

We refer the reader to Fig. 1.1 for a pictorial presentation of our model. This model has been formulated by Anton Wakolbinger in an oral presentation [7]. In contradiction with the models studied in [3–5], the evolution of the N first individuals $\eta_t(1), \ldots, \eta_t(N)$ depends upon the next ones, and $X_t^N = N^{-1}(\eta_t(1) + \cdots + \eta_t(N))$ is not a Markov process. We will show however that for each $t > 0$ the $\{\eta_t(k),\ k \geq 1\}$ are well defined (which is not obvious in our setup) and constitute an exchangeable sequence of $\{0,1\}$-valued random variables. We can then apply de Finetti's theorem and prove that $X_t^N \to X_t$ in probability, locally uniformly in $t \geq 0$, where X_t is a $[0,1]$-valued Markov process, solution of the Wright–Fisher SDE with selection (1.1).

In fact $\{X_t^N,\ t \geq 0\}$ is approximately Markovian, in a sense which will be clear below. It is possible, also no certain, that the techniques of proof from [3,5] and [4] might be adaptable in the present situation. We rather prefer to use a quite different approach. In particular, there is no mention of a generator in this paper. Rather, we use extensively de Finetti's theorem, tightness, and a duality argument between the Wright–Fisher diffusion with selection and a birth and death process which can be related to an ancestral recombination graph (in short ARG, see [6] for this notion). We do not claim any superiority of our method of proof over that of [3–5]. We just think that new approaches may be interesting in that they bring new insights into the problem.

Our paper is organized as follows. Section 1.2 presents the duality relation between a birth–death process and Wright–Fisher's diffusion with selection. We both construct our process and establish a crucial exchangeability property satisfied by our look-down model with selection in Sect. 1.3. We prove the convergence result in Sect. 1.4.

In this paper, we use $\mathbb{N}$ to denote the set of positive integers $\{1, 2, \ldots\}$.

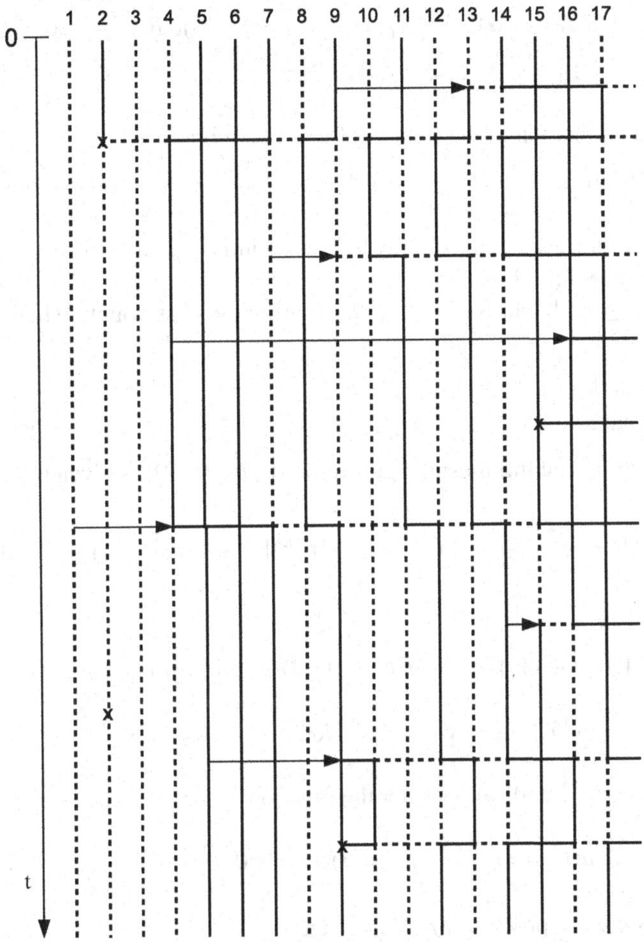

Fig. 1.1 The vertical time axis is shown on the *left*, time flows from *top* to *bottom*. *Solid lines* represent type b individuals, while *dotted lines* represent type B individuals

1.2 Wright–Fisher Diffusion with Selection and Duality

Let $(\Omega, \mathscr{F}, (\mathscr{F}_t)_{t\geq 0}, \mathbb{P})$ be a stochastic basis on which a d-dimensional Brownian motion $(B_t)_{t\geq 0}$ is defined. We assume that $\mathscr{F}_t = \sigma\{B_s, 0 \leq s \leq t\} \vee \mathscr{N}$, where $\mathscr{N}$ is the class of $\mathbb{P}$-null sets of $\mathscr{F}$.

Definition 1.1. A Wright–Fisher diffusion with selection is a [0, 1]-valued Markov process $Y = \{Y_t, t \geq 0\}$ with continuous paths, solution of the following stochastic differential equation:

$$\begin{cases} dY_t = -\alpha Y_t(1-Y_t)dt + \sqrt{Y_t(1-Y_t)}dB_t, & t \geq 0, \\ Y_0 = y, \quad 0 < y < 1, \end{cases} \tag{1.1}$$

where B is a realization of the standard Brownian motion, $\alpha \in \mathbb{R}$.

In all what follows, $\alpha > 0$. Y_t will denote the proportion of non-advantageous alleles.

In this section we study the duality between a jump Markov process and Wright–Fisher diffusion with selection.

Let $\{R_t, t \geq 0\}$ be an $\mathbb{N}$-valued jump Markov process which, when in state k, jumps to

1. $k-1$ at rate $\binom{k}{2}$
2. $k+1$ at rate αk, $\alpha > 0$

In other words, the infinitesimal generator of $\{R_t, t \geq 0\}$ is given by

$$Qf(k) = \frac{k(k-1)}{2}[f(k-1) - f(k)] + \alpha k[f(k+1) - f(k)]$$

for any $f : \mathbb{N} \to \mathbb{R}$.

Proposition 1.1. *Let* $(Y_t)_{t\geq 0}$ *given by* (1.1). *Then for any* $n \geq 1$ *and* $t \geq 0$ *we have*

$$\mathrm{e}[Y_t^n | Y_0 = y] = \mathrm{e}[y^{R_t} | R_0 = n], \quad 0 \leq y \leq 1.$$

Proof. We fix $n \geq 1$ and we consider the function $u : \mathbb{R}_+ \times [0,1] \to \mathbb{R}$ given by

$$u(t,y) = \mathrm{e}(y^{R_t} | R_0 = n), \quad t \geq 0, \quad 0 \leq y \leq 1.$$

Let $f : \mathbb{N} \to \mathbb{R}$. The process $(M_t^f)_{t\geq 0}$ given by

$$M_t^f = f(R_t) - f(R_0) - \int_0^t \left[\binom{R_s}{2}[f(R_s-1) - f(R_s)] + \alpha R_s[f(R_s+1) - f(R_s)]\right] ds \tag{1.2}$$

is a local martingale. Applying (1.2) with the particular choice $f(n) = y^n$ for each $y \in [0,1]$, there exists a local martingale $(M_t^{(1)})_{t\geq 0}$ such that $M_0^{(1)} = 0$ and

$$y^{R_t} = y^{R_0} + y(1-y)\int_0^t \left(\frac{1}{2}R_s(R_s-1)y^{R_s-2} - \alpha R_s y^{R_s-1}\right) ds + M_t, \quad t \geq 0. \tag{1.3}$$

Applying (1.2) with $f(n) = y^{2n}$ and comparing with Itô formula for the square of y^{R_t}, we deduce that

$$< M >_t = (y-1)^2 \int_0^t y^{2R_s-2}\left[\binom{R_s}{2} + \alpha R_s y^2\right] ds.$$

Moreover, M_t is in fact a square integrable martingale. Indeed, by a natural coupling, one may stochastically upper bound the birth and death process $\{R_t; t \ge 0\}$ by pure birth process $\{Z_t; t \ge 0\}$ issued from R_0, which jumps from k to $k+1$, at the birth times of $\{R_t; t \ge 0\}$. This is a Yule process. From this it is easy to show that each term in (1.2) is integrable.

Taking the conditional expectation $\mathrm{e}(\cdot|R_0 = n)$, we deduce from (1.3)

$$u(t,y) = u(0,y) + y(1-y)\int_0^t \mathrm{e}\left(\frac{1}{2}R_s(R_s-1)y^{R_s-2} - \alpha R_s y^{R_s-1} | R_0 = n\right) ds$$
$$= u(0,y) + y(1-y)\int_0^t \left(\frac{1}{2}\frac{\partial^2 u}{\partial y^2}(s,y) - \alpha\frac{\partial u}{\partial y}(s,y)\right) ds.$$

Hence u solves the following linear parabolic PDE:

$$\begin{cases} \partial_t u(t,y) = \frac{1}{2}y(1-y)\partial^2_{yy}u(t,y) - \alpha y(1-y)\partial_y u(t,y) & t \ge 0, \quad 0 < y < 1, \\ u(0,y) = y^n, \quad u(t,0) = 0, \quad u(t,1) = 1. \end{cases} \tag{1.4}$$

It is easy to check that u is of class $C^{1,2}(\mathbb{R} \times (0,1))$. Itô's formula applied to the function $(s,y) \longmapsto u(t-s,y)$ yields

$$u(0,Y_t) = u(t,Y_0) + \int_0^t \frac{\partial u}{\partial x}(t-r,Y_r)\sqrt{Y_r(1-Y_r)}\,dB_r$$
$$+ \int_0^t \left[-\frac{\partial u}{\partial s}(t-r,Y_r) - \alpha X_r(1-X_r)\frac{\partial u}{\partial x}(t-r,Y_r) + \frac{1}{2}Y_r(1-Y_r)\frac{\partial^2 u}{\partial x^2}(t-r,Y_r)\right] dr.$$

Using (1.4), we deduce that

$$u(0,Y_t) = u(t,Y_0) + N_t,$$

where $(N_t)_{t\ge 0}$ is a zero-mean martingale. It remains to take the expectation in the last identity to get the desired result.

Remark 1.1. Strong uniqueness of (1.1) is well known. Weak uniqueness follows from that result as well as from the duality argument in Proposition 1.1.

1.3 Look-Down with Selection, Exchangeability

1.3.1 Construction of Our Process

We consider the look-down model with selection defined in the introduction. We first need to give a construction of our $\{\eta_t(i),\ i \ge 1,\ t \ge 0\}$. For each N, consider

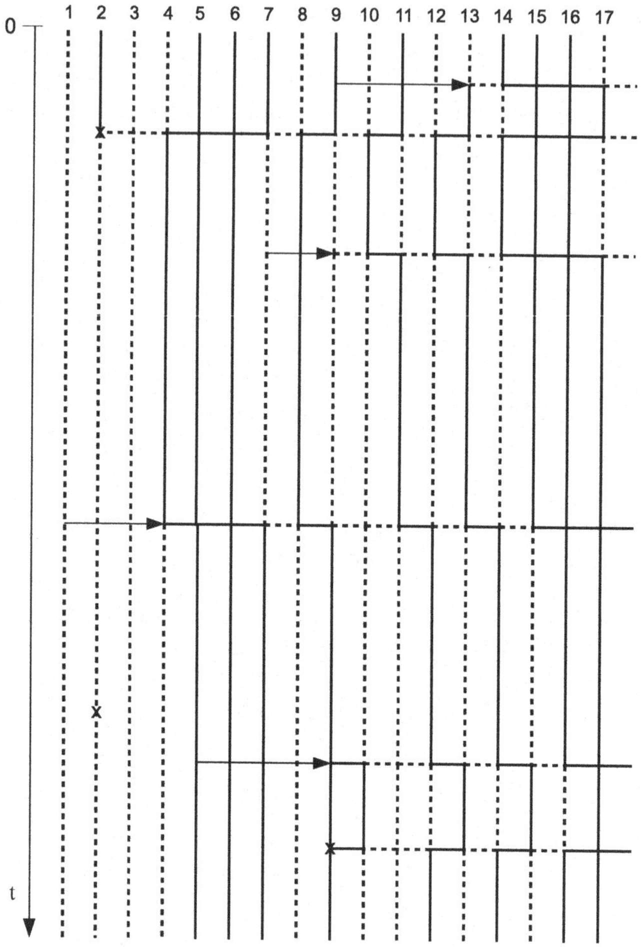

Fig. 1.2 $\eta^N(i)$ for $1 \le i \le 17$ and $N = 14$

the process $\{\eta_t^N(i),\ i \ge 1,\ t \ge 0\}$, obtained by applying only the arrows between $1 \le i < j \le N$ and the crosses on levels 1 to N. In other words, we disregard all the arrows pointing to levels above N, as well as all the crosses on levels above N. We then have a finite number of arrows and crosses on any finite time interval, and $\{\eta_t^N(i),\ i \ge 1,\ t \ge 0\}$ is constructed in an obvious way, by implementing the effect of the arrows and crosses, in the order in which they are met. We show in Fig. 1.2 the same realization of the look-down with selection as in Fig. 1.1, but where the arrows and crosses above $N = 14$ have been erased.

In the rest of this section, we refer to η^N as the just defined process. It follows from the Borel–Cantelli Lemma and the next proposition that for N large enough (depending upon ω)$\{(\eta_t^{2N+k}(1), \dots, \eta_t^{2N+k}(N)),\ t \ge 0\}$ does not depend upon $k \ge 1$, hence η^N converges to a limit η as $N \to \infty$.

Proposition 1.2. *For each $N > 32\alpha$,*

$$\mathbb{P}\left(\exists 1 \le i \le N, k \ge 1, t > 0 \text{ such that } \eta_t^{2N}(i) \ne \eta_t^{2N+k}(i)\right) \le 24\alpha \left(\frac{8\alpha}{N}\right)^{N-1}.$$

Proof. For each $i \ge 1, t > 0$, let $\xi_t^{i,2N}$ denote the level on which the individual who was sitting on level i at time $t = 0$ sits at time t, where the evolution corresponds to the "$2N$-model," i.e., all arrows pointing to levels above $2N$ and all crosses on levels above $2N$ have been erased. Each time there is a birth on a level smaller than or equal to $\xi_{t-}^{i,2N}$, $\xi_t^{i,2N}$ has a jump of size $+1$. Each time there is a death on a level smaller than or equal to $\xi_{t-}^{i,2N}$, $\xi_t^{i,2N}$ has a jump of size -1. In other words, $\xi_t^{i,2N}$ follows the position of the individual who was sitting on level i at time $t = 0$ until his possible death, then follows the position of his left neighbor, etc. We insist upon the rule that when this individual is killed, he is replaced by his immediate left neighbor. We have

$$\begin{aligned}
&\{\exists 1 \le i \le N, k \ge 1, t > 0 \text{ such that } \eta_t^{2N}(i) \ne \eta_t^{2N+k}(i)\} \\
&\quad \subset \left\{\exists i \ge 1, 0 \le t < t' \text{ such that } \xi_t^{i,2N} > 2N,\ \xi_{t'}^{i,2N} = N\right\} \\
&\quad = \left\{\exists 1 \le i \le 2N+1, 0 \le t < t' \text{ such that } \xi_t^{i,2N} > 2N,\ \xi_{t'}^{i,2N} = N\right\}.
\end{aligned}$$

In other words, for the crosses and arrows on levels higher than $2N$ to interfere with the behavior of the population at levels 1 to N, we need that at least one individual visit the level N, after having visited the level $2N+1$, and the identity follows from the following monotonicity property: $i < j \Rightarrow \xi_t^{i,2N} \le \xi_t^{j,2N}$ a.s. for all $t > 0$. Consequently

$$\begin{aligned}
&\mathbb{P}\left(\exists 1 \le i \le N, k \ge 1, t > 0 \text{ such that } \eta_t^{2N}(i) \ne \eta_t^{2N+k}(i)\right) \\
&\quad \le \sum_{i=1}^{2N+1} \mathbb{P}\left(\exists 0 \le t < t' \text{ such that } \xi_t^{i,2N} > 2N,\ \xi_{t'}^{i,2N} = N\right).
\end{aligned} \tag{1.5}$$

We first show that for each $N > 32\alpha$,

$$\mathbb{P}\left(\exists t > 0 \text{ such that } \xi_t^{2N+1,2N} = N\right) \le \left(\frac{8\alpha}{N}\right)^N. \tag{1.6}$$

We can couple the process $\xi_t^{2N+1,2N}$ with a birth and death process ρ_t^N, with birth rate $N(N+1)/2$ and death rate $\alpha(2N+1)$, with the properties

$$\rho_0^N = 2N+1,\ \rho_t^N \le \xi_t^{2N+1,2N},\ 0 \le t \le \tau_N,$$

where

$$\tau_N = \inf\{t > 0,\ \rho_t^N = N\}.$$

Clearly

$$\mathbb{P}(\exists t > 0 \text{ such that } \xi_t^{2N+1,2N} = N) \leq \mathbb{P}(\tau_N < \infty),$$

hence (1.6) follows from

Lemma 1.1. *For each $N \geq 32\alpha$,*

$$\mathbb{P}(\tau_N < \infty) \leq \left(\frac{8\alpha}{N}\right)^N.$$

Proof. Let $\{X_n,\ n \geq 1\}$ and $\{Y_n,\ n \geq 1\}$ be two mutually independent sequences of i.i.d. r.v.'s, the X_n's being exponential with parameter $N(N+1)/2$, the Y_n's being exponential with parameter $\alpha(2N+1)$. We have

$$\mathbb{P}(\tau_N < \infty) \leq \sum_{n=1}^{\infty} \mathbb{P}(X_1 + \cdots + X_n > Y_1 + \cdots + Y_{n+N}). \tag{1.7}$$

Now for each $0 < c_N < N(N+1)/2$,

$$\begin{aligned}\mathbb{P}(X_1 + \cdots + X_n > Y_1 + \cdots + Y_{n+N}) &= \mathbb{P}\left(\exp[c_N(X_1 + \cdots + X_n - Y_1 - \cdots - Y_{n+N})] > 1\right)\\ &\leq \left(\mathrm{e}e^{c_N X_1}\right)^n \left(\mathrm{e}e^{-c_N Y_1}\right)^{n+N}\\ &= \left(\frac{N(N+1)/2}{N(N+1)/2 - c_N}\right)^n \left(\frac{\alpha(2N+1)}{\alpha(2N+1) + c_N}\right)^{n+N}.\end{aligned}$$

We choose $c_N = N(N+1)/4$, and deduce

$$\mathbb{P}(X_1 + \cdots + X_n > Y_1 + \cdots + Y_{n+N}) \leq \left(\frac{16\alpha}{N}\right)^n \left(\frac{8\alpha}{N}\right)^N.$$

Summing from $n = 1$ to ∞ yields the result, since

$$\sum_{n=1}^{\infty} \left(\frac{16\alpha}{N}\right)^n \leq 1 \quad \text{if } N > 32\alpha.$$

We can now conclude the proof of Proposition 1.2. We note that for $N > 32\alpha$, any $1 \leq i \leq 2N$,

$$\mathbb{P}\left(\exists 0 \leq t < t' \text{ such that } \xi_t^{i,2N} > 2N, \xi_{t'}^{i,2N} = N\right) \leq \left(\frac{8\alpha}{N}\right)^N.$$

Indeed, wait until $\theta_{i,2N} = \inf\{t > 0,\ \xi_t^{i,2N} = 2N + 1\}$, which is a stopping time at which the Markov process $\{\eta_t^{2N}(j),\ j \geq 1\}_{t \geq 0}$ starts afresh, and then use the same argument as that of Lemma 1.1. Consequently

$$\begin{aligned}
&\mathbb{P}\left(\exists 1 \leq i \leq 2N+1, k \geq 1, t > 0 \text{ such that } \eta_t^{2N}(i) \neq \eta_t^{2N+k}(i)\right) \\
&\quad \leq \sum_{i=1}^{2N+1} \mathbb{P}\left(\exists 0 \leq t < t' \text{ such that } \xi_t^{i,2N} > 2N, \xi_{t'}^{i,2N} = N\right) \\
&\quad \leq (2N+1)\left(\frac{8\alpha}{N}\right)^N \\
&\quad \leq 3\frac{(8\alpha)^N}{N^{N-1}}.
\end{aligned}$$

From now on, we equip the probability space $(\Omega, \mathscr{F}, \mathbb{P})$ with the filtration defined by $\mathscr{F}_t = \sigma\{\eta_s(i),\ i \geq 1,\ 0 \leq s \leq t\}$. Any stopping time will be defined with respect to that filtration.

1.3.2 *Exchangeability*

Our goal in this section is to show that for all $t > 0$, the sequence $\{\eta_t(i),\ i \geq 1\}$ is exchangeable. It in fact suffices to show that for all $t > 0$, any $n \geq 1$, $\eta_t^{[n]} := (\eta_t(1), \ldots, \eta_t(n))$ is an exchangeable sequence of $\{0,1\}$-valued r.v.'s.

For any $t \geq 0$, $n \geq 1$, $\eta_t^{[n]}$ is a $\{0,1\}^n$-valued random vector. Let S_n denote the group of permutations of $\{1, 2, \ldots, n\}$.

For $\pi \in S_n$ and $a^{[n]} = (a_i)_{1 \leq i \leq n} \in \{0,1\}^n$, we define the vectors

$$\begin{aligned}
\pi^{-1}(a^{[n]}) &= (a_{\pi^{-1}(1)}, \ldots, a_{\pi^{-1}(n)}) = (a_i^{\pi})_{1 \leq i \leq n}, \\
\pi(\eta_t^{[n]}) &= (\eta_t(\pi(1)), \ldots, \eta_t(\pi(n))).
\end{aligned}$$

We should point out that $\pi(\eta_t^{[n]})$ is a permutation of $(\eta_t(1), \ldots, \eta_t(n))$ and it is clear from the definitions that

$$\{\pi(\eta_t^{[n]}) = a^{[n]}\} = \{\eta_t^{[n]} = \pi^{-1}(a^{[n]})\} \quad \text{for any} \quad \pi \in S_n. \tag{1.8}$$

We want to prove

Proposition 1.3. *Suppose that $\{\eta_0(i), i \geq 1\}$ are exchangeable random variables. Then for all $t > 0$, $\{\eta_t(i), i \geq 1\}$ is an exchangeable sequence of $\{0,1\}$-valued random variables.*

We first establish two lemmas.

Lemma 1.2. *For any stopping time S, any $\mathbb{N}$-valued $\mathscr{F}_S$-measurable random variable $\mathbf{n}$, if the random vector $\eta_S^{[\mathbf{n}]} = (\eta_S(1), \ldots, \eta_S(\mathbf{n}))$ is exchangeable, and T is the first time after S of an arrow pointing to a level $\leq \mathbf{n}$ or a death at a level $\leq \mathbf{n}$, then conditionally upon the fact that T is the time of an arrow, the random vector $\eta_T^{[\mathbf{n}+1]} = (\eta_T(1), \ldots, \eta_T(\mathbf{n}), \eta_T(\mathbf{n}+1))$ is exchangeable.*

Proof. For the sake of simplifying the notations, we condition upon $\mathbf{n} = n$ and $T = t$. We start with some notation

$$A_t^{i,j} := \{\text{The arrow at time } t \text{ is drawn from level } i \text{ to level } j\},\ 1 \leq i < j \leq n.$$

We define

$$\widehat{\mathbb{P}}_{t,n}[.] = \mathbb{P}(.|T = t, \mathbf{n} = n).$$

Thanks to (1.8), we deduce that, for $\pi \in S_{n+1}$

$$\begin{aligned}\widehat{\mathbb{P}}_{t,n}(\pi(\eta_t^{[n+1]}) = a^{[n+1]}) &= \sum_{1\leq i<j\leq n} \widehat{\mathbb{P}}_{t,n}\left(\eta_t^{[n+1]} = \pi^{-1}(a^{[n+1]}), A_t^{i,j}\right) \\ &= \sum_{1\leq i<j\leq n} \widehat{\mathbb{P}}_{t,n}\left(\eta_t(1) = a_1^\pi, \ldots, \eta_t(n+1) = a_{n+1}^\pi, A_t^{i,j}\right),\end{aligned} \tag{1.9}$$

On the event $A_t^{i,j}$, we have

$$\eta_t(k) = \begin{cases} \eta_{t-}(k) & \text{if } 1 \leq k < j, \\ \eta_{t-}(i) & \text{if } k = j, \\ \eta_{t-}(k-1) & \text{if } j < k \leq n+1. \end{cases}$$

This implies that

$$A_t^{i,j} \cap \{\eta_t^{[n+1]} = (a_1^\pi, \ldots, a_{n+1}^\pi)\} \subset \{a_i^\pi = a_j^\pi\}.$$

For $1 \leq j \leq n$, define the mapping $\rho_j : \{0,1\}^{n+1} \longrightarrow \{0,1\}^n$ by

$$\rho_j(b_1, \ldots, b_{n+1}) = (b_1, \ldots, b_{j-1}, b_{j+1}, \ldots, b_{n+1}).$$

The right-hand side of (1.9) is equal to

$$\sum_{1\leq i<j\leq n} \mathbf{1}_{\{a_i^\pi = a_j^\pi\}} \widehat{\mathbb{P}}_{t,n}\left(\eta_{t-}^{[n]} = \rho_j(\pi^{-1}(a^{[n+1]})), A_t^{i,j}\right).$$

It is easy to see that the events $(\eta_{t-}^{[n]} = \rho_j(\pi^{-1}(a^{[n+1]})))$ and $A_t^{i,j}$ are independent. Thus

$$\widehat{\mathbb{P}}_{t,n}(\pi(\eta_t^{[n+1]}) = a^{[n+1]}) = \sum_{1\le i<j\le n} \mathbf{1}_{\{a_i^\pi = a_j^\pi\}} \widehat{\mathbb{P}}_{t,n}\left(\eta_{t-}^{[n]} = \rho_j(\pi^{-1}(a^{[n+1]}))\right) \widehat{\mathbb{P}}_{t,n}(A_t^{i,j})$$
$$= \frac{2}{n(n-1)} \sum_{1\le i<j\le n} \mathbf{1}_{\{a_i^\pi = a_j^\pi\}} \mathbb{P}\left(\eta_{t-}^{[n]} = \rho_j(\pi^{-1}(a^{[n+1]}))\right).$$

If $a_j = a_{\pi(j)}$, $\rho_j(a^{[n+1]})$ and $\rho_j(\pi^{-1}(a^{[n+1]}))$ contain the same number of $0'$s and $1'$s. Since $\eta_{t-}^{[n]}$ is exchangeable, this implies that

$$\mathbf{1}_{\{a_i^\pi = a_j^\pi = \gamma\}} \mathbb{P}\left(\eta_{t-}^{[n]} = \rho_j(\pi^{-1}(a^{[n+1]}))\right) = \mathbf{1}_{\{a_i = a_j = \gamma\}} \mathbb{P}\left(\eta_{t-}^{[n]} = \rho_j(a^{[n+1]})\right), \quad \gamma \in \{0,1\}.$$

On the other hand, we have $\#\{1 \le i < j \le n : a_i^\pi = a_j^\pi\} = \#\{1 \le i < j \le n : a_i = a_j\}$. Let $k = \inf(\pi(i), \pi(j))$ and $\ell = \sup(\pi(i), \pi(j))$. If $a_i = a_j$, then we have $a_k^\pi = a_\ell^\pi = a_i = a_j$. Finally, we obtain

$$\widehat{\mathbb{P}}_{t,n}(\pi(\eta_t^{[n+1]}) = a^{[n+1]}) = \frac{2}{n(n-1)} \sum_{1\le k<\ell\le n} \mathbf{1}_{\{a_k^\pi = a_\ell^\pi\}} \mathbb{P}\left(\eta_{t-}^{[n]} = \rho_\ell(\pi^{-1}(a^{[n+1]})\right)$$
$$= \frac{2}{n(n-1)} \sum_{1\le i<j\le n} \mathbf{1}_{\{a_i = a_j\}} \mathbb{P}\left(\eta_{t-}^{[n]} = \rho_j(a^{[n+1]})\right)$$
$$= \widehat{\mathbb{P}}_{t,n}(\eta_t^{[n+1]} = a^{[n+1]}).$$

We have proved that for any $\pi \in S_{n+1}$ and $t \ge 0$, $\widehat{\mathbb{P}}_{t,n}(\pi(\eta_t^{[n+1]}) = a^{[n+1]}) = \widehat{\mathbb{P}}_{t,n}(\eta_t^{[n+1]} = a^{[n+1]})$. The result follows.

Lemma 1.3. *For any stopping time S, any $\mathbb{N}$-valued $\mathcal{F}_S$-measurable random variable $\mathbf{n}$, if the random vector $\eta_S^{[\mathbf{n}]} = (\eta_S(1), \ldots, \eta_S(\mathbf{n}))$ is exchangeable, and T is the first time after S of an arrow pointing to a level $\le \mathbf{n}$ or a death at a level $\le \mathbf{n}$, then conditionally upon the fact that T is the time of a death, the random vector $\eta_T^{[\mathbf{n}-1]} = (\eta_T(1), \ldots, \eta_T(\mathbf{n}-1))$ is exchangeable.*

Proof. For the sake of simplifying the notations, we condition upon $\mathbf{n} = n$ and $T = t$. Let $\pi \in S_{n-1}$ be arbitrary. We consider the events

$$B_t^i := \{\text{the level of the dying individual at time } t \text{ is } i\}.$$

Let $\widehat{\mathbb{P}}_{t,n}[.] = \mathbb{P}(.|T = t, \mathbf{n} = n)$. Using (1.8) we deduce that

$$\widehat{\mathbb{P}}_{t,n}(\pi(\eta_t^{[n-1]}) = a^{[n-1]}) = \sum_{1\le i\le n} \widehat{\mathbb{P}}_{t,n}\left(\eta_t^{[n-1]} = \pi^{-1}(a^{[n-1]}), B_t^i\right)$$
$$= \sum_{1\le i\le n} \widehat{\mathbb{P}}\left(\eta_t(1) = a_1^\pi, \dots, \eta_t(n-1) = a_{n-1}^\pi, B_t^i\right).$$

Define

$$c_i^{\pi,n} = (a_1^\pi, \dots, a_{i-1}^\pi, 1, a_i^\pi, \dots, a_{n-1}^\pi), \qquad c_i^n = (a_1, \dots, a_{i-1}, 1, a_i, \dots, a_{n-1}).$$

Using the property of the look-down with selection, the last term in the previous relation is equal to

$$\sum_{1\le i\le n} \widehat{\mathbb{P}}_{t,n}\left(\eta_{t-}^{[n]} = c_i^{\pi,n}, B_t^i\right) = \sum_{1\le i\le n} \widehat{\mathbb{e}}_{t,n}\left(\mathbf{1}_{\{\eta_{t-}^{[n]} = c_i^{\pi,n}\}}\mathbf{1}_{B_t^i}\right)$$
$$= \sum_{1\le i\le n} \mathbb{P}\left(\eta_{t-}^{[n]} = c_i^{\pi,n}\right)\widehat{\mathbb{P}}_{t,n}\left(B_t^i \mid \eta_{t-}^{[n]} = c_i^{\pi,n}\right)$$
$$= \frac{1}{1+\sum_{j=1}^n a_j^\pi} \sum_{1\le i\le n} \mathbb{P}\left(\eta_{t-}^{[n]} = c_i^{\pi,n}\right).$$

Thanks to the exchangeability of $(\eta_{t-}(1), \dots, \eta_{t-}(n))$, we have

$$\widehat{\mathbb{P}}_{t,n}(\pi(\eta_t^{[n-1]}) = a^{[n-1]}) = \frac{1}{1+\sum_{j=1}^n a_j} \sum_{1\le i\le n} \mathbb{P}\left(\eta_{t-}^{[n]} = c_i^n\right),$$

since $\sum_{j=1}^{n-1} a_j^\pi = \sum_{j=1}^{[n-1]} a_j$ and $c_i^{\pi,n}$ is a permutation of c_i^π. The result follows.

We can now proceed with the

Proof of Proposition 1.3. For each $N \ge 1$, let $\{V_t^N,\ t \ge 0\}$ denote the $\mathbb{N}$-valued process which describes the position at time t of the individual sitting on level N at time 0, with the convention that, if that individual dies, we replace him by his left neighbor. When $V_t^N = k$, V_t^N is shifted to $k+1$ at rate $k(k-1)/2$ and shifted to $k-1$ at rate $\alpha(\eta_t(1) + \cdots + \eta_t(k))$. Lemma 1.1 shows that $\inf_{t\ge 0} V_t^N \to \infty$, as $N \to \infty$.

It follows from Lemmas 1.2 and 1.3 that for each $t > 0$, $N \ge 1$, $(\eta_t(1), \dots, \eta_t(V_t^N))$ is an exchangeable random vector.

Consequently, for any $t > 0$, $n \ge 1$, $\pi \in S_n$, $a^{[n]} \in \{0,1\}$,

$$|\mathbb{P}(\eta_t^{[n]} = a^{[n]}) - \mathbb{P}(\eta_t^{[n]} = \pi^{-1}(a^{[n]}))| \le \mathbb{P}(V_t^N < n),$$

which goes to zero, as $N \to \infty$. The result follows.

Remark 1.2. The collection of random process

$$\{\eta_t(i), t \geq 0\}_{i\geq 1} \text{ is not exchangeable.}$$

Indeed, $\eta_t(1)$ can jump from 1 to 0, but never from 0 to 1, while the other $\eta_t(i)$ do not have that property.

For $N \geq 1$ and $t \geq 0$, denote by X_t^N the proportion of type b individuals at time t among the first N individuals, i.e.,

$$X_t^N = \frac{1}{N}\sum_{i=1}^{N} \eta_t(i). \tag{1.10}$$

We are interested in the limit of $(X_t^N)_{t\geq 0}$ as N tends to infinity. For this, let us recall the following useful result due to de Finetti (see, e.g., [1]).

Theorem 1.1. *An exchangeable (countably infinite) sequence $\{X_n, n \geq 1\}$ of random variables is a mixture of i.i.d. sequences in the sense that conditionally upon $\mathcal{G}$ (the tail σ-field of the sequence $\{X_n, n \geq 1\}$) the X_n's are i.i.d.*

As a consequence, we have the following asymptotic property for fixed t of the sequence $(X_t^N)_{N\geq 1}$ defined by (1.10).

Corollary 1.1. *For each $t \geq 0$,*

$$X_t = \lim_{N\to\infty} X_t^N \quad \textit{exist a.s.} \tag{1.11}$$

Proof. Let $t \geq 0$ and $n \geq 1$. Let us introduce the filtration $\mathcal{F}_n = \sigma(\eta_t(n+1), \eta_t(n+2), \ldots)$. We have (here "converges" means "converges as $N \to \infty$")

$$\mathbb{P}\left(N^{-1}\sum_{i=1}^{N} \eta_t(i) \text{ converges}\right) = \mathrm{e}\left[\mathbb{P}\left(N^{-1}\sum_{i=1}^{N} \eta_t(i) \text{ converges}\Big| \bigcap_{n=0}^{\infty} \mathcal{F}_n\right)\right].$$

From Proposition 1.3 and Theorem 1.1, conditionally upon $\bigcap_{n=0}^{\infty} \mathcal{F}_n$, $\eta_t(i), i \geq 1$ are i.i.d. bounded random variables. Thanks to the law of large numbers, $N^{-1}\sum_{i=1}^{N} \eta_t(i)$ converge a.s. as $N \to \infty$. This implies

$$\mathbb{P}\left(N^{-1}\sum_{i=1}^{N} \eta_t(i) \text{ converges}\Big| \bigcap_{n=0}^{\infty} \mathcal{F}_n\right) = 1,$$

which establishes the desired result.

1.4 Convergence to the Wright–Fisher Diffusion with Selection

1.4.1 Preliminary Results

Before stating the main theorem of this section, let us establish some auxiliary results which we shall need in its proof.

Proposition 1.4. *Let $\{\xi_1, \xi_2, \dots, \}$ be a countable exchangeable sequence of $\{0, 1\}$-valued random variables and $\mathcal{G}$ denote its tail σ-field. Let $\mathcal{H}$ be some additional σ-algebra. If conditionally upon $\mathcal{G} \vee \mathcal{H}$, the ξ_i's are exchangeable, then their conditional law given $\mathcal{G} \vee \mathcal{H}$ is their conditional law given $\mathcal{G}$.*

Proof. Let $n \geq 1$ and $f : \{0, 1\}^n \to \mathbb{R}$ be an arbitrary mapping. It follows from the assumption that

$$\begin{aligned} e(f(\xi_1, \dots, \xi_n)|\mathcal{G} \vee \mathcal{H}) &= e\left(N^{-1} \sum_{k=1}^{N} f(\xi_{(k-1)n+1}, \dots, \xi_{kn})|\mathcal{G} \vee \mathcal{H}\right) \\ &= e[f(\xi_1, \dots, \xi_n)|\mathcal{G}], \end{aligned}$$

where the second equality follows from the fact that the quantity inside the previous conditionally expectation converges a.s. to $e[f(\xi_1, \dots, \xi_n)|\mathcal{G}]$ as $N \to \infty$, as a consequence of exchangeability and de Finetti's theorem.

Let us look backward from time s to time 0. For each $0 \leq r \leq s$, we denote by $Z_r^{N,s}$ the highest level occupied by the ancestors at time r of the N first individuals at time s. We show in Fig. 1.3 how to find $Z_r^{16,s}$ on the same realization as shown in Fig. 1.1.

We have, with the notations $\alpha' = 3\alpha/2$, $\beta = e\alpha'$, $\gamma = \beta e$, $\rho = 8 \vee [12 + 16\beta(1 - e^{-1})^{-1}]$,

Lemma 1.4. *For any $0 < r - h < r$, $h \leq \gamma^{-1}$, $N \geq \rho$,*

$$\mathbb{P}(Z_{r-h}^{N,r} > N) \leq \beta N h e^{-chN^2} + (\gamma h)^{N/e},$$

with $c = (1 - e^{-1})/16$.

Proof. We have

$$\{Z_{r-h}^{N,r} > N\} = A_1^{N,r,h} \cup A_2^{N,r,h},$$

where

$$A_1^{N,r,h} = \{Z_{r-h}^{N,r} > N\} \cap \left(\left\{\inf_{r-h<s<r} Z_s^{N,r} \leq \frac{N}{2}\right\} \cup \left\{\sup_{r-h<s<r} Z_s^{N,r} \geq \frac{3N}{2}\right\}\right)$$

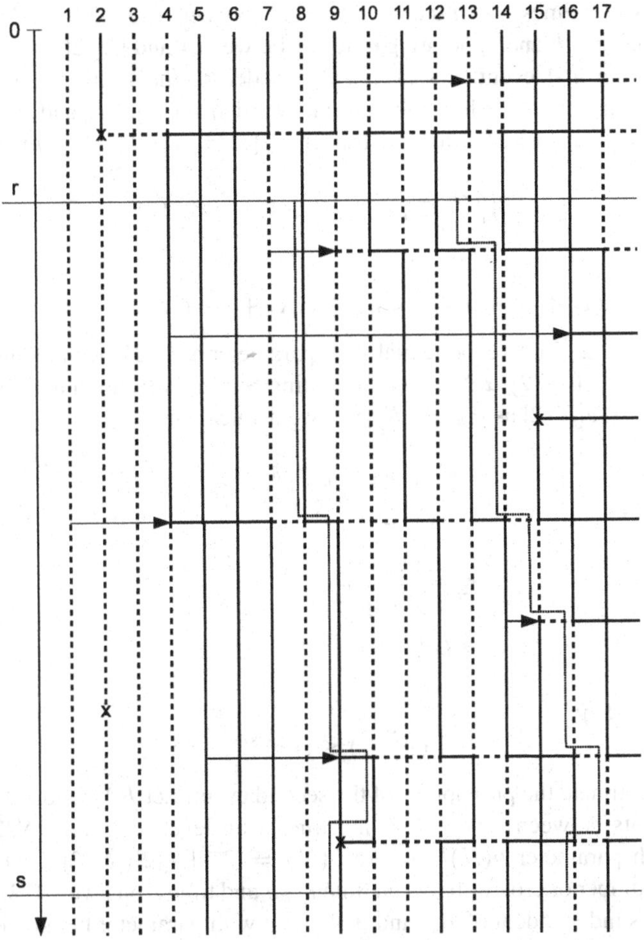

Fig. 1.3 We have represented the genealogies of the individuals sitting on levels 9 and 16 at time s. The reader can easily check that the ancestors at time r of the individuals sitting on levels $1, 2, \ldots, 16$ at time s are respectively $1, 2, 3, 1, 4, 5, 6, 7, 8, 7, 9, 10, 11, 12, 12, 13$. Here $Z_r^{16,s} = 13$. Note that all individuals sitting at time r between levels 1 and 13 have descendants (also there has been one death); this is not always the case

$$A_2^{N,r,h} = \{Z_{r-h}^{N,r} > N\} \cap \left\{ \inf_{r-h<s<r} Z_s^{N,r} > \frac{N}{2} \right\} \cap \left\{ \sup_{r-h<s<r} Z_s^{N,r} < \frac{3N}{2} \right\}.$$

On the event $A_1^{N,r,h}$, there must have been at least $N/2$ death events on a time interval of length h, while on the event $A_2^{N,r,h}$, there must have been more death events than birth events on the left of the curve $Z_s^{N,r}$, between time $r-h$ and time r, which, given the restrictions on the inf and the sup of that curve, implies that there are more crosses, between times $r - h$ and r, on levels between 1 and $3N/2$, than arrows, between times $r - h$ and r, pointing to levels between 1 and $[N/2]$.

Consider the first event. Let $\{Y_n,\ n \geq 1\}$ denote the waiting times between successive death events, when looking backward from time r, and we take into account those events which affect the trajectory $\{Z_s^{N,r},\ r - h \leq s \leq r\}$. Then

$$A_1^{N,r,h} \subset \{Y_1 + \cdots + Y_{[N/2]} < h\}.$$

Clearly

$$\mathbb{P}(Y_1 + \cdots + Y_{[N/2]} < h) \leq \mathbb{P}(X_1 + \cdots + X_{[N/2]} < h),$$

where the X_k's are i.i.d., exponential with parameter $\alpha' N$. Hence the law of $X_1 + \cdots + X_{[N/2]}$ is $\Gamma([N/2], \alpha' N)$, and, exploiting Stirling's formula and the obvious inequality $N \leq e[N/2]$ as soon as $N \geq 8$, we deduce that

$$\begin{aligned}
\mathbb{P}(A_1^{N,r,h}) &\leq \frac{(\alpha' N)^{[N/2]}}{([N/2]-1)!} \int_0^h e^{-\alpha' N r} r^{[N/2]-1} dr \\
&\leq \frac{(\alpha' N)^{[N/2]}}{[N/2]!} h^{[N/2]} \\
&\leq (\gamma h)^{[N/2]}.
\end{aligned}$$

Then, for $h \leq 1/\gamma$

$$\mathbb{P}(A_1^{N,r,h}) \leq (\gamma h)^{N/e}.$$

We now estimate the probability of the second event. Let B_h^N denote the number of birth events between time $r - h$ and time r on levels $1, 2, \ldots, [N/2]$. B_h^N is Poisson with parameter $b(N)h$, where $b(N) = 2^{-1}[N/2]([N/2] - 1)$. Let D_h^N denote the number of crosses between time $r - h$ and time r on levels $1, 2, \ldots, N + [N/2]$. D_h^N is independent of B_h^N and is Poisson with parameter bounded above by $\alpha' N h$. We have

$$\begin{aligned}
\mathbb{P}(A_2^{N,r,h}) &\leq \mathbb{P}(B_h^N < D_h^N) \\
&= \sum_{k=0}^{\infty} \mathbb{P}(D_h^N > k)\mathbb{P}(B_h^N = k) \\
&\leq \mathrm{e}[e^{D_h^N}; D_h^N > 0] \sum_{k=0}^{\infty} \frac{b(N)h/e)^k}{k!} e^{-b(N)h} \\
&\leq \beta N h \exp\left(-b(N)h(1 - e^{-1}) + \beta N h\right) \\
&\leq \beta N h e^{-cN^2 h}
\end{aligned}$$

for $N \geq 12 + 16\beta(1 - e^{-1})^{-1}$, if we choose $c = (1 - e^{-1})/16$.

If $\pi \in S_n, a \in \{0,1\}^n$, we shall write as above $\pi(a) = (a_{\pi(1)}, \ldots, a_{\pi(n)})$. Recall that a partition P of $\{1, \ldots, n\}$ induces an equivalence relation, whose equivalence classes are the blocks of the partition. Hence we shall write $i \simeq_P j$ whenever i and j are in the same block of P. Finally we write $|P|$ for the number of blocks of the partition P.

Recall the definition of X^N stated in (1.10). We have

Proposition 1.5. *Assume that the random sequence $\{\eta_0(1), \eta_0(2), \ldots\}$ is exchangeable. For all $N \geq 1$, $k \geq 1$, $0 \leq r_k < r_{k-1} < \cdots < r_1 < r_0 = s$, $a \in \{0,1\}^N$, $p_1, p_2, \ldots, p_k$ such that $0 \leq Np_1, Np_2, \ldots, Np_k \leq N$ are integers, $\pi \in S_N$,*

$$\mathbb{P}\left(\eta_s^{[N]} = a, \bigcap_{i=1}^{k}\{X_{r_i}^N = p_i\}, \bigcap_{i=1}^{k}\{Z_{r_i}^{N,r_{i-1}} \leq N\}\right)$$
$$= \mathbb{P}\left(\eta_s^{[N]} = \pi(a), \bigcap_{i=1}^{k}\{X_{r_i}^N = p_i\}, \bigcap_{i=1}^{k}\{Z_{r_i}^{N,r_{i-1}} \leq N\}\right).$$

Proof. We prove this in the case $k = 2$, the case $k > 2$ is similar.

For all $a^1 \in \{0,1\}^N$, we denote by $\mathscr{P}_{a^1}$ the set of partitions P of $\{1, 2, \ldots, N\}$ which are such that $i \simeq_P j \Rightarrow a_i^1 = a_j^1$. For any $i = 1, 2$, let $\eta_{r_{i-1}}^N = a^1$, where $a^1 \in \{0,1\}^N$. Let $P^i \in \mathscr{P}_{a^1}$ be the genealogy at time r_i of the individuals sitting on positions $\{1, \ldots, N\}$ at time r_{i-1}. $|P^i|$ is the number of ancestors at time r_i of the individuals $\{1, \ldots, N\}$ at time r_{i-1}. Each block of the partition P^i is a subset of $\{1, \ldots, N\}$ consisting of those individuals who have the same ancestor at time r_i. We assume that the blocks of P^i are arranged in increasing order of their smallest element.

For $1 \leq j \leq |P^i|$, let us define

$$c_j^{P^i} = \begin{cases} 1, & \text{if the } j\text{-th block of } P^i \text{ consists of type b individuals,} \\ 0, & \text{otherwise.} \end{cases}$$

Let

$$I^i = \{\ell_1^i, \ldots, \ell_{|P^i|}^i\}, \qquad J(I^i) = \{1 \leq j < \ell_{|P^i|}^i;\ j \notin \{\ell_1^i, \ldots, \ell_{|P^i|}^i\}\},$$

where $i = 1, 2$ and $1 \leq \ell_1^i < \ell_2^i < \cdots < \ell_{|P^i|}^i$ denote the levels of the $|P^i|$ ancestors at time r_i. Note that on the set $\{Z_{r_{i-1}}^{N,r_i} \leq N\}$, $|P^i| \leq \ell_{|P^i|}^i \leq N$.

For $i = 1, 2$, we define

$$\mathscr{H}_{P^i} = \left\{(\ell_1, \ldots, \ell_{|P^i|});\ 1 \leq \ell_1 < \cdots < \ell_{|P^i|} \leq N\right\},$$

$$\mathscr{A}_{r_i, r_{i-1}}(a, P^i, I^i, p_i) = \Big\{b \in \{0,1\}^N : \sum_{k=1}^{N} b_k = Np_i, \forall 1 \leq j \leq |P^i|, b_{\ell_j^i} = c_j^{P^i},$$

$$\times \forall j \in J(I^i), b_j = 1\}.$$

Note that the set $\mathscr{H}_{P^i}$ depends only upon $|P^i|$. Consider the event

$$A_{I^i}^{P^i} := \{\text{the succession of births and deaths between } r_{i-1} \text{ and } r_i \text{ produces } P^i \text{ and } I^i\}.$$

For $i = 1, 2$ and $a^1 \in \{0,1\}^N$, we have

$$\{\eta_{r_{i-1}}^{[N]} = a^1, X_{r_i}^N = p_i, Z_{r_{i-1}}^{N,r_i} \le N\} = \bigcup_{P^i \in \mathscr{P}_{a^1}} \bigcup_{I^i \in \mathscr{H}_{P^i}} \bigcup_{b \in \mathscr{A}_{r_i, r_{i-1}}(a^1, P^i, I^i, p_i)} \{A_{I^i}^{P^i}, \eta_{r_i}^{[N]} = b\},$$

from which we deduce that

$$\begin{aligned}
&\{\eta_{r_0}^{[N]} = a, X_{r_1}^N = p_1, X_{r_2}^N = p_2, Z_{r_1}^{N,r_0} \le N, Z_{r_2}^{N,r_1} \le N\} \\
&\quad = \bigcup_{P^1 \in \mathscr{P}_a} \bigcup_{I^1 \in \mathscr{H}_{P^1}} \bigcup_{a^1 \in \mathscr{A}_{r_1,r_0}(a, P^1, I^1, p_1)} \bigcup_{P^2 \in \mathscr{P}_{a^1}} \bigcup_{I^2 \in \mathscr{H}_{P^2}} \bigcup_{a^2 \in \mathscr{A}_{r_2,r_1}(a^1, P^2, I^2, p_2)} \\
&\qquad \times \{A_{I^1}^{P^1}, A_{I^2}^{P^2}, \eta_{r_2}^{[N]} = a^2\}
\end{aligned}$$

and from the independence of $A_{I^1}^{P^1}$, $A_{I^2}^{P^2}$ and $\eta_{r_2}^{[N]}$

$$\begin{aligned}
&\mathbb{P}(\eta_{r_0}^{[N]} = a, X_{r_1}^N = p_1, X_{r_2}^N = p_2, Z_{r_1}^{N,r_0} \le N, Z_{r_2}^{N,r_1} \le N) \\
&\quad = \sum_{P^1 \in \mathscr{P}_a} \sum_{I^1 \in \mathscr{H}_{P^1}} \sum_{a^1 \in \mathscr{A}_{r_1,r_0}(a, P^1, I^1, p_1)} \sum_{P^2 \in \mathscr{P}_{a^1}} \sum_{I^2 \in \mathscr{H}_{P^2}} \sum_{a^2 \in \mathscr{A}_{r_2,r_1}(a^1, P^2, I^2, p_2)} \\
&\qquad \times \mathbb{P}(A_{I^1}^{P^1}) \mathbb{P}(A_{I^2}^{P^2}) \mathbb{P}(\eta_{r_2}^{[N]} = a^2).
\end{aligned}$$

Similarly, for any $\pi \in S_N$ and $1 \le i \le 2$, if $\pi(P^i)$ is defined by $k \simeq_{P^i} j \Leftrightarrow \pi(k) \simeq_{\pi(P^i)} \pi(j)$

$$\begin{aligned}
&\mathbb{P}(\eta_{r_0}^{[N]} = \pi(a), X_{r_1}^N = p_1, X_{r_2}^N = p_2, Z_{r_1}^{N,r_0} \le N, Z_{r_2}^{N,r_1} \le N) \\
&= \sum_{P^1 \in \mathscr{P}_{\pi(a)}} \sum_{I^1 \in \mathscr{H}_{P^1}} \sum_{a^1 \in \mathscr{A}_{r_1,r_0}(\pi(a), P^1, I^1, p_1)} \sum_{P^2 \in \mathscr{P}_{a^1}} \sum_{I^2 \in \mathscr{H}_{P^2}} \sum_{a^2 \in \mathscr{A}_{r_2,r_1}(a^1, P^2, I^2, p_2)} \\
&\qquad \times \mathbb{P}(A_{I^1}^{P^1}) \mathbb{P}(A_{I^2}^{P^2}) \mathbb{P}(\eta_{r_2}^{[N]} = a^2) \\
&= \sum_{P^1 \in \mathscr{P}_a} \sum_{I^1 \in \mathscr{H}_{P^1}} \sum_{a^1 \in \mathscr{A}_{r_1,r_0}(\pi(a), \pi(P^1), I^1, p_1)} \sum_{P^2 \in \mathscr{P}_{a^1}} \sum_{I^2 \in \mathscr{H}_{P^2}} \sum_{a^2 \in \mathscr{A}_{r_2,r_1}(a^1, P^2, I^2, p_2)} \\
&\qquad \times \mathbb{P}(A_{I^1}^{\pi(P^1)}) \mathbb{P}(A_{I^2}^{P^2}) \mathbb{P}(\eta_{r_2}^{[N]} = a^2)
\end{aligned}$$

For $a^1 \in \{0,1\}^N$, we now describe a one-to-one correspondence ρ_π between $\mathscr{A}_{r_1,r_0}(a^1, P^i, I^i, p_i)$ and $\mathscr{A}_{r_1,r_0}(\pi(a^1), \pi(P^i), I^i, p_i)$. Let $b \in \mathscr{A}_{r_1,r_0}(a^1, P^i, I^i, p_i)$. We define $b' = \rho_\pi(b)$ as follows:

$$b'_j = \begin{cases} 1 & \text{if } j \in J(I^i), \\ b_j & \text{if } j > l_k^{I^i}, \end{cases}$$

where $k = |P^i|$ and

$$b'_{\ell_j^{I^i}} = c_j^{\pi(P^i)}, \quad \text{for all } 1 \leq j \leq |P^i|.$$

It is plain that $\sum_{j=1}^N b_j = \sum_{j=1}^N b'_j$. Clearly there exists $\pi' \in S_N$ such that $b' = \pi'(b)$. Moreover, for any $\pi \in S_N$ and $i = 1, 2$, $\mathbb{P}(A_{I^i}^{P^i}) = \mathbb{P}(A_{I^i}^{\pi(P^i)})$

Consequently

$$\begin{aligned}
&\sum_{a^1 \in \mathscr{A}_{r_1,s}(\pi(a),\pi(P^1),I^1,p_1)} \sum_{P^2 \in \mathscr{P}_{a^1}} \sum_{I^2 \in \mathscr{H}_{p2}} \sum_{a^2 \in \mathscr{A}_{r_2,r_1}(a^1,P^2,I^2,p_2)} \mathbb{P}(A_{I^1}^{\pi(P^1)})\mathbb{P}(A_{I^2}^{P^2})\mathbb{P}(\eta_{r_2}^{[N]} = a^2) \\
&= \sum_{a^1 \in \mathscr{A}_{r_1,s}(a,P^1,I^1,p_1)} \sum_{P^2 \in \mathscr{P}_{a^1}} \sum_{I^2 \in \mathscr{H}_{p2}} \sum_{a^2 \in \mathscr{A}_{r_2,r_1}(\pi'(a^1),\pi'(P^2),I^2,p_2)} \mathbb{P}(A_{I^1}^{\pi(P^1)})\mathbb{P}(A_{I^2}^{\pi'(P^2)})\mathbb{P}(\eta_{r_2}^{[N]} = a^2) \\
&= \sum_{a^1 \in \mathscr{A}_{r_1,s}(a,P^1,I^1,p_1)} \sum_{P^2 \in \mathscr{P}_{a^1}} \sum_{I^2 \in \mathscr{H}_{p2}} \sum_{a^2 \in \mathscr{A}_{r_2,r_1}(a^1,P^2,I^2,p_2)} \mathbb{P}(A_{I^1}^{P^1})\mathbb{P}(A_{I^2}^{P^2})\mathbb{P}(\eta_{r_2}^{[N]} = \pi'(a^2)) \\
&= \sum_{a^1 \in \mathscr{A}_{r_1,s}(a,P^1,I^1,p_1)} \sum_{P^2 \in \mathscr{P}_{a^1}} \sum_{I^2 \in \mathscr{H}_{p2}} \sum_{a^2 \in \mathscr{A}_{r_2,r_1}(a^1,P^2,I^2,p_2)} \mathbb{P}(A_{I^1}^{P^1})\mathbb{P}(A_{I^2}^{P^2})\mathbb{P}(\eta_{r_2}^{[N]} = a^2),
\end{aligned}$$

where the last identity follows from the fact $\eta_r^{[N]}$ is exchangeable. The result follows.

1.4.2 Tightness of $(X_t^N)_{t \geq 0}$

Before we establish tightness, we collect some results which will be required for its proof.

Lemma 1.5. *For any* $0 < r \leq h$, $N \geq 1$, $\varphi : [0,1]^3 \to \mathbb{R}$ *Borel measurable, any* $1 \leq i < N$,

$$\begin{aligned}
&\Big| \mathrm{e}(\varphi(X_{t-h}^N, X_t^N, X_{t+r}^N); \eta_{t+r}(i) = 0, \eta_{t+r}(N) = 1) \\
&\qquad - \mathrm{e}(\varphi(X_{t-h}^N, X_t^N, X_{t+r}^N); \eta_{t+r}(i) = 1, \eta_{t+r}(N) = 0) \Big| \\
&\qquad \leq \mathrm{e}(|\varphi(X_{t-h}^N, X_t^N, X_{t+r}^N)|; \{Z_t^{N,t+r} > N\} \cup \{Z_{t-h}^{N,t} > N\}).
\end{aligned}$$

Proof. Define

$$
\begin{aligned}
Z &= \varphi(X^N_{t-h}, X^N_t, X^N_{t+r}),\\
A &= \{\eta_{t+r}(i) = 0, \eta_{t+r}(N) = 1\},\\
B &= \{\eta_{t+r}(i) = 1, \eta_{t+r}(N) = 0\},\\
C &= \{Z^{N,t+r}_t > N\} \cup \{Z^{N,t}_{t-h} > N\}.
\end{aligned}
$$

We have shown in Proposition 1.5 that whenever φ is an indicator function,

$$
\mathrm{e}[Z(\mathbf{1}_A - \mathbf{1}_B)] = \mathrm{e}[Z(\mathbf{1}_A - \mathbf{1}_B)\mathbf{1}_C].
$$

The same result clearly follows for a general φ as in our statement by linearity of the expectation. But

$$
|\mathrm{e}[Z(\mathbf{1}_A - \mathbf{1}_B)\mathbf{1}_C]| \le \mathrm{e}[|Z|; C].
$$

The result follows.

It follows readily from Lemma 1.4 that with β and c as in that lemma.

Lemma 1.6. *For $N \ge \rho$, $h, r \le \gamma^{-1}$,*

$$
\mathbb{P}(\{Z^{N,t+r}_t > N\} \cup \{Z^{N,t}_{t-h} > N\}) \le \beta\left(Nre^{-cN^2r} + Nhe^{-cN^2h}\right) + (\gamma h)^{N/e} + (\gamma r)^{N/e}.
$$

We first deduce from the above estimates with $h = 0$, taking into account the obvious inequality $|X^N_{t+r} - X^N_t| \le 1$,

Corollary 1.2. *For any $t > 0$, $0 < r \le \gamma^{-1}$, $N \ge \rho$, $1 \le i < N$,*

$$
\begin{aligned}
&\Big|\mathrm{e}\left((X^N_{t+r} - X^N_t); \eta_{t+r}(i) = 0, \eta_{t+r}(N) = 1\right)\\
&\qquad - \mathrm{e}\left((X^N_{t+r} - X^N_t); \eta_{t+r}(i) = 1, \eta_{t+r}(N) = 0\right)\Big|\\
&\quad \le \beta N r\, e^{-cN^2r} + (\gamma r)^{N/e}.
\end{aligned}
$$

We now deduce

Corollary 1.3. *If $\beta' = \sqrt{\beta}$, $c' = c/2$, $N \ge \rho$, $h, r \le \gamma^{-1}$,*

$$
\begin{aligned}
&\Big|\mathrm{e}\left((X^N_t - X^N_{t-h})^2(X^N_{t+r} - X^N_t); \eta_{t+r}(i) = 0, \eta_{t+r}(N) = 1\right)\\
&\qquad - \mathrm{e}\left((X^N_t - X^N_{t-h})^2(X^N_{t+r} - X^N_t); \eta_{t+r}(i) = 1, \eta_{t+r}(N) = 0\right)\Big|\\
&\quad \le \beta'\left(\sqrt{Nr}\, e^{-c'N^2r} + \sqrt{Nh}\, e^{-c'N^2h} + (\gamma h)^{N/2e} + (\gamma r)^{N/2e}\right)\sqrt{\mathrm{e}\left[(X^N_t - X^N_{t-h})^4\right]}.
\end{aligned}
$$

Proof. Combining the two above lemmas and again $|X^N_{t+r} - X^N_t| \le 1$ with Schwarz's inequality in the form

$$e(Z;C) \le \sqrt{e(Z^2) \times \mathbb{P}(C)}$$

yields the result.

We are going to invoke a tightness criterium which involves an estimate of $e\left[(X_t^N - X_{t-h}^N)^2 (X_{t+h}^N - X_t^N)^2\right]$. More precisely, we have

Proposition 1.6. *For any $T > 0$, there exists a constant $K > 0$ which depends only upon α and T such that for all $N \ge \rho$, $0 < h < t \le T$,*

$$e\left[\left(X_t^N - X_{t-h}^N\right)^2 \left(X_{t+h}^N - X_t^N\right)^2\right] \le K h^{5/4}.$$

Proof. We note that it suffices to prove the result for h sufficiently small. Hence we may and do assume from now on that $h \le (2\gamma)^{-1}$. We have

$$dX_r^N = \frac{1}{N}\left[\sum_{1\le i<j\le N} \xi_{r-}^i dP_r^{i,j} + \sum_{1\le i\le N} \theta_{r-}^i dP_r^i\right],$$

where $P^{i,j}$, $1 \le i < j$, P^i, $i \ge 1$ are mutually independent Poisson processes, the $P^{i,j}$'s being standard, and the P^i's having intensity α,

$$\xi_r^i = \mathbf{1}_{\{\eta_r(i)=1,\eta_r(N)=0\}} - \mathbf{1}_{\{\eta_r(i)=0,\eta_r(N)=1\}},$$
$$\theta_r^i = -\mathbf{1}_{\{\eta_r(i)=1,\eta_r(N+1)=0\}}.$$

Let

$$\gamma_t^N := \left(X_t^N - X_{t-h}^N\right)^2.$$

It now follows from Corollary 1.3

$$\gamma_t^N \left(X_{t+h}^N - X_t^N\right)^2 = \frac{2}{N}\left[\sum_{1\le i<j\le N}\int_t^{t+h} \gamma_t^N \left(X_{r-}^N - X_t^N\right) \xi_{r-}^i dP_r^{i,j}\right.$$
$$\left. + \sum_{1\le i\le N}\int_t^{t+h} \gamma_t^N \left(X_{r-}^N - X_t^N\right) \theta_{r-}^i dP_r^i\right]$$
$$+ \frac{1}{N^2}\left[\sum_{1\le i<j\le N}\int_t^{t+h} \gamma_t^N (\xi_{r-}^i)^2 dP_r^{i,j} + \sum_{1\le i\le N}\int_t^{t+h} \gamma_t^N (\theta_{r-}^i)^2 dP_r^i\right]$$

$$e\left[\gamma_t^N \left(X_{t+h}^N - X_t^N\right)^2\right] = \frac{2}{N} e\left[\sum_{1\le i<j\le N}\int_t^{t+h} \gamma_t^N \left(X_r^N - X_t^N\right) \xi_r^i dr\right.$$
$$\left. + \alpha \sum_{1\le i\le N}\int_t^{t+h} \gamma_t^N \left(X_r^N - X_t^N\right) \theta_r^i dr\right]$$

$$+\frac{1}{N^2}\mathrm{e}\left[\sum_{1\le i<j\le N}\int_t^{t+h}\gamma_t^N(\xi_r^i)^2dr+\alpha\sum_{1\le i\le N}\int_t^{t+h}\gamma_t^N(\theta_r^i)^2dr\right]$$

$$\le\beta'\left[N^{3/2}\int_0^h\sqrt{r}e^{-c'N^2r}dr+N^{3/2}h^{3/2}e^{-c'N^2h}\right.$$

$$\left.+N\int_0^h(\gamma r)^{N/2e}dr+N(\gamma h)^{N/2e}h\right]$$

$$\times\sqrt{\mathrm{e}\left[(\gamma_t^N)^2\right]}+C_\alpha h\mathrm{e}(\gamma_t^N),$$

with $C_\alpha=3\alpha+1/2$. But from Hölder's inequality with $p=4, q=4/3$,

$$N^{3/2}\int_0^h\sqrt{r}e^{-c'N^2r}dr\le N^{3/2}\left(\int_0^h r^2\right)^{1/4}\left(\int_0^h e^{-4c'N^2r/3}dr\right)^{3/4}$$

$$\le C\frac{N^{3/2}}{N^{3/2}}h^{3/4}\le Ch^{3/4},$$

for some $C>0$, while

$$N^{3/2}h^{3/2}e^{-c'N^2h}\le C'h^{3/4},$$

with $C'=\sup_{x>0}x^{3/4}e^{-c'x}<\infty$. Next

$$N\int_0^h(\gamma r)^{N/2e}dr=\frac{N}{1+N/2e}(\gamma h)^{1+N/2e}$$

$$\le 2e\gamma^{3/4}h^{3/4},$$

since $\gamma h\le 1$. Finally

$$N(\gamma h)^{N/2e}h\le Ch^{3/4},$$

where

$$C=\sup_{N\ge1}\sup_{h\le(2\gamma)^{-1}}\gamma^{1/4}N(\gamma h)^{N/2e-1/4}<\infty.$$

We have shown that

$$\mathrm{e}\left[\gamma_t^N\left(X_{t+h}^N-X_t^N\right)^2\right]\le C''h^{3/4}\sqrt{\mathrm{e}\left[(\gamma_t^N)^2\right]}+C_\alpha h\mathrm{e}(\gamma_t^N).\tag{1.12}$$

It remains to estimate $\mathrm{e}(\gamma_t^N)$. The computations are quite similar to the previous ones, but simpler. We use Corollary 1.2, but with the interval $[t,t+h]$ replaced by the interval $[t-h,t]$.

$$
\begin{aligned}
(X_t^N - X_{t-h}^N)^2 &= \frac{2}{N}\Bigg[\sum_{1\le i<j\le N}\int_{t-h}^{t}(X_{r-}^N - X_{t-h}^N)\,\xi_{r-}^i\,dP_r^{i,j} \\
&\quad + \sum_{1\le i\le N}\int_{t-h}^{t}(X_{r-}^N - X_{t-h}^N)\,\theta_{r-}^i\,dP_r^i\Bigg] \\
&\quad + \frac{1}{N^2}\Bigg[\sum_{1\le i<j\le N}\int_{t-h}^{t}(\xi_{r-}^i)^2 dP_r^{i,j} + \sum_{1\le i\le N}\int_{t-h}^{t}(\theta_{r-}^i)^2 dP_r^i\Bigg] \\
\mathrm{e}\left[(X_t^N - X_{t-h}^N)^2\right] &= \frac{2}{N}\mathrm{e}\Bigg[\sum_{1\le i<j\le N}\int_{t-h}^{t}(X_r^N - X_{t-h}^N)\,\xi_{r-}^i\,dr \\
&\quad + \alpha\sum_{1\le i\le N}\int_{t-h}^{t}(X_r^N - X_{t-h}^N)\,\theta_{r-}^i\,dr\Bigg] \\
&\quad + \frac{1}{N^2}\mathrm{e}\Bigg[\sum_{1\le i<j\le N}\int_{t-h}^{t}(\xi_r^i)^2 dr + \alpha\sum_{1\le i\le N}\int_{t-h}^{t}(\theta_r^i)^2 dr\Bigg] \\
&\le \beta N^2\int_0^h r e^{-cN^2 r}dr + C_\alpha h \\
&\le \frac{\beta}{c}h + C_\alpha h \\
&\le C'''h.
\end{aligned}
$$

Moreover,

$$\mathrm{e}\left[(\gamma_t^N)^2\right] \le \mathrm{e}[\gamma_t^N] \le C'''h.$$

The result follows if we combine this last estimate with (1.12), keeping in mind that $h \le T$, and K may depend upon T.

It now follows from Proposition 1.6 and Theorem 13.5 in [2] that the collection of random processes $\{X_t^N,\ t \ge 0\}_{N\ge 1}$ is tight in $D([0,\infty))$. Since we already know that for all $k \ge 1$, all $0 \le t_1, t_2, \dots, t_k < \infty$,

$$(X_{t_1}^N, X_{t_2}^N, \dots, X_{t_k}^N) \to (X_{t_1}, X_{t_2}, \dots, X_{t_k}) \quad a.s., as\ N \to \infty,$$

we have that $X^N \Rightarrow X$ weakly in $D([0,\infty))$. Moreover, since $\sup_t |X_t^N - X_{t-}^N| = 1/N$, it follows from Theorem 13.4 in [2] that X possesses an a.s. continuous modification, and the weak convergence holds for the topology of locally uniform convergence in $[0,+\infty)$.

We have in fact a slightly stronger result.

Corollary 1.4. *The process X possesses an a.s. continuous modification, and for all $T > 0$,*

$$\sup_{0 \le t \le T} |X_t^N - X_t| \to 0 \text{ in probability, as } N \to \infty.$$

Proof. To each $\delta > 0$, we associate $n \ge 1$ and $0 = t_0 < t_1 < \cdots < t_n = T$, such that $\sup_{1 \le i \le n}(t_i - t_{i-1}) \le \delta$. We have, with the notation $y \wedge z = \inf(y, z)$,

$$\begin{aligned}\sup_{0 \le t \le T} |X_t^N - X_t| &\le \sup_i \sup_{t_{i-1} \le t \le t_i} |X_t^N - X_{t_{i-1}}^N| \wedge |X_t^N - X_{t_i}^N| + \sup_i |X_{t_i}^N - X_{t_i}| \\ &\quad + \sup_i \sup_{t_{i-1} \le t \le t_i} |X_t - X_{t_i}| \\ &\le w_T''(X^N, \delta) + \sup_i |X_{t_i}^N - X_{t_i}| + w_T(X, \delta),\end{aligned}$$

where

$$\begin{aligned} w_T(x, \delta) &= \sup_{0 \le s,t \le T, |s-t| \le \delta} |x(t) - x(s)|, \\ w_T''(x, \delta) &= \sup_{0 \le t_1 < t < t_2 \le T, t_2 - t_1 \le \delta} |x(t) - x(t_1)| \wedge |x(t) - x(t_2)|. \end{aligned}$$

From the proof of Theorem 13.5 in [2], we know that Proposition 1.6 implies that

$$\mathbb{P}(w_T''(X^N, \delta) > \varepsilon) \le \varepsilon^{-4} C_T (2\delta)^{1/4}.$$

Since X is continuous a.s., for each $\varepsilon > 0$,

$$\mathbb{P}(w_T(X, \delta) > \varepsilon) \to 0, \quad \text{as } \delta \to 0.$$

Moreover,

$$\sup_i |X_{t_i}^N - X_{t_i}| \to 0 \quad \text{a. s., as } N \to \infty.$$

The result follows.

1.4.3 The Main Result

In this section, we prove our main result. Before let us establish

Lemma 1.7. $\forall s > 0, k \ge 1, 0 \le r_k < r_{k-1} < \cdots < r_1 < r_0 = s$, $\forall N \ge 1$, $\forall a \in \{0, 1\}^N, \forall \pi \in S_N, \forall A_{r_j} \in \sigma(X_{r_j}), 0 \le j \le k$,

$$\mathbb{P}\left(\{\eta_s^{[N]} = a\}, \bigcap_{j=1}^{k} A_{r_j}\right) = \mathbb{P}\left(\{\eta_s^{[N]} = \pi(a)\}, \bigcap_{j=1}^{k} A_{r_j}\right).$$

Proof. $\forall m \geq N$, $p_0, p_1, \ldots, p_k \in [0,1]$ such that $mp_j \in \mathbb{N}$ for $0 \leq j \leq k$,

$$\mathbb{P}\left(\{\eta_s^{[N]} = \pi(a)\}, \bigcap_{j=0}^{k}\{X_{r_j}^m = p_j\}, \bigcap_{i=1}^{k}\{Z_{r_i}^{m,r_i-1} \leq m\}\right)$$

$$= \sum_{b\in\mathscr{A}(a,p_0)} \mathbb{P}\left(\eta_s^m = (\pi(a), b), \bigcap_{j=1}^{k}\{X_{r_j}^m = p_j\}, \bigcap_{i=1}^{k}\{Z_{r_i}^{N,r_i-1} \leq m\}\right)$$

$$= \sum_{b\in\mathscr{A}(a,p_0)} \mathbb{P}\left(\eta_s^m = (\pi'(a,b), \bigcap_{j=1}^{k}\{X_{r_j}^m = p_j\}, \bigcap_{i=1}^{k}\{Z_{r_i}^{N,r_i-1} \leq m\}\right),$$

where $\mathscr{A}(a, p_0) = \{b \in \{0,1\}^{m-N} : \sum_{i=1}^{N} a_i + \sum_{j=1}^{m-N} b_j = mp_0\}$, $\pi' \in S_m$. Thanks to Proposition 1.5, we deduce that

$$\mathbb{P}\left(\{\eta_s^{[N]} = \pi(a)\}, \bigcap_{j=0}^{k}\{X_{r_j}^m = p_j\}, \bigcap_{i=1}^{k}\{Z_{r_i}^{m,r_i-1} \leq m\}\right)$$

$$= \mathbb{P}\left(\{\eta_s^{[N]} = a\}, \bigcap_{j=0}^{k}\{X_{r_j}^m = p_j\}, \bigcap_{i=1}^{k}\{Z_{r_i}^{m,r_i-1} \leq m\}\right),$$

which implies that for all $f_j \in C_b([0,1])$, $1 \leq j \leq k$,

$$\mathrm{e}\left[\prod_{j=1}^{k} f_j(X_{r_j}^m); \{\eta_s^{[N]} = a\}; \bigcap_{i=1}^{k}\{Z_{r_i}^{m,r_i-1} \leq m\}\right]$$

$$= \mathrm{e}\left[\prod_{j=1}^{k} f_j(X_{r_j}^m); \{\eta_s^{[N]} = \pi(a)\}; \bigcap_{i=1}^{k}\{Z_{r_i}^{m,r_i-1} \leq m\}\right]$$

from which we deduce

$$\left|\mathrm{e}\left(\prod_{j=1}^{k} f_j(X_{r_j}^m); \{\eta_s^{[N]} = a\}\right) - \mathrm{e}\left(\prod_{j=1}^{k} f_j(X_{r_j}^m); \{\eta_s^{[N]} = \pi(a)\}\right)\right|$$

$$\leq 2\left(\prod_{j=1}^{k}||f_j||_\infty\right)\mathbb{P}(\cup_{i=1}^{k}\{Z_{r_i}^{m,r_{i-1}} > m\})$$

$$\leq 2k\beta hm\, e^{-chm^2}\prod_{j=1}^{k}||f_j||_\infty,$$

where $h = \inf_{1\leq j\leq k}\{r_j - r_{j-1}\}$, and the last line follows from Lemma 1.4. Letting $m \to \infty$, we deduce that

$$\mathrm{e}\left[\prod_{j=1}^{k} f_j(X_{r_j}); \{\eta_s^{[N]} = a\}\right] = \mathrm{e}\left[\prod_{j=1}^{k} f_j(X_{r_j}); \{\eta_s^{[N]} = \pi(a)\}\right].$$

The lemma has been established.

We are now in position to prove our main result.

Theorem 1.2. *The* $[0, 1]$*-valued process* $\{X_t, t \geq 0\}$ *defined by* (1.11) *admits a continuous version which is a weak solution of the Wright–Fisher equation* (1.1).

Proof. We already know from Corollary 1.4 that $\{X_t, t \geq 0\}$ defined by (1.11) possesses a continuous modification. The proof of Theorem 1.2 is structured as follows. In step 1 we show that $\{X_t, t \geq 0\}$ is a Markov process. In step 2 we show that X_t is a weak solution of the Wright–Fisher equation (1.1) .

STEP 1: We want to show that $\{X_t, t \geq 0\}$ defined by (3.3) is a Markov process. For $0 \leq s < t$, let $H_{s,t}$ denote the history between s and t which affects the vector $\{\eta_s(i), i \geq 1\}$ defined as follows. For all $N \geq 1$, the history $H_{s,t}^N$ is described by the time ordered sequence of all birth and death events affecting the levels between 1 and N, from time s to time t. $H_{s,t}$ is the union over $N \in \mathbb{N}$ of the $H_{s,t}^N$'s. X_t is a function of $\{\eta_s(i), i \geq 1\}$ and $H_{s,t}$. But $H_{s,t}$ is independent of $\sigma(X_r,\ 0 \leq r \leq s) \vee \sigma(\eta_s(i), i \geq 1)$. Consequently, for any $0 < x \leq 1$, there exists a measurable function $G_x : \{0, 1\}^{\mathbb{N}} \to [0, 1]$ such that

$$\mathbb{P}(X_t \leq x|\sigma(X_r,\ 0 \leq r \leq s)) = \mathrm{e}(G_x(\eta_s(i), i \geq 1)|\sigma(X_r,\ 0 \leq r \leq s)).$$

We know that conditionally upon $X_s = x$, the $\eta_s = \{\eta_s(i), i \geq 1\}$ are i.i.d. Bernoulli with parameter x. So all we need to show is that conditionally upon $\sigma(X_r,\ 0 \leq r \leq s\}$, the $\{\eta_s(i), i \geq 1\}$ are i.i.d Bernoulli with parameter X_s. In view of Proposition 1.4, it suffices to prove that conditionally upon $\sigma(X_r,\ 0 \leq r \leq s\}$, the $\eta_s(i)$ are exchangeable. This will follow from the fact that the same is true conditionally upon $\sigma(X_{r_1}, \ldots, X_{r_k}, X_s\}$, for all $k \geq 1, 0 \leq r_k < r_{k-1} < \ldots r_1 < r_0 = s$.

Hence it suffices to show that for all $N \geq 1$, $\eta_s^{[N]} = (\eta_s(1), \ldots, \eta_s(N))$ is conditionally exchangeable, given $\sigma(X_{r_1}, \ldots, X_{r_k}, X_s\}$. This is established in Lemma 1.7. The Markov property of the process $\{X_t, t \geq 0\}$ follows.

STEP 2: We now finally show that the time-homogeneous Markov process $\{X_t, t \geq 0\}$ has the right transition probability. Since X_t takes values in the compact set $[0, 1]$, the conditional law of X_t, given that $X_0 = x$ is determined by its moments. Hence all we have to show is that for all $t > 0, x \in [0, 1], n \geq 1$,

$$\mathrm{e}[X_t^n | X_0 = x] = \mathrm{e}[Y_t^n | Y_0 = x], \tag{1.13}$$

where $\{Y_t,\ t \geq 0\}$ solves (1.1).

We known that conditionally upon X_t, the $\{\eta_s(i), i \geq 1\}$ are i.i.d. Bernoulli with parameter X_t. Consequently, for all $n \geq 1$,

$$X_t^n = \mathbb{P}(\eta_t(1) = \cdots = \eta_t(n) = 1 \mid X_t).$$

This implies that

$$\begin{aligned}
\mathrm{e}_x[X_t^n] &= \mathrm{e}_x[\mathbb{P}(\eta_t(1) = \cdots = \eta_t(n) = 1 | X_t)] \\
&= \mathbb{P}_x(\eta_t(1) = \cdots = \eta_t(n) = 1) \\
&= \mathbb{P}_x(\text{the } 1 \ldots \widetilde{R}(t) \text{ individuals at time 0 are } \mathtt{b}) \\
&= \mathrm{e}_n[x^{\widetilde{R}_t}],
\end{aligned}$$

where $\widetilde{R}_s = Z_{t-s}^{N,t}$, for all $0 \leq s \leq t$. And it is easy to see that $\widetilde{R}_t$ and R_t defined in Sect. 1.2 have the same law. Equation (1.13) then follows from Proposition 1.1. The result is proved.

Remark 1.3. For any $N \geq 1$, the process $\{X_t^N, t \geq 0\}$ is not a Markov process. Indeed, the past values $\{X_s^N, 0 \leq s < t\}$ give us some clue as to what the values of $\eta_t(N+1), \eta_t(N+2), \ldots$ may be, and this influences the law of the future values $\{X_{t+r}^N, r > 0\}$.

Acknowledgements This work was partially supported by the ANR MANEGE (contract ANR-09-BLANC-0215-03). The authors wish to thank an anonymous referee, whose excellent and very detailed report permitted us to correct two errors and some imprecisions in an earlier version of this paper.

References

1. Aldous, D.: Exchangeability and related topics. In: Ecole d'été St Flour 1983. Lectures Notes in Math., vol. 1117, pp. 1–198 (1985)
2. Billingsley, P.: Convergence of Probability Measures, 2nd edn. Wiley, New York (1999)
3. Donnelly, P., Kurtz, T.G.: A countable representation of the Fleming Viot measure-valued diffusion. Ann. Probab. **24**, 698–742 (1996)
4. Donnelly, P., Kurtz, T.G.: Genealogical processes for Fleming-Viot models with selection and recombination. Ann. Appl. Probab. **9**, 1091–1148 (1999)
5. Donnelly, P., Kurtz, T.G.: Particle representations for measure-valued population models. Ann. Probab. **27**, 166–205 (1999)
6. Pardoux, É., Salamat, M.: On the height and length of the ancestral recombination graph. J. Appl. Prob. **46**, 669–689 (2009)
7. Wakolbinger, A.: Lectures at the Evolutionary Biology and Probabilistic Models. Summer School ANR MAEV, La Londe Les Maures (2008, unpublished)

Chapter 2
Control of Inventories with Markov Demand

Alain Bensoussan

Abstract We consider inventory control problems in discrete time. The horizon is infinite, and we consider discounted payoffs as well nondiscounted payoffs (ergodic control). We may have backlog or not. We may have set-up costs or not. In the traditional framework, the demand is modeled as a sequence of i.i.d. random variables. The ordering strategy is given by a base stock policy or an s, S policy, whether or not there is a set-up cost. We consider here the situation when the demand is modeled by a Markov chain. We show how the base stock policy and the s, S policy can be extended.

2.1 Introduction

We consider inventory control problems, with nonindependent demands. In real problems, the successive demands are linked for a lot of reasons, and the assumption of independence is too limited. The simplest way to model the linkage is to assume that the demands form a Markov process as such or are derived from a Markov process. Our objective is to show how the methods used in the case of independent demands can be adapted to this situation. In a recent book by Beyer et al. [1] a comprehensive presentation of these problems is given. We refer also to the references of this book for the related literature [2, 3]. In this work, the authors consider that the demand comes from an underlying state of demand, which is modeled as a Markov chain with a finite number of states. The fact that the number of states is finite simplifies mathematical arguments. We will here discuss the

A. Bensoussan (✉)
University of Texas at Dallas, Box 830688, Richardson, TX 75083-0688, USA

The Hong Kong Polytechnic Institute, Graduate School of Business, M921, Li Ka Shing Tower, Hung Hom, Kowloon, Hong Kong
e-mail: Alain.Bensoussan@utdallas.edu

L. Decreusefond and J. Najim (eds.), *Stochastic Analysis and Related Topics*, Springer Proceedings in Mathematics & Statistics 22, DOI 10.1007/978-3-642-29982-7_2,

situation in which the demand itself is a Markov process. We will consider backlog and non-backlog situations (but develop the no backlog case only) and set-up and non-set-up situations. The model is on an infinite horizon, and we study discounted as well as non-discounted (ergodic control) situations.

2.2 No Backlog and No Set-Up Cost

2.2.1 The Model

Let $\Omega, \mathscr{A}, P$ be a probability space, on which is defined a Markov chain z_n. This Markov chain represents the demand. Its state space is R^+ and its transition probability is $f(\zeta|z)$. We shall assume that

$$f(\zeta|z) \text{ is uniformly continuous in both variables and bounded} \tag{2.1}$$

$$\int_0^{+\infty} \zeta f(\zeta|z) d\zeta \leq c_0 z + c_1 \tag{2.2}$$

We can assume that $z_1 = z$, a fixed constant or more generally a random varible with given probability distribution. We define the filtration

$$\mathscr{F}^n = \sigma(z_1, \cdots, z_n)$$

A control policy, denoted by V, is a sequence of random variables v_n such that v_n is $\mathscr{F}^n$ measurable. When $z_1 = z$, then v_1 is deterministic. Also, we assume as usual that $v_n \geq 0$. We next define the inventory as the sequence

$$y_{n+1} = (y_n + v_n - z_{n+1})^+, \quad y_1 = x \tag{2.3}$$

The process y_n is adapted to the filtration $\mathscr{F}^n$. The joint process y_n, z_n is also a Markov chain. We can write, for a test function $\varphi(x, z)$ (bounded continuous on $R^+ \times R^+$)

$$E[\varphi(y_2, z_2)|y_1 = x, z_1 = z, v_1 = v] = E[\varphi((x + v - z_2)^+, z_2)|z_1 = z]$$

This defines a Markov chain to which is associated the operator

$$\Phi^v \varphi(x, z) = \int_{x+v}^{+\infty} \varphi(0, \zeta) f(\zeta|z) d\zeta + \int_0^{x+v} \varphi(x + v - \zeta, \zeta) f(\zeta|z) d\varsigma \tag{2.4}$$

We notice the property

$$\begin{aligned}&\Phi^v \varphi_v(x,z) \text{ is uniformly continuous in } v, x, z \text{ if}\\&\varphi_v(x,z) = \phi(x,z,v) \text{ is uniformly continuous and bounded in } x, z, v\end{aligned} \tag{2.5}$$

We next define the function

$$l(x,z,v) = cv + hx + pE[(x+v-z_2)^-|z_1 = z] \tag{2.6}$$

which will model the one period cost. The cost function is then

$$J_{x,z}(V) = E\sum_{n=1}^{\infty} \alpha^{n-1} l(y_n, z_n, v_n) \tag{2.7}$$

We are interested in the value function

$$u(x,z) = \inf_V J_{x,z}(V) \tag{2.8}$$

Set $l_v(x,z) = l(x,z,v)$. Note the inequalities

$$cv + hx \leq l_v(x,z) \leq cv + hx + p(c_0 z + c_1) \tag{2.9}$$

and

$$cv \leq \Phi^v l_v(x,z) \leq cv + h(x+v) + p(c_0^2 z + c_0 c_1 + c_1) \tag{2.10}$$

Consider then the function

$$w_0(x,z) = \sum_{n=1}^{\infty} \alpha^{n-1} (\Phi^0)^{n-1} l_0(x,z) \tag{2.11}$$

which corresponds to the payoff, when the control is identically 0.

Lemma 2.1. *We assume that*

$$\alpha c_0 < 1 \tag{2.12}$$

then the series $w_0(x,z) < \infty$ *and more precisely*

$$w_0(x,z) \leq \frac{hx}{1-\alpha} + \frac{pc_0 z}{1-c_0\alpha} + \frac{pc_1}{(1-\alpha)(1-c_0\alpha)} \tag{2.13}$$

Proof. It is an immediate consequence of formulas (2.11) and (2.10). □

We can now write the Bellman equation

$$u(x,z) = \inf_{v \geq 0} [l(x,z,v) + \alpha \Phi^v u(x,z)] \tag{2.14}$$

We state the following:

Theorem 2.1. *We assume* (2.1), (2.2), (2.6), (2.12). *Then there exists one and only one solution of equation* (2.14), *such that* $0 \leq u \leq w_0$. *It is continuous and coincides with the value function* (2.8). *There exists an optimal feedback* $\hat{v}(x,z)$ *and an optimal control policy* $\hat{V}$.

Proof. We use a standard monotonicity argument to prove that on the interval $(0, w_0)$, the set of solutions of (2.14) is not empty. It has a minimum and a maximum solution. The minimum solution is l.s.c. and the maximum solution is u.s.c. The minimum solution coincides with the value function. There exists an optimal feedback $\hat{v}(x,z)$and an optimal control policy $\hat{V}$. To prove uniqueness, we have to prove that the minimum and maximum solution coincide. We begin by proving a bound on $\hat{v}(x,z)$. It will be first convenient to mention a slightly better estimate for w_0. Indeed, we can write

$$w_0(x,z) \leq h \sum_{n=1}^{\infty} (\alpha \Phi^0)^{n-1} x(x,z) + \frac{pc_0 z}{1 - c_0\alpha} + \frac{pc_1}{(1-\alpha)(1-c_0\alpha)} \tag{2.15}$$

and (2.13) was simply derived from (2.15) by using $\Phi^0 x(x,z) \leq x$. Next, from (2.14), considering $\underline{u}$, the minimum solution, we can state that

$$\underline{u}(x,z) \geq h \sum_{n=1}^{\infty} (\alpha \Phi^0)^{n-1} x(x.z)$$

which follows from $l(x,z,v) \geq hx$ and

$$\Phi^v \varphi(x,z) \geq \Phi^0 \varphi(x,z), \forall v \geq 0, \forall \varphi \text{ increasing in } x \tag{2.16}$$

Therefore, we can write

$$l(x,z,v) + \alpha \Phi^v \underline{u}(x,z) \geq cv + h \sum_{n=1}^{\infty} (\alpha \Phi^0)^{n-1} x(x,z)$$

Therefore, in minimizing in v, we can bound from above the range of v. More precisely, we get

$$\hat{v}(x,z) \leq \frac{pc_0 z}{c(1-c_0\alpha)} + \frac{pc_1}{c(1-\alpha)(1-c_0\alpha)} \tag{2.17}$$

We consider the optimal trajectory $\hat{y}_n, z_n$ obtained from the optimal feedback namely

$$\hat{y}_{n+1} = (\hat{y}_n + \hat{v}_n - z_{n+1})^+, \ \hat{v}_n = \hat{v}(\hat{y}_n, z_n)$$

with $\hat{y}_1 = x, z_1 = z$. It can be shown that the maximum solution will coincide with the minimum solution, if we can check that $\hat{V}$ satisfies the property

$$\alpha^n E\bar{u}(\hat{y}_{n+1}, z_{n+1}) \to 0, \text{ as } n \to \infty$$

It is sufficient to replace $\bar{u}$ by w_0 and by estimate (2.13), it is sufficient to show that

$$\alpha^n E\hat{y}_{n+1}, \quad \alpha^n E z_{n+1} \to 0, \text{ as } n \to \infty \tag{2.18}$$

However, by standard Markov arguments, $\alpha^n E z_{n+1} \leq (\alpha c_0)^n z$. From the Assumption (2.12), the second part of (2.18) follows immediately. We next use

$$\begin{aligned}\hat{y}_{n+1} &\leq \hat{y}_n + \hat{v}_n \\ &\leq \hat{y}_n + \frac{pc_0 z_n}{c(1-c_0\alpha)} + \frac{pc_1}{c(1-\alpha)(1-c_0\alpha)}\end{aligned}$$

Therefore,

$$E\hat{y}_{n+1} \leq E\hat{y}_n + \frac{pc_0^n z}{c(1-c_0\alpha)} + \frac{pc_1}{c(1-\alpha)(1-c_0\alpha)}$$

and we deduce the estimate

$$E\hat{y}_{n+1} \leq x + \frac{pzc_0}{c(1-\alpha c_0)}\frac{1-c_0^n}{1-c_0} + \frac{npc_1}{c(1-\alpha)(1-c_0\alpha)}$$

Using again the Assumption (2.12), we deduce the first part of (2.18). This completes the proof. □

2.2.2 *Base-Stock Policy*

We want now to check that the optimal feedback $\hat{v}(x,z)$ can be obtained by a base stock policy, with a base stock depending on the value of z. We have

Theorem 2.2. *We make the assumptions of Theorem 2.1 and $p > c$. We assume also*

$$f(x|z) \geq a_0(M) > 0, \ \forall x, z \leq M \tag{2.19}$$

Then the function $u(x,z)$ is convex and C^1in the argument x. Moreover the optimal feedback is given by

$$\hat{v}(x,z) = \left| \begin{array}{ll} S(z) - x & \text{if } x \leq S(z) \\ 0 & \text{if } x \geq S(z) \end{array} \right. \tag{2.20}$$

The function $S(z)$ is uniformly continuous and the derivative in x, $u'(x,z)$ is uniformly continuous.

Proof. We consider the increasing process

$$u_{n+1}(x,z) = \inf_{v \geq 0}[l(x,z,v) + \alpha \Phi^v u_n(x,z)], \qquad u_0(x) = 0$$

which we write also as

$$u_{n+1}(x,z) = (h-c)x + \inf_{\eta \geq x}\{c\eta + pE[(\eta-z_2)^-|z_1 = z] + \alpha E[u_n((\eta-z_2)^+, z_2)|z_1 = z]\}$$

and we are going to show recursively that the sequence $u_n(x,z)$ is convex and C^1 in x. Define

$$G_n(x,z) = cx + pE[(x-z_2)^-|z_1 = z] + \alpha E[u_n((x-z_2)^+, z_2)|z_1 = z]$$

then the function $G_n(x,z)$ attains its minimum in $S_n(z)$, which can be uniquely defined by taking the smallest minimum. We prove by induction that $u_n(x,z)$, $G_n(x,z)$ are convex and C^1in x. Moreover

$$G_n'(x,z) = c - p\bar{F}(x|z) + \alpha E[u_n'(x-z_2, z_2)\mathbf{1}_{x \geq z_2}|z_1 = z]$$

We see that $G_n'(0,z) = c - p$ and we can check that

$$G_n'(+\infty, z) = c + h\alpha\frac{1-\alpha^n}{1-\alpha}$$

Therefore, there exists a point $S_n(z)$ such that $G_n'(S_n(z), z) = 0$. Note that $S_n(z) > 0$, since $G_n'(0,z) = c - p < 0$. We have

$$u_{n+1}(x,z) = \left| \begin{array}{ll} (h-c)x + G_n(S_n(z), z) & \text{if } x \leq S_n(z) \\ (h-c)x + G_n(x,z) & \text{if } x \geq S_n(z) \end{array} \right.$$

It follows that the limit $u(x,z)$ is convex in x. Clearly $u(x,z) \to +\infty$, as $x \to +\infty$. Also

$$G(x,z) = cx + pE[(x-z_2)^-|z_1 = z] + \alpha E[u((x-z_2)^+, z_2)|z_1 = z]$$

is convex and $\to +\infty$ as $x \to +\infty$. So the minimum is attained in $S(z)$, which can be defined in a unique way, by taking the smallest minimum.

Since $h-c \leq u_n'(x,z) \leq \frac{h}{1-\alpha}$, we can assert that $u(x,z)$ is Lipschitz continuous in x. The same is true for $G(x,z)$. But

$$G'(x,z) = c - p\bar{F}(x|z) + \alpha E[u'(x-z_2, z_2)\mathbf{1}_{x > z_2}|z_1 = z] \to c - p \text{ as } x \to 0$$

therefore also $S(z) > 0$. Then, from convexity

$$u(x,z) = \left| \begin{array}{ll} (h-c)x + G(S(z), z) & \text{if } x \leq S(z) \\ (h-c)x + G(x,z) & \text{if } x \geq S(z) \end{array} \right.$$

Next, from

$$c - p\bar{F}(S_n(z)|z) + \alpha E[u_n'(S_n(z) - z_2, z_2)\mathbf{1}_{S_n(z)>z_2}|z_1 = z] = 0$$

we deduce, using the estimate on u_n' and the property

$$\bar{F}(S_n(z)|z) \leq \frac{c_0 z + c_1}{S_n(z)}$$

that

$$S_n(z) \leq \frac{(c_0 z + c_1)(p + \alpha(h - c))}{c + \alpha(h - c)} \tag{2.21}$$

The same estimate holds for $S(z)$. It is easy to check that $S_n(z) \to S(z)$, and $S(z)$ is the smallest minimum of $G(x, z)$ in x. Furthermore, from the continuity properties of $G(x, z)$ in both arguments, we can check that $S(z)$ is a continuous function. The feedback $\hat{v}(x, z)$ defined by (2.20) is also continuous in both arguments. Define

$$\chi(x, z) = u'(x, z) - h + c$$

as an element of B (space of measurable bounded functions on $R^+ \times R^+$), then χ is the unique solution in B of the equation

$$\chi(x, z) = g(x+\hat{v}(x, z), z)+\alpha E[\chi(x+\hat{v}(x, z)-z_2, z_2)\mathbf{1}_{x+\hat{v}(x,z)>z_2}|z_1 = z], \ x, z \in R^+ \tag{2.22}$$

where

$$g(x, z) = c + \alpha(h - c) - (p + \alpha(h - c))\bar{F}(x|z)$$

Since the function $g(x + \hat{v}(x, z), z)$ is continuous in the pair x, z, the solution $\chi(x, z)$ is also continuous. Let us check that $S(z)$ is uniformly continuous. Let us first check that

$$(G'(x_1, z)-G'(x_2, z))(x_1-x_2) \geq (p-\alpha(c-h))(F(x_1|z)-F(x_2|z))(x_1-x_2) \tag{2.23}$$

Assume to fix the ideas that $x_1 > x_2$. We have

$$(G'(x_1, z) - G'(x_2, z))(x_1 - x_2) = p(F(x_1|z) - F(x_2|z))(x_1 - x_2)+$$

$$\alpha E[(u'(x_1 - z_2, z_2) - u'(x_2 - z_2, z_2))(x_1 - x_2)\mathbf{1}_{z_2<x_2}|z_1 = z]+$$

$$+\alpha E[u'(x_1 - z_2, z_2)\mathbf{1}_{x_2<z_2<x_1}(x_1 - x_2)|z_1 = z]$$

The second term is positive, from the convexity of u. Using the left estimate on u' for the last term, we deduce (2.23). We then obtain

$$(S(z) - S(z'))(G'(S(z), z') - G'(S(z'), z)) \geq$$

$$(p - \alpha(c - h))(S(z) - S(z'))\left[\int_{S(z')}^{S(z)} (f(\xi|z) + f(\xi|z'))d\xi\right]$$

If $z, z' < m$, then from (2.21) we can find $M_m > m$ such that $S(z), S(z') < M_m$. From the Assumption (2.19), we deduce

$$(S(z)-S(z'))(G'(S(z),z')-G'(S(z'),z)) \geq 2a_0(M_m)(p-\alpha(c-h))(S(z)-S(z'))^2$$

Next we have

$$G'(x,z') - G'(x,z) = \int_0^x (p + \alpha u'(x-\zeta,\zeta))(f(\zeta|z') - f(\zeta|z))d\zeta$$

hence

$$|G'(x,z') - G(x,z)| \leq xC \sup_{0<\zeta<x} |f(\zeta|z') - f(\zeta|z)|$$

Applying this estimate with $x = S(z)$ and $x = S(z')$ and combining estimates, we obtain easily that $S(z)$ is uniformly continuous. It follows that the feedback $\hat{v}(x,z)$ is uniformly continuous. Then from (2.22), we obtain that $\chi(x,z)$ is uniformly continuous. The proof has been completed. □

Remark 2.1. We have $\chi(x,z) = 0, \forall x \leq S(z)$, and

$$\chi(x,z) = g(x,z) + \alpha E[\chi(x - z_2, z_2)\mathbf{1}_{x>z_2}|z_1 = z], \ \forall x \geq S(z) \tag{2.24}$$

Also

$$0 = g(S(z),z) + \alpha E[\chi(S(z) - z_2, z_2)\mathbf{1}_{S(z)>z_2}|z_1 = z] \tag{2.25}$$

So $S(z)$ is not the solution of $g(S(z),z) = 0$. If we consider the function

$$G^*(x,z) = (p + \alpha(h-c))E[(x-z_2)^+|z_1 = z] - (p-c)x + pE[z_2|z_1 = z]$$

then the solution of $g(S,z) = 0$ is denoted by $S^*(z)$ and minimizes $G^*(x,z)$. We have $G(x,z) \geq G^*(x,z)$.

2.2.3 Ergodic Theory

We turn now to the case when $\alpha \to 1$. We write $u_\alpha(x,z)$ to satisfy

$$u_\alpha(x,z) = \left| \begin{array}{ll} (h-c)x + cS_\alpha(z) + pE[(S_\alpha(z)-z_2)^-|z_1 = z]+ & \\ \alpha E[u_\alpha((S_\alpha(z)-z_2)^+, z_2)|z_1 = z] & \text{if } x \leq S_\alpha(z) \\ hx + pE[(x-z_2)^-|z_1 = z] + \alpha E[u_\alpha((x-z_2)^+, z_2)|z_1 = z] & \text{if } x \geq S_\alpha(z) \end{array} \right. \tag{2.26}$$

We shall make the assumptions

$$c_0 = 0 \tag{2.27}$$

$$\inf_{\substack{0 \leq \zeta \leq a \\ z}} f(\zeta|z) \geq \gamma(a) > 0, \forall a$$

$$f(\zeta|z) \text{ is ergodic} \tag{2.28}$$

$$\int |f(\zeta|z) - f(\zeta|z')|d\zeta \leq \delta|z - z'| \tag{2.29}$$

$$\sup_z F(x|z) = \delta_0(x) < 1, \forall x$$

We denote by $\varpi(z)$ the invariant probability density corresponding to the Markov chain $f(\zeta|z)$. We state

Theorem 2.3. *We assume* (2.1), (2.2) *with* $c_0 = 0$, (2.6) *with* $p > c$ *and* (2.27)–(2.29). *Then, for a subsequence (still denoted* α*) converging to* 1, *we have, for any compact* K *of* R^+

$$\sup_{z \in K} |S_\alpha(z) - S(z)| \leq \epsilon(\alpha, K), \quad \epsilon(\alpha, K) \to 0, \text{ as } \alpha \uparrow 1 \tag{2.30}$$

$$\sup_{\substack{x \leq M \\ z \leq N}} |u_\alpha(x,z) - \frac{\rho_\alpha}{1-\alpha} - u(x,z)| \to 0, \forall M, N \tag{2.31}$$

with $\rho_\alpha \to \rho$ *and*

$$u(x,z) + \rho = \begin{vmatrix} (h-c)x + cS(z) + pE[(S(z) - z_2)^-|z_1 = z] + \\ E[u((S(z) - z_2)^+, z_2)|z_1 = z] \quad \text{if } x \leq S(z) \\ hx + pE[(x - z_2)^-|z_1 = z] + E[u((x - z_2)^+, z_2)|z_1 = z] \text{ if } x \geq S(z) \end{vmatrix} \tag{2.32}$$

The function $u(x,z)$ *satisfies the growth condition*

$$\sup_{\substack{\xi \leq x \\ z}} |u(\xi, z)| \leq C_0 + \frac{C_1 x}{1 - \delta_0(x)} \tag{2.33}$$

It is C^1 *in* x *and Lipschitz continuous in* z. *The following estimates hold:*

$$\sup_{\substack{\xi \leq x \\ z}} |u_x(\xi, z)| \leq \frac{C}{1 - \delta_0(x)}, \tag{2.34}$$

$$|u(x,z) - u(x,z')| \leq [C_0(m_0) + \frac{C_1(m_0)x}{1 - \delta_0(x)}]\delta|z - z'| \tag{2.35}$$

$$(1-\delta_0(x)) \sup_{\substack{\xi \leq x \\ z}} |u_{xx}(\xi, z)| \leq C, \ (1-\delta_0(x)) \sup_{\substack{\xi \leq x \\ z}} |u_{xz}(\xi, z)| \leq C, \ a.e. \tag{2.36}$$

Given $S(z)$, the pair $u(x,z), \rho$ satisfying the above conditions and $\int u(0,z)\varpi(z)dz = 0$ is uniquely defined. Also

$$u(x,z) + \rho = \inf_{v \geq 0}[l(x,z,v) + \Phi^v u(x,z)] \tag{2.37}$$

Proof. We begin with (2.30). We first note that, thanks to $c_0 = 0$, we have

$$S_\alpha(z) \leq \frac{c_1(p+h)}{\min(h,c)} = m_0 \tag{2.38}$$

We pursue the estimates obtained in Theorem 2.2. We have first

$$(S_\alpha(z) - S_\alpha(z'))(G'_\alpha(S_\alpha(z), z') - G'_\alpha(S_\alpha(z'), z)) \geq 2\min(h,c)\gamma(m_0)(S_\alpha(z) - S_\alpha(z'))^2$$

We know that $u_\alpha(x,z)$ is C^1 in x. Then, from (2.26), we have (denoting $u'_\alpha(x,z) = u'_{\alpha x}(x,z)$

$$\begin{aligned} u'_\alpha(x,z) &= h - c \quad \text{if} \quad x < S_\alpha(z) \\ &= h - p\bar{F}(x|z) + \alpha E[u'_\alpha(x - z_2, z_2)\mathbf{1}_{x > z_2}|z_1 = z], \text{ if } x > S_\alpha(z) \end{aligned}$$

from which we can assert that

$$\sup_{\substack{\xi \leq x \\ z}} |u'_\alpha(\xi, z)| \leq \frac{\max(h,p)}{1 - \delta_0(x)} \tag{2.39}$$

Therefore,

$$|G'_\alpha(S_\alpha(z), z') - G'_\alpha(S_\alpha(z), z)| = (p + \frac{\max(h,p)}{1 - \sup_z F(m_0|z)}) \int |f(\zeta|z') - f(\zeta|z)|d\zeta$$

$$|G'_\alpha(S_\alpha(z'), z') - G'_\alpha(S_\alpha(z'), z)| = (p + \frac{\max(h,p)}{1 - \sup_z F(m_0|z)}) \int |f(\zeta|z') - f(\zeta|z)|d\zeta$$

Collecting results we obtain the estimate

$$|S_\alpha(z) - S_\alpha(z')| \leq \frac{p + \dfrac{\max(h,p)}{1 - \delta_0(m_0)}}{\min(h,c)\gamma(m_0)} \int |f(\zeta|z') - f(\zeta|z)|d\zeta \tag{2.40}$$

and from the first Assumption (2.29), we finally obtain

$$|S_\alpha(z) - S_\alpha(z')| \leq \frac{p + \dfrac{\max(h,p)}{1-\delta_0(m_0)}}{\min(h,c)\gamma(m_0)}\delta|z - z'| \tag{2.41}$$

Therefore, the sequence $S_\alpha(z)$ is uniformly Lipschitz continuous. It is standard that one can extract a subsequence, which converges in the space of continuous functions on compact sets, for any compact set K, towards a function $S(z)$. Therefore, (2.30) is satisfied.

Define $\chi_\alpha(x,z) = u'_\alpha(x,z) - h + c$. We have

$$\chi_\alpha(x,z) = g_\alpha(x,z) + \tag{2.42}$$

$$+ \alpha E[\chi_\alpha(x - z_2, z_2)\mathbf{1}_{x>z_2}|z_1 = z], \; x > S_\alpha(z) \tag{2.43}$$

$$= 0, \qquad x \leq S_\alpha(z)$$

with

$$g_\alpha(x,z) = (c + \alpha(h-c) - (p + \alpha(h-c))\bar{F}(x|z) \tag{2.44}$$

As it has been done for $u'_\alpha(x,z)$, we can state

$$\sup_{\substack{0 \leq \xi \leq x \\ z}} |\chi_\alpha(\xi,z)| \leq \frac{\max(h,p)}{1-\delta_0(x)} \tag{2.45}$$

If we consider $\psi_\alpha(x,z) = \chi'_\alpha(x,z) = u''_\alpha(x,z)$, then using the fact that $\chi_\alpha(0,z) = 0$, we see that

$$\psi_\alpha(x,z) = (p + \alpha(h-c))f(x|z) + \tag{2.46}$$

$$+ \alpha E[\psi_\alpha(x - z_2, z_2)\mathbf{1}_{x>z_2}|z_1 = z], \quad x > S_\alpha(z)$$

$$\psi_\alpha(x,z) = 0, \quad x < S_\alpha(z)$$

The function $\psi_\alpha(x,z)$ is not continuous, however it is measurable and bounded. We have

$$\sup_{\substack{0 \leq \xi \leq x \\ z}} |\psi_\alpha(\xi,z)| \leq \frac{((h-c)^+ + p)\|f\|}{1-\delta_0(x)} \tag{2.47}$$

where $\|f\| = \sup_{x,z} f(x|z)$. We next obtain an estimate on $\chi_\alpha(x,z) - \chi_\alpha(x,z')$. Assume first $x > S_\alpha(z), x > S_\alpha(z')$. Then, from (2.42), we have

$$\chi_\alpha(x,z) - \chi_\alpha(x,z') = g_\alpha(x,z) - g_\alpha(x,z') +$$

$$+ \alpha \int_0^x \chi_\alpha(x - \zeta, \zeta)(f(\zeta|z) - f(\zeta|z'))d\zeta$$

From estimate (2.45) and the first Assumption (2.29), we deduce

$$|\chi_\alpha(x,z) - \chi_\alpha(x,z')| \le \left(p + (h-c)^+ + \frac{\max(h,p)}{1-\delta_0(x)}\right)\delta|z-z'|$$

$$x > S_\alpha(z), x > S_\alpha(z')$$

Assume now to fix ideas that $S_\alpha(z') > x > S_\alpha(z)$. Then $\chi_\alpha(x,z') = 0$ and

$$0 = g_\alpha(S_\alpha(z),z) + \alpha E[\chi_\alpha(S_\alpha(z) - z_2, z_2)\mathbf{1}_{S_\alpha(z)>z_2}|z_1 = z]$$

Therefore,

$$\begin{aligned}\chi_\alpha(x,z) - \chi_\alpha(x,z') &= g_\alpha(x,z) - g_\alpha(S_\alpha(z),z) + \alpha E[\chi_\alpha(x-z_2,z_2)\mathbf{1}_{x>z_2}|z_1 = z] - \\ &\quad - \alpha E[\chi_\alpha(S_\alpha(z)-z_2,z_2)\mathbf{1}_{S_\alpha(z)>z_2}|z_1 = z] \\ &= (p + \alpha(h-c))(F(x|z) - F(S_\alpha(z)|z)) + \\ &\quad + \alpha E[\chi_\alpha(x-z_2,z_2)\mathbf{1}_{x>z_2>S_\alpha(z)}|z_1 = z] + \\ &\quad + \alpha E[(\chi_\alpha(x-z_2,z_2) - \chi_\alpha(S_\alpha(z)-z_2,z_2))\mathbf{1}_{S_\alpha(z)>z_2}|z_1 = z]\end{aligned}$$

It follows

$$|\chi_\alpha(x,z) - \chi_\alpha(x,z')| \le \left[p + (h-c)^+ + \frac{\max(h,p)}{1-\delta_0(x)} + \frac{(h-c)^+ + p}{1-\delta_0(x)}\right]\|f\|(x - S_\alpha(z))$$

Finally, we can state the estimate

$$|\chi_\alpha(x,z) - \chi_\alpha(x,z')| \le \frac{C(m_0)}{1-\delta_0(x)}\delta|z-z'| \tag{2.48}$$

where $C(m_0)$ depends only on constants and on m_0. Therefore, considering the gradient of χ_α in both variables, we have obtained the estimate

$$|D\chi_\alpha(x,z)| \le \frac{C}{1-\delta_0(x)} \tag{2.49}$$

From this estimate, we can assert that, for a subsequence (still denoted α),

$$\sup_{\substack{x \le M \\ z \le N}} |\chi_\alpha(x,z) - \chi(x,z)| \to 0, \text{ as } \alpha \to 0, \forall M, N \tag{2.50}$$

Therefore, also

$$\sup_{\substack{x \le M \\ z}} |\alpha E[\chi_\alpha(x - z_2, z_2)\mathbf{1}_{x>z_2} - E[\chi(x - z_2, z_2)\mathbf{1}_{x>z_2}|z_1 = z]| \le$$

$$\sup_{\substack{x \le M \\ z}} |E[\chi_\alpha(x-z_2, z_2)\mathbf{1}_{x>z_2} - E[\chi(x-z_2, z_2)\mathbf{1}_{x>z_2}|z_1 = z]| + (1-\alpha)\frac{\max(h, p)}{1 - \delta_0(M)}$$

and

$$\sup_{\substack{x \le M \\ z}} |E[\chi_\alpha(x - z_2, z_2)\mathbf{1}_{x>z_2} - E[\chi(x - z_2, z_2)\mathbf{1}_{x>z_2}|z_1 = z]| \le$$

$$\sup_{\substack{x \le M \\ z \le N}} |\chi_\alpha(x, z) - \chi(x, z)| + 2\frac{\max(h, p)}{1 - \delta_0(M)}\frac{c_1}{N}$$

from which we deduce easily that

$$\sup_{\substack{x \le M \\ z}} |\alpha E[\chi_\alpha(x - z_2, z_2)\mathbf{1}_{x>z_2} - E[\chi(x - z_2, z_2)\mathbf{1}_{x>z_2}|z_1 = z]| \to 0, \forall M$$

From (2.42), it follows that

$$\chi(x, z) = g(x, z) + E[\chi(x - z_2, z_2)\mathbf{1}_{x>z_2}|z_1 = z], \ \forall x > S(z)$$

where

$$g(x, z) = h - (p + h - c)\bar{F}(x|z) \tag{2.51}$$

Also χ(x,z) = 0, if $x < S(z)$. Moreover from

$$0 = g_\alpha(S_\alpha(z), z) + \alpha E[\chi_\alpha(S_\alpha(z) - z_2, z_2)\mathbf{1}_{S_\alpha(z)>z_2}|z_1 = z]$$

and

$$|\alpha E[\chi_\alpha(S_\alpha(z) - z_2, z_2)\mathbf{1}_{S_\alpha(z)>z_2}|z_1 = z] - E[\chi(S_\alpha(z) - z_2, z_2)\mathbf{1}_{S_\alpha(z)>z_2}|z_1 = z]| \le$$

$$\sup_{\substack{x \le m_0 \\ z}} |\alpha E[\chi_\alpha(x - z_2, z_2)\mathbf{1}_{x>z_2} - E[\chi(x - z_2, z_2)\mathbf{1}_{x>z_2}|z_1 = z]|$$

$$|E[\chi(S_\alpha(z)-z_2,z_2)\mathbf{1}_{S_\alpha(z)>z_2}|z_1=z]-E[\chi(S(z)-z_2,z_2)\mathbf{1}_{S(z)>z_2}|z_1=z]|\leq$$
$$\frac{((h-c)^++p)\|f\|}{1-\delta_0(m_0)}|S_\alpha(z)-S(z)|+\frac{2\max(h,p)}{1-\delta_0(m_0)}|F(S_\alpha(z)|z)-F(S(z)|z)|$$

we obtain easily

$$0=g(S(z),z)+E[\chi((S(z)-z_2)^+,z_2)|z_1=z]$$

and the function $\chi(x,z)$ is continuous in x. Let us next set

$$\Gamma_\alpha(z)=u_\alpha(0,z)$$
$$G_\alpha(z)=E[(S_\alpha(z)-z_2)^-|z_1=z]$$

then from the first equation (2.26) one can check

$$\begin{aligned}\Gamma_\alpha(z)&=cS_\alpha(z)+pG_\alpha(z)+\alpha E[u_\alpha((S_\alpha(z)-z_2)^+,z_2)|z_1=z]\\&=\Psi_\alpha(z)+\alpha E[\Gamma_\alpha(z_2)|z_1=z]\end{aligned}$$

with

$$\Psi_\alpha(z)=cS_\alpha(z)+pG_\alpha(z)+\alpha E\left[\int_0^{(S_\alpha(z)-z_2)^+}(h-c+\chi_\alpha(\xi,z_2))d\xi|z_1=z\right]$$

and

$$0\leq\Psi_\alpha(z)\leq\left[\max(h,c)+\frac{\max(h,p)}{1-\delta(m_0)}\right]m_0+pc_1$$

This estimate also holds for the limit

$$\Psi(z)=cS(z)+pG(z)+E\left[\int_0^{(S(z)-z_2)^+}(h-c+\chi(\xi,z_2))d\xi|z_1=z\right]$$

with

$$G(z)=E[(S(z)-z_2)^-|z_1=z]$$

Define

$$\rho=\int\Psi(z)\varpi(z)dz$$

Consider now the equation

$$\Gamma(z)+\rho=\Psi(z)+E[\Gamma(z_2)|z_1=z],\quad\int\Gamma(z)\varpi(z)dz=0$$

From ergodic theory, we can assert that

$$\sup_z |\Gamma(z)| \le \sup_z |\Psi(z) - \rho| \frac{3-\beta}{1-\beta}$$

where $0 < \beta < 1$ depends only on the Markov chain. Similarly, if we set

$$\rho_\alpha = \int \Psi_\alpha(z)\varpi(z)dz, \qquad \tilde{\Gamma}_\alpha(z) = \Gamma_\alpha(z) - \frac{\rho_\alpha}{1-\alpha}$$

we can write

$$\tilde{\Gamma}_\alpha(z) + \rho_\alpha = \Psi_\alpha(z) + \alpha E[\tilde{\Gamma}_\alpha(z_2)|z_1 = z]$$

we can also assert that

$$\begin{aligned}\sup_z |\tilde{\Gamma}_\alpha(z)| &\le \sup_z |\Psi_\alpha(z) - \rho_\alpha| \frac{3-\beta}{1-\beta} \\ &\le 2\left[(\max(h,c) + \frac{\max(h,p)}{1-\delta(m_0)})m_0 + pc_1\right]\frac{3-\beta}{1-\beta}\end{aligned}$$

Moreover

$$\tilde{\Gamma}_\alpha(z) - \tilde{\Gamma}_\alpha(z') = \Psi_\alpha(z) - \Psi_\alpha(z') + \alpha \int \tilde{\Gamma}_\alpha(\zeta)(f(\zeta|z) - f(\zeta|z'))d\zeta$$

Using properties (2.41), (2.45) and the Assumption (2.29), we can check that

$$|\Psi_\alpha(z) - \Psi_\alpha(z')| \le C(m_0)\delta|z - z'|$$

and thus also

$$|\tilde{\Gamma}_\alpha(z) - \tilde{\Gamma}_\alpha(z')| \le C_1(m_0)\delta|z - z'|$$

Therefore, the functions $\tilde{\Gamma}_\alpha(z)$ are uniformly Lipschitz continuous and bounded. It follows that for a subsequence we obtain

$$\sup_{0 \le z \le N} \left|\Gamma_\alpha(z) - \frac{\rho_\alpha}{1-\alpha} - \Gamma(z)\right| \to 0, \ \forall M \tag{2.52}$$

Therefore, also

$$\sup_{\substack{0 \le x \le M \\ 0 \le z \le N}} \left|u_\alpha(x,z) - \frac{\rho_\alpha}{1-\alpha} - u(x,z)\right| \to 0, \forall x, \forall M, N \tag{2.53}$$

with

$$u(x,z) = (h-c)x + + \int_0^x \chi(\xi,z)d\xi + \Gamma(z) \tag{2.54}$$

We deduce

$$u(x,z) = (h-c)x + \Gamma(z), \forall x \leq S(z) \tag{2.55}$$

From (2.52), (2.53) it is clear that

$$\Gamma(z) = u(0,z) \tag{2.56}$$

However

$$E[u((S(z)-z_2)^+, z_2|z_1 = z] = E[u(0,z_2)|z_1 = z] + E\left[\int_0^{(S(z)-z_2)^+} (h-c+\chi(\xi,z_2))d\xi|z_1 = z\right]$$

hence

$$\Gamma(z) = -\rho + cS(z) + pG(z) + E[u((S(z)-z_2)^+, z_2|z_1 = z]$$

and thus the first relation (2.32) is proven. Consider now the situation with $x \geq S(z)$. Define the function

$$\tilde{u}(x,z) = hx + pE[(x-z_2)^-|z_1 = z] + E[u((x-z_2)^+, z_2)|z_1 = z]$$

We obtain

$$\begin{aligned} \tilde{u}'(x,z) &= h - p\bar{F}(x|z) + E[u'(x-z_2,z_2)\mathbf{1}_{x>z_2}|z_1 = z] \\ &= h - c + \chi(x,z), \quad x \geq S(z) \end{aligned}$$

Also

$$\begin{aligned} \tilde{u}(S(z),z) &= hS(z) + pE[(S(z)-z_2)^-|z_1 = z] + \Gamma(z) \\ &= u(S(z),z) + \rho \end{aligned}$$

From these two relations we get $\tilde{u}(x,z) = u(x,z) + \rho, \forall x \geq S(z)$. This concludes the second part of (2.32). Note also that

$$u_\alpha(x,z) - u_\alpha(x,z') = \Gamma_\alpha(z) - \Gamma_\alpha(z') + \int_0^x (\chi_\alpha(\xi,z) - \chi_\alpha(\xi,z')d\xi$$

Using already proven estimates we obtain

$$|u_\alpha(x,z) - u_\alpha(x,z')| \leq \left[C_0(m_0) + \frac{C_1(m_0)x}{1-\delta_0(x)}\right]\delta|z-z'|$$

The limit $u(x,z)$ satisfies all estimates (2.34), (2.36). To prove (2.37), we first check that

$$u_\alpha(x,z) \leq l(x,z,v) + \alpha E[u_\alpha((x+v-z_2)^+, z_2)|z_1 = z], \forall x,z,v$$

Therefore, it easily follows that

$$u(x,z) + \rho \leq l(x,z,v) + E[u((x+v-z_2)^+, z_2)|z_1 = z], \forall x,z,v$$

However (2.37) can be read as

$$u(x,z) + \rho = l(x,z,\hat{v}(x,z)) + E[u((x+\hat{v}(x,z)-z_2)^+, z_2)|z_1 = z], \forall x,z$$

where

$$\hat{v}(x,z) = \left| \begin{array}{ll} S(z) - x & \text{if } x \leq S(z) \\ 0 & \text{if } x \geq S(z) \end{array} \right.$$

Combining we get equation (2.37). Let us prove uniqueness, for $S(z)$ given. We first prove that $\chi(x,z)$ is uniquely defined. To prove this it is sufficient to prove that if

$$\begin{aligned} \chi(x,z) &= E[\chi(x-z_2,z_2)\mathbf{1}_{x>z_2}|z_1 = z], \ \forall x > S(z) \\ &= 0, \ \forall x \leq S(z) \end{aligned}$$

and

$$(1-\delta_0(x)) \sup_{\substack{0 \leq \xi \leq x \\ z}} |\chi(\xi,z)| < \infty$$

then $\chi(x,z)=0$. This is clear. The function $\Psi(z)$ is thus uniquely defined. It follows that the pair ρ, $\Gamma(z)=u(0,z)$ is also uniquely defined, with the condition $\int \Gamma(z)\varpi(z)dz=0$. Therefore, $u(x,z)$ is also uniquely defined. The proof of the theorem has been completed. □

Example 2.1. Consider the situation of independent demands, then $f(x|z) = f(x)$. In that case $\varpi(x) = f(x)$. Then $S(z) = S$, and

$$\rho = (p+h-c)(S-D)^+ - (p-c)S + p\bar{D}$$

We see also that $\Psi(z) = \rho$, and thus $\Gamma(z) = 0$. □

Consider next the situation

$$f(x|z) = \lambda(z)\exp -\lambda(z)x$$

with the assumption $0 < \lambda_0 \leq \lambda(z) \leq \lambda_1$. We also assume that $\lambda(z)$ is Lipschitz continuous. Then all assumptions of Theorem 2.3 are satisfied. □

We turn to the interpretation of the number ρ. Consider the feedback $\hat{v}(x,z)$ associated with the base stock $S(z)$. Define the controlled process

$$\hat{y}_{n+1} = (\hat{y}_n + \hat{v}_n - z_{n+1})^+, \ \hat{v}_n = \hat{v}(\hat{y}_n, z_n)$$

with $\hat{y}_1 = x, \ z_1 = z$. We define the policy $\hat{V} = (\hat{v}_1, \cdots, \hat{v}_n, \cdots)$. We consider the averaged cost

$$J^n_{x,z}(\hat{V}) = \frac{\sum_{j=1}^n El(\hat{y}_j, z_j, \hat{v}_j)}{n}$$

Similarly, for any policy $V = (v_1, \cdots, v_n \cdots)$ we define the averaged cost

$$J^n_{x,z}(V) = \frac{\sum_{j=1}^n El(y_j, z_j, v_j)}{n}$$

with

$$y_{n+1} = (y_n + v_n - z_{n+1})^+, \quad y_1 = x, \ z_1 = z$$

We state

Proposition 2.1. *We have the property*

$$\rho = \lim_{n \to \infty} J^n_{x,z}(\hat{V}) \tag{2.57}$$

Furthermore, consider the set of policies

$$\mathscr{U} = \{V \mid |u(y_n, z_n)| \leq C_x\}$$

then we have

$$\rho = \inf_{V \in \mathscr{U}} \lim_{n \to \infty} J^n_{x,z}(V) \tag{2.58}$$

Proof. We first notice that

$$\hat{y}_n \leq \max(x, m_0)$$

Therefore, from estimate (2.33), we get

$$|u(\hat{y}_n, z_n)| \leq C_0 + \frac{C_1 \max(x, m_0)}{1 - \delta_0(\max(x, m_0))}$$

Therefore, $\hat{V}$ belongs to $\mathscr{U}$. From (2.37), we can write

$$u(\hat{y}_n, z_n) + \rho = l(\hat{y}_n, z_n, \hat{v}_n) + E[u(\hat{y}_{n+1}, z_{n+1}) | \hat{y}_n, z_n]$$

Taking the expectation and adding up we obtain

$$\rho = J^n_{x,z}(\hat{V}) + \frac{Eu(\hat{y}_{n+1}, z_{n+1}) - u(x,z)}{n}$$

and thus the property (2.57) follows immediately. Similarly, for any policy we can write

$$\rho \leq J^n_{x,z}(V) + \frac{Eu(y_{n+1}, z_{n+1}) - u(x,z)}{n}$$

Therefore, if $V \in \mathscr{U}$, we have $\rho \leq J^n_{x,z}(V)$. This implies (2.58). The proof has been completed. □

Remark 2.2. We cannot state that the process $\hat{y}_n, z_n$ is ergodic. Consequently, we cannot give an interpretation of the function $u(x,z)$ itself.

2.3 No Backlog and Set-Up Cost

2.3.1 *Model*

We now study the situation of set-up cost, and we consider the no shortage model. We have to study the Bellman equation

$$u(x,z) = \inf_{v \geq 0}[K\mathbf{1}_{v>0} + l(x,z,v) + \alpha \Phi^v u(x,z)] \tag{2.59}$$

where

$$\Phi^v \varphi(x,z) = E[\varphi((x+v-z_2)^+, z_2)|z_1 = z] \tag{2.60}$$

and

$$f(\zeta|z) \text{ is uniformly continuous in both variables and bounded} \tag{2.61}$$

$$\int_0^{+\infty} \zeta f(\zeta|z) d\zeta \leq c_0 z + c_1 \tag{2.62}$$

$$l(x,z,v) = cv + hx + pE[(x+v-z_2)^-|z_1 = z] \tag{2.63}$$

We look for solutions of (2.59) in the interval $[0, w_0(x,z)]$ with

$$w_0(x,z) = l_0(x,z) + \alpha E[w_0(x-z_2, z_2)|z_1 = z] \tag{2.64}$$

In Lemma 2.1, we have proven the estimate

$$w_0(x,z) \le \frac{hx}{1-\alpha} + \frac{pc_0 z}{1-c_0\alpha} + \frac{pc_1}{(1-\alpha)(1-c_0\alpha)} \tag{2.65}$$

with the assumption

$$c_0\alpha < 1 \tag{2.66}$$

As usual we consider the payoff function

$$J_{x,z}(V) = E\sum_{n=1}^{\infty} \alpha^{n-1}[K\mathbf{1}_{v_n>0} + l(y_n, z_n, v_n)] \tag{2.67}$$

with $V = (v_1, \cdots, v_n, \cdots)$ adapted process with positive values and

$$y_{n+1} = (y_n + v_n - z_{n+1})^+, \; y_1 = x \tag{2.68}$$

We define the value function

$$u(x,z) = \inf_V J_{x,z}(V) \tag{2.69}$$

We state

Theorem 2.4. *We assume* (2.60)–(2.63), (2.66). *The value function defined in* (2.69) *is the unique l.s.c. solution of the Bellman equation* (2.59) *in the interval* $[0, w_0]$. *There exists an optimal feedback* $\hat{v}(x,z)$.

2.3.2 $s(z), S(z)$ Policy

We now prove the following result.

Theorem 2.5. *We make the assumptions of Theorem 2.4 and* $p > c$. *Then the function* $u(x,z)$ *is* K*-convex and continuous. It tends to* $+\infty$ *as* $x \to +\infty$. *Considering the numbers* $s(z), S(z)$ *associated to* $u(x,z)$, *the optimal feedback is given by*

$$\hat{v}(x,z) = \left| \begin{array}{ll} S(z) - x & \text{if } x \le s(z) \\ 0 & \text{if } x > s(z) \end{array} \right. \tag{2.70}$$

The functions $s(z)$ $S(z)$ *are continuous.*

Proof. We consider the increasing sequence

$$u_{n+1}(x,z) = \inf_{v\ge 0}\{K\mathbf{1}_{v>0} + cv + hx + pE[(x+v-z_2)^-|z_1 = z] + \alpha E[u_n((x+v-z_2)^+, z_2)|z_1 = z]\} \tag{2.71}$$

with $u_0(x,z) = 0$. Define

$$G_n(x,z) = cx + pE[(x - z_2)^-|z_1 = z] + \alpha E[u_n((x - z_2)^+, z_2)|z_1 = z] \quad (2.72)$$

We can write

$$u_{n+1}(x,z) = (h - c)x + \inf_{\eta \geq x}[K\mathbf{1}_{\eta \geq x} + G_n(\eta, z)] \quad (2.73)$$

We are going to show, by induction, that both $u_n(x,z), G_n(x,z)$ are $K-$convex in x, continuous, and $\rightarrow +\infty$ as $x \rightarrow +\infty$, for $n \geq 1$. The properties are clear for $n = 1$. We assume they are verified for n, we prove them for $n + 1$. Since $G_n(x,z)$ is $K-$convex in x, continuous, and $\rightarrow +\infty$ as $x \rightarrow +\infty$, we can define $s_n(z)$, $S_n(z)$ with

$$G_n(S_n(z), z) = \inf_\eta G_n(\eta, z) \quad (2.74)$$

$$s_n(z) = \left| \begin{array}{ll} 0 \quad \text{if} & G_n(0,z) \leq K + \inf_\eta G_n(\eta, z) \\ G_n(s_n(z), z) = K + \inf_\eta G_n(\eta, z) \text{ if} & G_n(0,z) > K + \inf_\eta G_n(\eta, z) \end{array} \right. \quad (2.75)$$

As usual we take the smallest minimum to define $S_n(z)$in a unique way. Since $G_n(\eta, z)$ is continuous, it is easy to check that $S_n(z)$ is continuous. Also $s_n(z)$ is continuous. We can write

$$u_{n+1}(x,z) = (h - c)x + G_n(\max(x, s_n(z)), z)$$

which shows immediately that $u_{n+1}(x,z)$ is K-convex and continuous. Furthermore $u_{n+1}(x,z) \rightarrow +\infty$, as $x \rightarrow +\infty$. We then have

$$\begin{aligned} G_{n+1}(x,z) = cx + pE[(x - z_2)^-|z_1 = z] + \alpha(h - c)E[(x - z_2)^+|z_1 = z] + \\ + \alpha E[G_n(\max(x - z_2, s_n(z_2)), z_2)|z_1 = z] \end{aligned}$$

It is the sum of a convex function and a K-convex function, hence $K-$convex. It is continuous and $\rightarrow +\infty$ as $x \rightarrow +\infty$. So the recurrence is proven. If we write (formally)

$$u_n'(x,z) = h - c + \chi_n(x,z)$$

then one has the recurrence

$$\begin{aligned} \chi_{n+1}(x,z) &= 0 \quad \text{if} \quad x < s_n(z) \\ &= \mu(x,z) + \alpha E[\chi_n(x - z_2, z_2)|z_1 = z] \end{aligned}$$

with

$$\mu(x,z) = c + \alpha(h - c) - (p + \alpha(h - c))\bar{F}(x|z) \quad (2.76)$$

The function $\chi_{n+1}(x, z)$ is discontinuous in $s_n(z)$. However one has the bound

$$-\frac{(p-c)}{1-\alpha} \leq \chi_{n+1}(x, z) \leq \frac{c+\alpha(h-c)}{1-\alpha} \quad \text{a.e.}$$

Therefore, the limit $u(x, z)$ is K-convex and satisfies

$$u'(x, z) = h - c + \chi(x, z)$$

with

$$-\frac{(p-c)}{1-\alpha} \leq \chi(x, z) \leq \frac{c+\alpha(h-c)}{1-\alpha} \quad \text{a.e.}$$

Hence $u(x, z)$ is continuous in x and $\to +\infty$ as $x \to +\infty$. Therefore, one defines uniquely $s(z), S(z)$ where $S(z)$ minimizes in x the function

$$G(x, z) = cx + pE[(x - z_2)^- | z_1 = z] + \alpha E[u((x - z_2)^+, z_2) | z_1 = z]$$

From this formula and the Lipschitz continuity of u in x, using the Assumption (2.61) one can see that $G(x, z)$ is continuous in both arguments. Hence $S(z)$ and $s(z)$ are continuous. The proof has been completed. □

2.3.3 Ergodic Theory

We now study the behavior of $u(x, z)$ as $\alpha \to +\infty$. We denote it by $u_\alpha(x, z)$ and we write the relations

$$u_\alpha(x, z) = (h - c)x + G_\alpha(\max(x, s_\alpha(z)), z) \tag{2.77}$$

$$G_\alpha(x, z) = g_\alpha(x, z) + \alpha E[G_\alpha(\max((x - z_2)^+, s_\alpha(z_2)), z_2) | z_1 = z] \tag{2.78}$$

with

$$g_\alpha(x, z) = cx + pE[(x - z_2)^- | z_1 = z] + \alpha(h - c)E[(x - z_2)^+ | z_1 = z] \tag{2.79}$$

We will use an approach different from that of the base stock case, since we cannot prove uniform Lipschitz properties for the function $s_\alpha(z)$. The present method will use less assumptions. We shall assume $c_0 = 0$ and

$$z_n \quad \text{is ergodic} \tag{2.80}$$

We denote by $\varpi(z)$ the invariant measure. We also assume

$$\int |f(\zeta|z) - f(\zeta|z')| d\zeta \leq \delta |z - z'|, \tag{2.81}$$

$$\int \zeta|f(\zeta|z) - f(\zeta|z')|d\zeta \leq \delta|z - z'|$$

$$\sup_z F(x|z) = \delta_0(x) < 1, \forall x \tag{2.82}$$

$$\sup_z \bar{F}(x|z) \to 0 \text{ as } x \to +\infty \tag{2.83}$$

We begin with

Lemma 2.2. *If* $s(z) > 0$, *then*

$$S(z) \leq \frac{p + \alpha(h - c)}{c + \alpha(h - c)} E[z_2|z]$$

Proof. If $s(z) > 0$ then we have

$$u(0, z) = K + G(S(z), z)$$

Set

$$\hat{y}_2 = (S(z) - z_2)^+, \quad \hat{v}_2 = \hat{v}(\hat{y}_2, z_2)$$

then

$$G(S(z), z) = cS(z) + pE[(S(z) - z_2)^-|z_1 = z] + \alpha E[u(\hat{y}_2, z_2)|z_1 = z]$$

Also we can write

$$u(\hat{y}_2, z_2) = (h - c)\hat{y}_2 + K\mathbf{1}_{\hat{v}_2>0} + G(\hat{y}_2 + \hat{v}_2, z_2)$$

$$= h\hat{y}_2 + K\mathbf{1}_{\hat{v}_2>0} + c\hat{v}_2 + pE[(\hat{y}_2 + \hat{v}_2 - z_3)^-|z_2] + \alpha E[u((\hat{y}_2 + \hat{v}_2 - z_3)^+, z_3)|z_2]$$

Therefore, we can write

$$u(0, z) = K + cS(z) + pE[(S(z) - z_2)^-|z_1 = z]+$$

$$+\alpha E[h\hat{y}_2 + K\mathbf{1}_{\hat{v}_2>0} + c\hat{v}_2 + p(\hat{y}_2 + \hat{v}_2 - z_3)^- + +\alpha u((\hat{y}_2 + \hat{v}_2 - z_3)^+, z_3)|z_1 = z] \tag{2.84}$$

We next have

$$u(0, z) \leq G(0, z)$$
$$= E[pz_2 + \alpha u(0, z_2)|z_1 = z]$$

Furthermore

$$u(0, z_2) \leq K + G(\hat{y}_2 + \hat{v}_2, z_2)$$

Replacing G and combining the two inequalities we get

$$u(0,z) \leq pE[z_2|z_1 = z] + \alpha K +$$
$$+ \alpha E[c(\hat{y}_2 + \hat{v}_2) + p(\hat{y}_2 + \hat{v}_2 - z_3)^- + \alpha u((\hat{y}_2 + \hat{v}_2 - z_3)^+, z_3)|z_1 = z]$$

Comparing with (2.84) we obtain easily the desired inequality. □

We deduce from the lemma that, whenever $c_0 = 0$

$$s_\alpha(z) \leq \frac{p + (h-c)^+}{\min(c,h)} c_1 \tag{2.85}$$

We now state

Theorem 2.6. *We assume* (2.61), (2.62), *with* $c_0 = 0$, (2.80)–(2.82). *Then there exists a number* ρ_α *such that, for a subsequence, still denoted* $\alpha \to 1$

$$\sup_{\substack{x \leq M \\ z}} |u_\alpha(x,z) - \frac{\rho_\alpha}{1-\alpha} - u(x,z)| \to 0, \forall M \tag{2.86}$$

and $\rho_\alpha \to \rho$. *The function* $u(x,z)$ *is Lipschitz continuous,* K*-convex and satisfies*

$$|u(x,z)| \leq C_0 + \frac{C_1 x}{1 - \delta_0(x)} \tag{2.87}$$

$$|u_x(x,z)| \leq \frac{C}{1-\delta_0(x)}, \quad |u_z(x,z)| \leq C_0 + \frac{C_1 x}{1-\delta_0(x)} \ a.e. \tag{2.88}$$

The pair $u(x,z), \rho$ *is the solution of*

$$u(x,z) + \rho = \inf_{v \geq 0}[K \mathbf{1}_{v>0} + l(x,z,v) + \Phi^v u(x,z)] \tag{2.89}$$

Proof. We set

$$\chi_\alpha(x,z) = u'_\alpha(x,z) - h + c$$

then we can write

$$\chi_\alpha(x,z) = \mathbf{1}_{x>s_\alpha(z)}\{\mu_\alpha(x,z) + \alpha E[\chi_\alpha(x-z_2,z_2)\mathbf{1}_{x-z_2>0}|z_1 = z] \tag{2.90}$$

with

$$\mu_\alpha(x,z) = g'_\alpha(x,z)$$
$$= c + \alpha(h-c) - (p + \alpha(h-c))\bar{F}(x|z)$$

The function $\chi_\alpha(x,z)$ is not continuous, but satisfies

$$\sup_{\substack{0\leq\xi\leq x\\ z}} |\chi_\alpha(\xi,z)| \leq \frac{\max(h,p)}{1-\delta_0(x)} \tag{2.91}$$

Let us next define $\Gamma_\alpha(z) = u_\alpha(0,z)$. We have

$$\begin{aligned}\Gamma_\alpha(z) &= G_\alpha(s_\alpha(z),z)\\ &= cs_\alpha(z) + pE[(s_\alpha(z)-z_2)^-|z_1=z] + \alpha E[u_\alpha((s_\alpha(z)-z_2)^+,z_2)|z_1=z]\\ &= g_\alpha(s_\alpha(z),z) + \alpha E\left[\int_0^{(s_\alpha(z)-z_2)^+} \chi_\alpha(\xi,z_2)d\xi|z_1=z\right] + \alpha E[\Gamma_\alpha(z_2)|z_1=z]\end{aligned}$$

Therefore, we can write

$$\Gamma_\alpha(z) = \Psi_\alpha(z) + \alpha E[\Gamma_\alpha(z_2)|z_1=z] \tag{2.92}$$

with

$$\Psi_\alpha(z) = g_\alpha(s_\alpha(z),z) + \alpha E\left[\int_0^{(s_\alpha(z)-z_2)^+} \chi_\alpha(\xi,z_2)d\xi|z_1=z\right]$$

Thanks to Lemma 2.2, we have estimate (2.85), so $s_\alpha(z) \leq m_0$. Using estimate (2.91), we see that $|\Psi_\alpha(z)| \leq C$. Let us then define

$$\rho_\alpha = \int \Psi_\alpha(z)\varpi(z)dz$$

and $\tilde{\Gamma}_\alpha(z) = \Gamma_\alpha(z) - \dfrac{\rho_\alpha}{1-\alpha}$. From ergodic theory, we can assert that

$$\sup_z |\tilde{\Gamma}_\alpha(z)| \leq \sup_z |\Psi_\alpha(z) - \rho_\alpha| \frac{3-\beta}{1-\beta} \leq C$$

and we can state the estimate

$$|u_\alpha(x,z) - \frac{\rho_\alpha}{1-\alpha}| \leq C + \frac{Cx}{1-\delta_0(x)}$$

Define $\tilde{u}_\alpha(x,z) = u_\alpha(x,z) - \dfrac{\rho_\alpha}{1-\alpha}$. We write equation (2.59) as

$$\tilde{u}_\alpha(x,z) + \rho_\alpha = \inf_{v\geq 0}[K\mathbf{1}_{v>0} + l(x,z,v) + \alpha\Phi^v\tilde{u}_\alpha(x,z)] \tag{2.93}$$

From Lemma 2.2 we can assert that the optimal feedback satisfies $x+\hat{v}_\alpha(x,z) \leq \max(x,m_0)$. Therefore, we can replace (2.93) by

$$\begin{aligned}\tilde{u}_\alpha(x,z)+\rho_\alpha &= \inf_{x+v\leq\max(x,m_0)}[K\mathbf{1}_{v>0}+l(x,z,v)+\alpha\Phi^v\tilde{u}_\alpha(x,z)]\\ &= \inf_{0\leq v\leq\max(x,m_0)-x} L_\alpha(x,z,v)\end{aligned}$$

Next

$$L_\alpha(x,z,v)-L_\alpha(x,z',v)=\int[p(x+v-\zeta)^-+\alpha\tilde{u}_\alpha((x+v-\zeta)^+,\zeta)](f(\zeta|z)-f(\zeta|z'))d\zeta$$

For $v \leq \max(x,m_0)-x$ we can write

$$\begin{aligned}|\tilde{u}_\alpha((x+v-\zeta)^+,\zeta)| &\leq C+\frac{C\max(x,m_0)}{1-\delta_0(\max(x,m_0))}\\ &\leq C_1'+\frac{C_2x}{1-\delta_0(x)}\end{aligned}$$

Using the Assumption (2.81) we get easily

$$|\tilde{u}_\alpha(x,z)-\tilde{u}_\alpha(x,z')| \leq \left(C_1+\frac{C_2x}{1-\delta_0(x)}\right)\delta|z-z'|$$

From the estimates obtained we can assert that $\tilde{u}_\alpha(x,z)$ has a converging subsequence (still denoted α) in the sense

$$\sup_{\substack{x\leq M\\ z\leq N}} |\tilde{u}_\alpha(x,z)-u(x,z)| \to 0, \text{ as } \alpha\to 0 \tag{2.94}$$

Since ρ_α is bounded, we can always assume that $\rho_\alpha \to \rho$. Denote

$$L(x,z,v)=K\mathbf{1}_{v>0}+l(x,z,v)+\Phi^v u(x,z)$$

then

$$L_\alpha(x,z,v)-L(x,z,v)=\alpha\Phi^v(\tilde{u}_\alpha-u)(x,z)-(1-\alpha)\Phi^v u(x,z)$$

For $v \leq \max(x,m_0)-x$, we have assuming $M>m_0$

$$\sup_{\substack{x\leq M\\ z}} |\Phi^v u(x,z)| \leq C+\frac{CM}{1-\delta_0(M)}$$

We also have

$$\sup_{\substack{x \le M \\ z}} |\Phi^v(\tilde{u}_\alpha - u)(x,z)| \le \sup_{\substack{x \le M \\ z}} |\tilde{u}_\alpha - u|(x,z) \sup_z \bar{F}(N,z) + $$
$$+ \sup_{\substack{x \le M \\ z \le N}} |\tilde{u}_\alpha - u|(x,z)$$

Using the Assumption (2.83) and (2.94), letting first $\alpha \to 1$, then $N \to \infty$, we deduce

$$\sup_{\substack{x \le M \\ v \le \max(x, m_0) - x \\ z}} |L_\alpha(x,z,v) - L(x,z,v)| \to 0, \text{ as } \alpha \to 1$$

Therefore we deduce easily (2.86) and also that the pair $u(x,z), \rho$ satisfies (2.89). Estimates (2.87), (2.88) follow immediately from the corresponding ones on $\tilde{u}_\alpha(x,z)$. The $K-$convexity of u follows from the $K-$convexity of $\tilde{u}_\alpha$. The proof has been completed. □

References

1. Beyer, D., Cheng, F., Sethi, S.P., Taksar M.: Markovian Demand Inventory Models. Springer, Berlin (2009)
2. Porteus, E.L.: Foundations of Stochastic Inventory Theory. Stanford University Press, Stanford (2002)
3. Scarf, H.: The optimality of (s, S) policies in the dynamic inventory problem. In: Arrow, K., Karlin, S., Suppes, P. (eds.) Mathematical Methods in the Social Sciences, chap. 13, pp. 196–202. Stanford University Press, Stanford (1960)

Chapter 3
On the Splitting Method for Some Complex-Valued Quasilinear Evolution Equations

Zdzisław Brzeźniak and Annie Millet

Abstract Using the approach of the splitting method developed by I. Gyöngy and N. Krylov for parabolic quasilinear equations, we study the speed of convergence for general complex-valued stochastic evolution equations. The approximation is given in general Sobolev spaces and the model considered here contains both the parabolic quasi-linear equations under some (non-strict) stochastic parabolicity condition as well as semi-linear Schrödinger equations.

3.1 Introduction

Once the well-posedness of a stochastic differential equation is proved, an important issue is to provide an efficient way to approximate the unique solution. The aim of this paper is to propose a fast converging scheme which gives a simulation of the trajectories of the solution on a discrete time grid and in terms of some spatial approximation. The first results in this direction were obtained for stochastic differential equations and it is well known that the limit is sensitive to the approximation. For example, the Stratonovich integral is the limit of Riemann sums with the mid-point approximation and the Wong Zakai approximation also leads to Stratonovich stochastic integrals and not to the Itô ones in this finite-dimensional framework.

Z. Brzeźniak
Department of Mathematics, University of York, Heslington, York YO10 5DD, UK
e-mail: zb500@york.ac.uk

A. Millet (✉)
SAMM, EA 4543, Université Paris 1 Panthéon Sorbonne, 90 Rue de Tolbiac, 75634 Paris Cedex 13, France

Laboratoire PMA (CNRS UMR 7599), Universités Paris 6-Paris 7, Boîte Courrier 188, 4 place Jussieu, 75252 Paris Cedex 05, France
e-mail: amillet@univ-paris1.fr; annie.millet@upmc.fr

L. Decreusefond and J. Najim (eds.), *Stochastic Analysis and Related Topics*, Springer Proceedings in Mathematics & Statistics 22, DOI 10.1007/978-3-642-29982-7_3,

There is a huge literature on this topic for stochastic PDEs, mainly extending classical deterministic PDE methods to the stochastic framework. Most of the papers deal with parabolic PDEs and take advantage of the smoothing effect of the second-order operator; see, e.g., [7,8,10–12,16,19] and the references therein. The methods used in these papers are explicit, implicit or Crank–Nicolson time approximations and the space discretization is made in terms of finite differences, finite elements or wavelets. The corresponding speeds of convergence are the "strong" ones, which is uniform in time on some bounded interval $[0, T]$ and with various functional norms for the space variable. Some papers also study numerical schemes in other "hyperbolic" situations, such as the KDV or Schrödinger equations as in [2–4]. Let us also mention the weak speed of convergence, which is of an approximation of the expected value of a functional of the solution by the similar one for the scheme obtained by [3,5]. These references extend to the infinite-dimensional setting, a very crucial problem for finite-dimensional diffusion processes.

Another popular approach in the deterministic setting, based on semi-group theory, is the splitting method which solves successively several evolution equations. This technique has been used in a stochastic case in a series of papers by Gyöngy and Krylov. Let us especially mention reference [9] which uses tools from [14, 18] and [13] and provides a very elegant approach to study quasilinear evolution equations under (non-strict) stochastic parabolicity conditions. In their framework, the smoothing effect of the second operator is exactly balanced by the quadratic variation of the stochastic integrals, which implies that there is no increase of space regularity with respect to that of the initial condition. Depending on the number of steps of the splitting, the speed of convergence is at least twice that of the classical finite differences or finite element methods. A series of papers has been using the splitting technique in the linear and non-linear cases for the deterministic Schrödinger equation; see, e.g., [1, 6, 17] and the references therein.

The stochastic Schrödinger equation studies complex-valued processes where the second-order operator $i\Delta$ does not improve (nor decay) the space regularity of the solution with respect to that of the initial condition. Well-posedness of this equation has been proven in a non-linear setting by de Bouard and Debussche [3]; these authors have also studied finite element discretization schemes for the corresponding solution under conditions stronger than that in [2].

The aim of this paper is to transpose the approach from [9] to general quasilinear complex-valued equations including both the "classical degenerate" parabolic setting as well as the quasilinear Schödinger equation. Indeed, the method used in [9] consists in replacing the usual splitting via semi-group arguments by the study of pth moments of $Z^0 - Z^1$ where Z^0 and Z^1 are solutions of two stochastic evolution equations with the same driving noise and different families of increasing processes V_0^r and V_1^r for $r = 0, 1, \dots, d_1$ driving the drift term. It does not extend easily to non-linear drift terms because it is based on some linear interpolation between the two cases V_0^r and V_1^r. Instead of getting an upper estimate of the pth moment of $Z^0 - Z^1$ in terms of the total variation of the measures defined by the differences $V_0^r - V_1^r$, using integration by parts, they obtain an upper estimate in terms of the sup norm of the process $(V_0^r(t) - V_1^r(t), t \in [0,T])$.

We extend this model as follows: given second-order linear differential operators $L^r, r = 0, \dots, d_1$ with complex coefficients, a finite number of sequences of first-order linear operators $S^l, l \geq 1$ with complex coefficients, a sequence of real-valued martingales M^l, $l \geq 1$ and a finite number of families of real-valued increasing processes V_i^r, $i \in \{0,1\}$, $r = 0, \dots, d_1$, we consider the following system of stochastic evolution equations:

$$dZ_i(t) = \sum_r L_r(t,\cdot)Z_i(t)dV_i^r(t) + \sum_l S_l(t,\cdot)Z_i(t)dM^l(t), \quad i = 1, \dots, d,$$

with an initial condition $Z_i(0)$ belonging to the Sobolev space $H^{m,2}$ for a certain $m \geq 0$. Then under proper assumptions on the various coefficients and processes, under which a stochastic parabolicity condition (see Assumptions **(A1)–(A4(m,p))** in Sect. 3.2), we prove that for $p \in [2, \infty)$, we have

$$\mathbb{E}\Big(\sup_{t\in[0,T]} \|Z_1(t) - Z_0(t)\|_m^p\Big) \leq C\Big(\mathbb{E}\|Z_1(0) - Z_0(0)\|_m^p + A^p\Big), \tag{3.1}$$

where $A = \sup_\omega \sup_{t\in[0,T]} \max_r |V_1^r(t) - V_0^r(t)|$. When the operator $L_r = i\Delta + \tilde{L}^r$ for certain first-order differential operator $\tilde{L}_r$, we obtain the quasilinear Schrödinger equation. Note that in this case, the diffusion operators S^l are linear and cannot contain first-order derivatives.

As in [9], this abstract result yields the speed of convergence of the following splitting method. Let $\tau_n = \{iT/n, i = 0, \dots, n\}$ denote a time grid on $[0,T]$ with constant mesh $\delta = T/n$ and define the increasing processes $A_t(n)$ and $B_t(n) = A_{t+\delta}(n)$, where

$$A_t(n) = \begin{cases} k\delta & \text{for } t \in [2k\delta, (2k+1)\delta], \\ t - (k+1)\delta & \text{for } t \in [(2k+1)\delta, (2k+2)\delta]. \end{cases}$$

Given a time-independent second-order differential operator L, first-order time-independent operators S^l and a sequence $(W^l, l \geq 1)$ of independent one-dimensional Brownian motions, let Z, Z_n and ζ_n be solutions to the evolution equations

$$dZ(t) = LZ(t)dt + \sum_l S_l Z(t) \circ dW_t^l,$$

$$dZ_n(t) = LZ_n(t)dA_t(n) + \sum_l S_l Z_n(t) \circ dW_{B_t(n)}^l,$$

$$d\zeta_n(t) = L\zeta_n(t)dB_t(n) + \sum_l S_l(t,\cdot)\zeta_n(t) \circ dW_{B_t(n)}^l,$$

where $\circ d W_t^l$ denotes the Stratonovich integral. The Stratonovich integral is known to be the right one to ensure stochastic parabolicity when the differential operators S_l contain first-order partial derivatives (see, e.g., [9]). Then $\zeta_n(2k\delta, x) = Z(k\delta, x)$, while the values of $Z_n(2k\delta, x)$ are those of the process $\tilde{Z}_n$ obtained by the following spitting method: one solves successively the correction and prediction equations on each time interval $[iT/n, (i+1)T/n)]$: $dv_t = Lv_t dt$ and then $d\tilde{v}_t = \sum_l S^l(t, \cdot)\tilde{v}_t \circ dW_t^l$. Then, one has $A = CT/n$, and we deduce that $\mathbb{E}\Big(\sup_{t\in[0;T]} \|Z(t) - \tilde{Z}_n(t)\|_m^p\Big) \le Cn^{-p}$. As in [9], a k-step splitting would yield a rate of convergence of Cn^{-kp}.

The paper is organized as follows. Section 3.2 states the model, describes the evolution equation, proves well-posedness as well as a priori estimates. In Sect. 3.3 we prove (3.1) first in the case of time-independent coefficients of the differential operators, then in the general case under more regularity conditions. As explained above, in Sect. 3.4 we deduce the rate of convergence of the splitting method for evolution equations generalizing the quasilinear Schrödinger equation. The speed of convergence of the non-linear Schrödinger equation will be addressed in a forthcoming paper. As usual, unless specified otherwise, we will denote by C a constant which may change from one line to the next.

3.2 Well-Posedness and First A Priori Estimates

3.2.1 Well-Posedness Results

Fix $T > 0$, $\mathbb{F} = (\mathscr{F}_t, t \in [0, T])$ be a filtration on the probability space $(\Omega, \mathscr{F}, \mathbf{P})$ and consider the following $\mathbb{C}$-valued evolution equation on the process $Z(t, x) = X(t, x) + iY(t, x)$ defined for $t \in [0, T]$ and $x \in \mathbb{R}^d$:

$$dZ(t,x) = \sum_{r=0}^{d_1} \big[L_r Z(t,x) + F_r(t,x)\big]\, dV_t^r + \sum_{l\ge 1} \big[S_l Z(t,x) + G_l(t,x)\big] dM_t^l, \tag{3.2}$$

$$Z(0,x) = Z_0(x) = X_0(x) + iY_0(x), \tag{3.3}$$

where d_1 is a positive integer, $(V_t^r, t \in [0, T])$, $r = 0, 1, \ldots, d_1$ are real-valued increasing processes, $\big(M_t^l, t \in [0, T]\big)$, $l \ge 1$, are independent real-valued $(\mathscr{F}_t, t \in [0, T])$-martingales, L_r (resp. S_l) are second (resp. first) order differential operators defined as follows:

$$L_r Z(t,x) = \sum_{j,k=1}^{d} D_k\Big(\big[a_r^{j,k}(t,x) + i b_r^{j,k}(t)\big] D_j Z(t,x)\Big) + \sum_{j=1}^{d} a_r^j(t,x) D_j Z(t,x) \\ + \big[a_r(t,x) + i b_r(t,x)\big] Z(t,x), \tag{3.4}$$

$$S_l Z(t,x) = \sum_{j=1}^d \sigma_l^j(t,x) D_j Z(t,x) + \big[\sigma_l(t,x) + i\tau_l(t,x)\big] Z(t,x). \tag{3.5}$$

Let $m \geq 0$ be an integer. Given $\mathbb{C}$-valued functions $Z(.) = X(.) + iY(.)$ and $\zeta(.) = \xi(.) + i\eta(.)$ which belong to $L^2(\mathbb{R}^d)$, let

$$(Z,\zeta) = (Z,\zeta)_0 := \int_{\mathbb{R}^d} Re\big(Z(x)\overline{\zeta(x)}\big)dx = \int_{\mathbb{R}^d} \big[X(x)\xi(x) + Y(x)\eta(x)\big]dx.$$

Thus, we have $(X,\xi) = \int_{\mathbb{R}^d} X(x)\,\xi(x)\,dx$, so that $(Z,\zeta) = (X,\xi) + (Y,\eta)$. For any multi-index $\alpha = (\alpha_1, \alpha_2, \ldots, \alpha_d)$ with non-negative integer components α_i, set $|\alpha| = \sum_j \alpha_j$ and for a regular enough function $f : \mathbb{R}^d \to \mathbb{R}$, let $D^\alpha f$ denote the partial derivative $(\frac{\partial}{\partial x_1})^{\alpha_1} \cdots (\frac{\partial}{\partial x_d})^{\alpha_d}(f)$. For $k = 1, \ldots, d$, let $D_k f$ denote the partial derivative $\frac{\partial f}{\partial x_k}$. For a $\mathbb{C}$-valued function $F = F_1 + iF_2$ defined on $\mathbb{R}^d$, let $D^\alpha F = D^\alpha F_1 + iD^\alpha F_2$ and $D_k F = D_k F_1 + iD_k F_2$. Finally, given a positive integer m, say that $F \in H^m$ if and only if F_1 and F_2 belong to the (usual) real Sobolev space $H^m = H^{m,2}$. Finally, given $Z = X + iY$ and $\zeta = \xi + i\eta$ which belong to H^m, set

$$\begin{aligned}(Z,\zeta)_m &= \sum_{\alpha:0\leq|\alpha|\leq m} \int_{\mathbb{R}^d} Re\big(D^\alpha Z(x)\overline{D^\alpha\zeta(x)}\big)dx \\ &= \sum_{\alpha:0\leq|\alpha|\leq m} \int_{\mathbb{R}^d} \big[D^\alpha X(x)D^\alpha\xi(x) + D^\alpha Y(x)D^\alpha\eta(x)\big]dx,\end{aligned} \tag{3.6}$$

$$\|Z\|_m^2 = (Z,Z)_m = \sum_{\alpha:|\alpha|\leq m} \int_{\mathbb{R}^d} Re\big(D^\alpha Z(x)\overline{D^\alpha Z(x)}\big)dx. \tag{3.7}$$

We suppose that the following assumptions are satisfied:

Assumption (A1) For $r = 0, \ldots, d_1$, $(V_t^r, t \in [0,T])$ are predictable increasing processes. There exists a positive constant $\tilde{K}$ and an increasing predictable process $(V_t, t \in [0,T])$ such that:

$$V_0 = V_0^r = 0,\ r = 0, \ldots, d_1, \quad V_T \leq \tilde{K} \text{ a.s.}, \tag{3.8}$$

$$\sum_{r=0}^{d_1} dV_t^r + \sum_{l\geq 1} d\langle M^l\rangle_t \leq dV_t \quad \text{a.s. in the sense of measures.} \tag{3.9}$$

Assumption (A2)

(i) For $r = 0, 1, \ldots, d_1$, the matrices $(a_r^{j,k}(t,x), j,k = 1, \ldots, d)$ and $(b_r^{j,k}(t)$, $j,k = 1, \ldots d)$ are $(\mathcal{F}_t)$-predictable real-valued symmetric for almost every ω, $t \in [0,T]$ and $x \in \mathbb{R}^d$.

(ii) For every $t \in (0,T]$ and $x, y \in \mathbb{R}^d$

$$\sum_{j,k=1}^{d} y^j y^k \Big[2 \sum_{r=0}^{d_1} a_r^{j,k}(t,x)\, dV_t^r - \sum_{l\geq 1} \sigma_l^j(t,x)\sigma_l^k(t,x) d\langle M^l\rangle_t \Big] \geq 0 \quad (3.10)$$

a.s. in the sense of measures.

Assumption (A3(m)) There exists a constant $\tilde{K}(m)$ such that for all $j,k = 1,\ldots,d$, $r = 0,\ldots,d_1$, $l \geq 1$, any multi-indices α (resp. β) of length $|\alpha| \leq m+1$ (resp. $|\beta| \leq m$), and for every $(t,x) \in (0,T]\times\mathbb{R}^d$ one has a.s.

$$|D^\alpha a_r^{j,k}(t,x)| + |b^{j,k}(t)| + |D^\alpha a_r^j(t,x)| + |D^\beta a_r(t,x)| + |D^\beta b_r(t,x)| \leq \tilde{K}(m), \quad (3.11)$$

$$|D^\alpha \sigma_l^j(t,x)| + |D^\alpha \sigma_j(t,x)| + |D^\alpha \tau_l(t,x)| \leq \tilde{K}(m). \quad (3.12)$$

Assumption (A4(m,p)) Let $p \in [2,+\infty)$; for any $r = 0,\ldots,d_1$, $l \geq 1$, the processes $F_r(t,x) = F_{r,1}(t,x) + iF_{r,2}(t,x)$ and $G_l(t,x) = G_{l,1}(t,x) + iG_{l,2}(t,x)$ are predictable, $F_r(t,\cdot) \in H^m$ and $G_r(t,\cdot) \in H^{m+1}$. Furthermore, if we denote

$$K_m(t) = \int_0^t \Big[\sum_{r=0}^{d_1} \|F_r(s)\|_m^2\, dV_s^r + \sum_{l\geq 1} \|G_l(s)\|_{m+1}^2 d\langle M^l\rangle_s \Big], \quad (3.13)$$

then

$$\mathbb{E}\big(\|Z_0\|_m^p + K_m^{\frac{p}{2}}(T)\big) < +\infty. \quad (3.14)$$

The following defines what is considered to be a (probabilistically strong) weak solution of the evolution equations (3.2) and (3.3).

Definition 3.1. A $\mathbb{C}$-valued $(\mathscr{F}_t)$-predictable process Z is a solution to the evolution equation (3.2) with initial condition Z_0 if $\mathrm{P}(\int_0^T \|Z(s)\|_1^2 ds < +\infty) = 1$, $\mathbb{E}\int_0^T |Z(s)|^2 dV_s < \infty$, and for every $t \in [0,T]$ and $\Phi = \phi + i\psi$, where ϕ and ψ are $\mathscr{C}^\infty$ functions with compact support from $\mathbb{R}^d$ to $\mathbb{R}$, one has a.s.

$$\begin{aligned}(Z(t),\Phi) = (Z(0),\Phi) &+ \sum_{r=0}^{d_1}\int_0^t \Big[-\sum_{j,k=1}^{d} \Big(\big[a_r^{j,k}(s,\cdot) + ib_r^{j,k}(s)\big] D_j Z(s,\cdot), D_k\Phi \Big) \\ &+ \sum_{j=1}^{d} \Big(a_r^j(s,\cdot) D_j Z(s,\cdot) + [a_r(s,\cdot) + ib_r(s,\cdot)] Z(s,\cdot) + F_r(s,\cdot), \Phi \Big) \Big] dV_r(s) \\ &+ \sum_{l\geq 1}\int_0^t \big(S_l(Z(s,\cdot)) + G_l(s,\cdot), \Phi\big)\, dM_s^l. \end{aligned} \quad (3.15)$$

Theorem 3.1. *Let $m \geq 1$ be an integer and suppose that Assumptions* **(A1), (A2), (A3(m))** *and* **(A4(m,2))** *(i.e., for $p = 2$) are satisfied.*

(i) Then Eqs. (3.2) *and* (3.3) *have a unique solution Z, such that*

$$\mathbb{E}\Big(\sup_{t\in[0,T]} \|Z(t,\cdot)\|_m^2\Big) \leq C\,\mathbb{E}\Big(\|Z_0\|_m^2 + K_m(T)\Big) < \infty, \tag{3.16}$$

for a constant C that only depends on the constants which appear in the above listed conditions. Almost surely, $Z \in \mathscr{C}([0,T], H^{m-1})$ and almost surely the map $[0,T] \ni t \mapsto Z(t,\cdot) \in H^m$ is weakly continuous.

(ii) Suppose furthermore that Assumption **(A4(m,p))** *holds for $p \in (2,+\infty)$. Then there exists a constant C_p as above such that*

$$\mathbb{E}\Big(\sup_{t\in[0,T]} \|Z(t,\cdot)\|_m^p\Big) \leq C_p\,\mathbb{E}\Big(\|Z_0\|_m^p + K_m^{p/2}(T)\Big). \tag{3.17}$$

Proof. Set $\mathscr{H} = H^m$, $\mathscr{V} = H^{m+1}$ and $\mathscr{V}' = H^{m-1}$. Then $\mathscr{V} \subset \mathscr{H} \subset \mathscr{V}'$ is a Gelfand triple for the equivalent norm $|(I-\Delta)^{m/2}u|_{L^2}$ on the space H_m. Given $Z = X + iY \in \mathscr{V}$ and $\zeta = \xi + i\eta \in \mathscr{V}'$ set

$$\langle Z,\zeta\rangle_m = \langle X,\xi\rangle_m + \langle Y,\eta\rangle_m, \quad \text{and } \langle Z,\zeta\rangle := \langle Z,\zeta\rangle_0,$$

where $\langle X,\xi\rangle_m$ and $\langle Y,\eta\rangle_m$ denote the duality between the (real) spaces H^{m+1} and H^{m-1}. For every multi-index α, let

$$\mathscr{I}(\alpha) = \{(\beta,\gamma)\,:\, \alpha = \beta+\gamma,\ |\beta|,|\gamma| \in \{0,\ldots,|\alpha|\}\}.$$

To ease notations, we skip the time parameter when writing the coefficients a_r, b_r, σ_l and τ_l. Then for $l \geq 1$, using the Assumption **(A3(m))**, we deduce that if $Z = X + iY \in H^{m+1}$, we have for $|\alpha| \leq m$,

$$D^\alpha[S_l(Z)] = \sum_{j=1}^d \sigma_l^j(x)\,D_j D^\alpha Z + \sum_{(\beta,\gamma)\in\mathscr{I}(\alpha)} C_l(\beta,\gamma) D^\beta Z, \tag{3.18}$$

with functions $C_l(\beta,\gamma)$ from $\mathbb{R}^d$ to $\mathbb{C}$ such that $\sup_{l\geq 1}\sup_{x\in\mathbb{R}^d} |C_l(\beta,\gamma)(x)| < +\infty$. A similar computation proves that for every multi-index α with $|\alpha| \leq m$, $r = 0,\ldots,d_1$

$$D^\alpha[L_r(Z)] = L_r(D^\alpha Z) + \sum_{j,k=1}^d \sum_{(\beta,\gamma)\in\mathscr{I}(\alpha):|\gamma|=1} D_k\Big(D^\gamma a_r^{j,k} D_j D^\beta Z\Big) + \sum_{|\beta|\leq m} C_r(\beta,\gamma) D^\beta Z, \tag{3.19}$$

for some bounded functions $C_r(\beta,\gamma)$ from $\mathbb{R}^d$ into $\mathbb{C}$. Hence for every $r=0,\dots,d_1$, one has a.s. $L_r:\mathscr{V}\times\Omega\to\mathscr{V}'$ and similarly, for every $l\geq 1$, a.s. $S_l:\mathscr{V}\times\Omega\to\mathscr{H}$.

For every $\lambda>0$ and $Z=X+iY\in\mathscr{H}$, let us set

$$L_{r,\lambda}Z := L_r Z + \lambda(\Delta X + i\Delta Y) = L_r Z + \lambda\Delta Z. \tag{3.20}$$

Consider the evolution equation for the process $Z^\lambda(t,x)=X^\lambda(t,x)+iY^\lambda(t,x)$,

$$dZ^\lambda(t,x) = \sum_{r=0}^{d_1}\big[L_{r,\lambda}Z^\lambda(t,\cdot)+F_r(t,x)\big]\,dV_t^r + \sum_{l\geq 1}\big[S_l Z^\lambda(t,x)+G_l(t,x)\big]dM_t^l, \tag{3.21}$$

$$Z^\lambda(0,x) = Z_0(x) = X_0(x) + iY_0(x). \tag{3.22}$$

In order to prove well-posedness of the problems (3.21) and (3.22), firstly we have to check the following stochastic parabolicity condition:

Condition (C1) There exists a constant $K>0$ such that for $Z\in H^{m+1}$, $t\in[0,T]$:

$$2\sum_{r=0}^{d_1}\langle L_r Z\,,\,Z\rangle_m\,dV_t^r + \sum_{l\geq 0}\|S_l(Z)\|_m^2\,d\langle M^l\rangle_t \leq K\|Z\|_m^2\,dV_t$$

a.s. in the sense of measures.

Let $Z=X+iY\in H^{m+1}$; using (3.19) and (3.18), we deduce that

$$2\sum_{|\alpha|=m}\sum_{r=0}^{d_1}\langle D^\alpha L^r Z\,,\,D^\alpha Z\rangle\,dV_t^r + \sum_{|\alpha|=m}\sum_{l\geq 1}|D^\alpha S_l Z|^2\,d\langle M^l\rangle_t = \sum_{\kappa=1}^{5} dT_\kappa(t), \tag{3.23}$$

where to ease notation we drop the time index in the right-hand sides and we set

$$\begin{aligned}
dT_1(t) = \sum_{|\alpha|=m}\sum_{j,k=1}^{d}\Big\{-2\sum_{r=1}^{d_1}\Big[&\big(a_r^{j,k}D_jD^\alpha X, D_kD^\alpha X\big) - \big(b_r^{j,k}D_jD^\alpha Y, D_kD^\alpha X\big)\\
&+\big(a_r^{j,k}D_jD^\alpha Y, D_kD^\alpha Y\big) + \big(b_r^{j,k}D_jD^\alpha X, D_kD^\alpha Y\big)\Big]dV_t^r\\
&+\sum_{l\geq 1}\Big[\big(\sigma_l^j D_jD^\alpha X, \sigma_l^k D_kD^\alpha X\big) + \big(\sigma_l^j D_jD^\alpha Y, \sigma_l^k D_kD^\alpha Y\big)\Big]d\langle M^l\rangle_t\Big\},
\end{aligned}$$

$$dT_2(t) = -2\sum_{r=0}^{d_1}\sum_{|\alpha|\le m}\Big\{\sum_{j,k=1}^{d}\sum_{(\beta,\gamma)\in\mathscr{I}(\alpha):|\gamma|=1}\Big[\big(D^\gamma a_r^{j,k} D_j D^\beta X, D_k D^\alpha X\big) + \big(D^\gamma a_r^{j,k} D_j D^\beta Y, D_k D^\alpha Y\big)\Big]$$
$$+\sum_{j=1}^{d}\Big[\big(a_r^j D_j D^\alpha X, D^\alpha X\big) + \big(a_r^j D_j D^\alpha Y, D^\alpha Y\big)\Big]\Big\}dV_t^r,$$

$$dT_3(t) = \sum_{l\ge 1}\sum_{j,k=1}^{d}\sum_{|\alpha|=m}\sum_{(\beta,\gamma)\in\mathscr{I}(\alpha):|\gamma|=1}\Big[\big(D^\gamma\sigma_l^j D_j D^\beta X, \sigma_l^k D_k D^\alpha X\big)$$
$$+ (D^\gamma\sigma_l^j D_j D^\beta Y, \sigma_l^k D_k D^\alpha Y)d\langle M^l\rangle_t,$$

$$dT_4(t) = \sum_{l\ge 1}\sum_{j,k=1}^{d}\sum_{|\alpha|=m}\Big[(\sigma_l^j D_j D^\alpha X, \sigma_l D^\alpha X) - (\sigma_l^j D_j D^\alpha X, \tau_l D^\alpha Y)$$
$$+ (\sigma_l^j D_j D^\alpha Y, \tau_l D^\alpha X) + (\sigma_l^j D_j D^\alpha Y, \sigma_l D^\alpha Y)\Big]d\langle M^l\rangle_t,$$

$$dT_5(t) = \sum_{|\alpha|\vee|\beta|\le m}\Big\{\sum_{r=0}^{d_1}\Big[\sum_{j,k=1}^{d}(C_r^{j,k}(.)D^\beta Z, D^\alpha Z)$$
$$+\sum_{j=1}^{d}(C_r^j(.)D^\beta Z, D^\alpha Z) + (C_r(.)D^\beta Z, D^\alpha Z)\Big]dV_t^r$$
$$+\sum_{l\ge 1}\Big[\sum_{j=1}^{d}\Big(\Big[\tilde C_l(.) + \sum_{l\ge 1}\tilde C_l^j(.)\Big]]D^\beta Z, D^\alpha Z\Big)\Big]d\langle M^l\rangle_t\Big\},$$

where $C_r^{j,k}$, C_r^j, C_r, $\tilde C_l^j$ and $\tilde C_l$ are bounded functions from $\mathbb{R}^d$ to $\mathbb{C}$ due to Assumption **(A4(m,p))** for any $p \in [2,\infty)$.

For every r the matrix b_r is symmetric; hence $\sum_{j,k}\big[(b_r^{j,k} D_j D^\alpha X, D_k D^\alpha Y) - (b_r^{j,k} D_j D^\alpha Y, D_k D^\alpha X)\big] = 0$. Hence, Assumption **(A2)** used with $y_j = D_j D^\alpha X$ and with $y_j = D_j D^\alpha Y$, $j = 1,\dots,d$, implies $T_1(t) \le 0$ for $t \in [0,T]$. Furthermore, Assumption **(A3(m))** yields the existence of a constant $C > 0$ such that $dT_5(t) \le C\,\|Z(t)\|_m^2 dV_t$ for all $t \in [0,T]$.

Integration by parts shows that for regular enough functions $f, g, h : \mathbb{R}^d \to \mathbb{R}$, $(\beta,\gamma) \in \mathscr{I}(\alpha)$ with $|\gamma| = 1$, we have

$$(fD^\beta g, D^\alpha h) = -(fD^\alpha g, D^\beta h\rangle - \langle D^\gamma f D^\beta g, D^\beta h\rangle. \tag{3.24}$$

Therefore, the symmetry of the matrices a_r implies that for $\phi \in \{X(t), Y(t)\}$ and $r = 0, \ldots, d_1$,

$$\sum_{j,k=1}^{d} \big(D^\gamma a_r^{j,k} D_j D^\beta \phi, D_k D^\alpha \phi\big) = -\frac{1}{2} \sum_{j,k=1}^{d} \big(D^\gamma (D^\gamma a_r^{j,k}) D_j D^\beta \phi, D_k D^\beta \phi\big).$$

A similar argument proves that for fixed $j = 1, \ldots, d$, $r = 0, \ldots, d_1$ and $\phi = X(t)$ or $\phi = Y(t)$,

$$(a_r^j D_j D^\alpha \phi, D^\alpha \phi) = -\frac{1}{2}(D_j a_r^j D^\alpha \phi, D^\alpha \phi).$$

Therefore, Assumption **(A3(0))** implies the existence of $K > 0$ such that $d T_2(t) \leq K \|Z(t)\|_m^2 d V_t$ for all $t \in [0, T]$. Furthermore, $d T_4(t)$ is the sum of terms $(\phi \psi, D_j \phi)$ $d \langle M^l \rangle_t$ where $\phi = D^\alpha X(t)$ or $\phi = D^\alpha Y(t)$, and $\psi = fg$, with $f \in \{\sigma_l^k\}$ and $g \in \{\sigma_l, \tau_l\}$. The identity $(\phi \psi, D_j \phi) = -\frac{1}{2}(\phi D_j \psi, \phi)$, which is easily deduced from integration by parts, and Assumptions **(A1)** and **(A3(m))** imply the existence of $K > 0$ such that $d T_4(t) \leq K \|Z(t)\|_m^2 d V_t$ for every $t \in [0, T]$. The term $d T_3(t)$ is the sum over $l \geq 1$ and multi-indices α with $|\alpha| = m$ of

$$A(l, \alpha) = \sum_{j,k=1}^{d} \sum_{(\beta,\gamma) \in \mathscr{I}(\alpha) : |\gamma| = 1} \big(D^\gamma f_l^j f_l^k D_j D^\beta \varphi, D_k D^\alpha \varphi\big),$$

with $\varphi = X(t)$ or $\varphi = Y(t)$ and $f_l^j = \sigma_l^j$ for every $j = 1, \ldots, d$. Then, $A(l, \alpha) = B(l, \alpha) - C(l, \alpha)$, where $B(l, \alpha) = \sum_{j,k=1}^{d} B_{j,k}(l, \alpha)$ and

$$B_{j,k}(l, \alpha) = \sum_{(\beta,\gamma) \in \mathscr{I}(\alpha) : |\gamma| = 1} \big(D^\gamma (f_l^j f_l^k) D_j D^\beta \varphi, D_k D^\alpha \varphi\big),$$

$$C(l, \alpha) = \sum_{j,k=1}^{d} \sum_{(\beta,\gamma) \in \mathscr{I}(\alpha) : |\gamma| = 1} \big(D^\gamma f_l^k f_l^j D_j D^\beta \varphi, D_k D^\alpha \varphi\big).$$

Integrating by parts twice and exchanging the partial derivatives D_j and D_k in each term of the sum in $C(l, \alpha)$, we deduce that

$$\begin{aligned}\big(D^\gamma f_l^k f_l^j D_j D^\beta \varphi, D_k D^\alpha \varphi\big) &= -\big([D_k [D^\gamma f_l^k f_l^j] D_j D^\beta \varphi + D^\gamma f_l^k f_l^j D_k D_j D^\beta \varphi], D^\alpha \varphi\big) \\ &= \big(- D_k [D^\gamma f_l^k f_l^j] D_j D^\beta \varphi + (D_j [D^\gamma f_l^k f_l^j] D_k D^\beta \varphi, D^\alpha \varphi\big) \\ &\quad + \big(D^\gamma f_l^k f_l^j D_k D^\beta \varphi, D_j D^\alpha \varphi\big).\end{aligned}$$

On the other hand, by symmetry we obviously have

$$\sum_{j,k} \big(D^\gamma f_l^j f_l^k D_k D^\beta \varphi, D_j D^\alpha \varphi\big) = \sum_{j,k} \big(D^\gamma f_l^k f_l^j D_j D^\beta \varphi, D_k D^\alpha \varphi\big).$$

Using Assumptions **(A1)** and **(A3(1))** we deduce that there exist bounded functions $\phi(\alpha, \tilde{\alpha}, l)$ defined for multi-indices $\tilde{\alpha}$ which have at most one component different from those of α, and such that

$$A(l,\alpha) = \frac{1}{2} B(l,\alpha) + \sum_{|\tilde{\alpha}|=m} \big(\phi(\alpha,\tilde{\alpha},l) D^{\tilde{\alpha}}\Phi,\, D^{\alpha}\Phi\big).$$

Furthermore, integration by parts yields

$$\sum_{j,k=1}^{d} B_{j,k}(l,\alpha) = -\frac{1}{2} \sum_{j,k=1}^{d} \sum_{(\beta,\gamma)\in\mathscr{I}(\alpha):|\gamma|=1} \big(D^{\gamma} D^{\gamma}[f_l^j f_l^k] D_j D^{\beta}\varphi,\, D_k D^{\beta}\varphi\big).$$

Thus, Assumption **(A3(1))** implies the existence of a constant $C > 0$ such that for the various choices of φ and f_l^k, $\sum_{l\geq 1}\sum_{|\alpha|=m} |B(l,\alpha)| d\langle M^l\rangle_t \leq C\|Z(t)\|_m^2 dV_t$ for every $t \in [0,T]$. Therefore, we deduce that we can find a constant $K > 0$ such that $dT_3(t) \leq K\|Z(t)\|_m^2 dV_t$. The above inequalities and (3.23) complete the proof of Condition **(C1)**.

Since L_r are linear operators, Condition **(C1)** implies the following classical Monotonicity, Coercivity and Hemicontinuity: for every $Z, \zeta \in H^{m+1}$ and $L_{r,\lambda}$ defined by (3.20),

$$2\sum_{r=0}^{d_1} \langle L_r Z - L_r\zeta\,,\, Z-\zeta\rangle_m dV_t^r + \sum_{l\geq 1} \|S_l(Z) - S_l(\zeta)\|_m^2 d\langle M^l\rangle_t \leq K\|Z-\zeta\|_m^2 dV_t,$$

$$2\sum_{r=0}^{d_1} \langle L_{r,\lambda} Z\,,\, Z\rangle_m dV_t^r + \sum_{l\geq 1} \|S_l(Z)\|_m^2 d\langle M^l\rangle_t + 2\lambda\|Z\|_{m+1}^2 \sum_{r=0}^{d_1} dV_t^r \leq K\|Z\|_m^2 dV_t$$

a.s. in the sense of measures, and for $Z_i \in H^{m+1}$, $i = 1,2,3$, $r = 0,\dots,d_1$ and $\lambda > 0$, the map $a \in \mathbb{C} \to \langle L_{r,\lambda}(Z_1 + aZ_2\,,\, Z_3\rangle_m$ is continuous.

The following condition **(C2)** gathers some useful bounds on the operators L_r and S_l for $0 \leq r \leq d_1$ and $l \geq 1$.

Condition (C2) There exist positive constants $K_i, i = 2,3,4$ such that for $Z \in H^{m+1}$, $\lambda \in [0,1]$, $r = 0,\dots,d_1$ and $l \geq 1$:

$$2\|L_{r,\lambda} Z\|_{m-1}^2 + \|S_l Z\|_m^2 \leq K_2\|Z\|_{m+1}^2 \text{ a.s.} \tag{3.25}$$

$$|(S_l Z, Z)_m| \leq K_3\|Z\|_m^2 \text{ and } |(S_l Z, G_l)_m| \leq K_4\|Z\|_m\|G_l\|_{m+1} \text{ a.s..} \tag{3.26}$$

The inequality (3.25) is a straightforward consequence of the Cauchy–Schwarz inequality and of Assumption **(A3(m))**. Using integration by parts and Assumptions **(A3(m))–(A4(m,p))**, we deduce that if $G_l(t) = G_{l,1}(t) + iG_{l,2}(t)$,

$$|(S_l Z, Z)_m| \leq \frac{1}{2} \sum_{|\alpha|=m} \sum_{j=1}^{d} \big|\big[\big(D_j \sigma_l^j D^\alpha X, D^\alpha X\big) + \big(D_j \sigma_l^j D^\alpha Y, D^\alpha Y\big)\big]\big| + \sum_{|\alpha|\vee|\beta|\leq m} \big|\big(C(\alpha,\beta,l) D^\alpha Z, D^\beta Z\big)\big|$$

$$|(S_l Z, G_l)_m| \leq \sum_{j=1}^{d} \sum_{|\alpha|=m} \big[\big|\big(\sigma_l^j D^\alpha X, D_j D^\alpha G_{l,1}\big)\big| + \big|\big(\sigma_l^j D^\alpha Y, D_j D^\alpha G_{l,2}\big)\big| + \sum_{|\alpha|\vee|\beta|\leq m} \big|\big(C(\alpha,\beta,l) D^\alpha Z, D^\beta G_l)\big)\big|,$$

for constants $C(\alpha,\beta,l)$, α,β,l such that $\sup_{\alpha,\beta,l} C(\alpha,\beta,l) \leq C < \infty$. Hence a simple application of the Cauchy–Schwarz and Young inequalities implies inequality (3.26).

We then proceed as in the proof of Theorem 3.1 in [13] for fixed $\lambda > 0$ (see also [9, 18]). To ease notations, we do not write the Galerkin approximation as the following estimates would be valid with constants which do not depend on the dimension of the Galerkin approximation, and hence would still be true for the weak and weak* limit in $L^2([0,T] \times \Omega; H_{m+1})$ and $L^2(\Omega; L^\infty(0,T; H_m))$. Let us fix a real number $N > 0$ and let $\tau_N = \inf\{t \geq 0 : \|Z^\lambda(t)\|_m \geq N\} \wedge T$. The Itô formula, the stochastic parabolicity condition **(C1)** and the Davis inequality imply that for any $t \in [0,T]$ and $\lambda \in (0,1]$,

$$\mathbb{E}\Big(\sup_{s\in[0,t]} \|Z^\lambda(s\wedge\tau_N)\|_m^2\Big) + 2\lambda\mathbb{E}\int_0^{t\wedge\tau_N} \|Z^\lambda(s)\|_{m+1}^2 ds \leq \mathbb{E}\|Z_0\|_m^2$$

$$+ 2\sum_{r=0}^{d_1} \mathbb{E}\int_0^t \big|\langle F_r(s\wedge\tau_N), Z^\lambda(s\wedge\tau_N)\rangle_m\big| dV_s^r$$

$$+ \sum_{l\geq 1} \mathbb{E}\int_0^t \big[2\big|\big(S_l Z^\lambda(s\wedge\tau_N), G_l(s\wedge\tau_N)\big)_m\big| + \|G_l(s\wedge\tau_N)\|_m^2\big] d\langle M^l\rangle_s$$

$$+ 6\,\mathbb{E}\Big(\sum_{l\geq 1}\Big\{\int_0^t \big(S_l Z^\lambda(s\wedge\tau_N) + G_l(s\wedge\tau_N), Z^\lambda(s\wedge\tau_N)\big)_m^2 d\langle M^l\rangle_s\Big\}^{\frac{1}{2}}\Big)$$

The Cauchy–Schwarz inequality, the upper estimate (3.26) in Condition **(C2)** and inequalities (3.8) and (3.9) in Assumption **(A1)** imply the existence of some constant $K > 0$ such that for any $\delta > 0$,

$$\mathbb{E}\Big(\sup_{s\in[0,t]}\|Z^\lambda(s\wedge\tau_N)\|_m^2\Big)+2\lambda\mathbb{E}\int_0^{t\wedge\tau_N}\|Z^\lambda(s)\|_{m+1}^2ds\le\mathbb{E}\|Z_0\|_m^2$$

$$+3\,\delta\mathbb{E}\Big(\sup_{s\in[0,t]}\|Z^\lambda(s\wedge\tau_N)\|_m^2\Big)+\tilde{K}\delta^{-1}\sum_{r=0}^{d_1}\mathbb{E}\int_0^t\|F_r(s)\|_m^2dV_s^r$$

$$+\mathbb{E}\int_0^t\sum_{l\ge1}K\big[\delta^{-1}+1\big]\|G_l(s)\|_{m+1}^2d\langle M^l\rangle_s+\mathbb{E}\int_0^t\|Z^\lambda(s\wedge\tau_N)\|_m^2dV_s.$$

For $\delta=\frac{1}{6}$, the Gronwall Lemma implies that for some constant C we have for all $N>0$ and $\lambda\in(0,1]$,

$$\mathbb{E}\Big(\sup_{s\in[0,t]}\|Z^\lambda(s\wedge\tau_N)\|_m^2\Big)+\lambda\mathbb{E}\int_0^{t\wedge\tau_N}\|Z^\lambda(s)\|_{m+1}^2ds\le C\mathbb{E}\big(\|Z_0\|_m^2+K_m(T)\big).$$

As $N\to\infty$, we deduce that $\tau_N\to\infty$ a.s. and by the monotone convergence theorem,

$$\mathbb{E}\Big(\sup_{s\in[0,T]}\|Z^\lambda(s)\|_m^2\Big)+\lambda\mathbb{E}\int_0^T\|Z^\lambda(s)\|_{m+1}^2ds\le C\mathbb{E}\big(\|Z_0\|_m^2+K_m(T)\big).$$

Furthermore, Z^λ belongs a.s. to $\mathscr{C}([0,T],H^m)$. As in [13], we deduce the existence of a sequence $\lambda_n\to0$ such that $Z^{\lambda_n}\to Z$ weakly in $L^2([0,T]\times\Omega;H^m)$. Furthermore, Z is a solution to (3.2) and (3.3) such that (3.16) holds and is a.s. weakly continuous from $[0,T]$ to H^m.

The uniqueness of the solution follows from the growth condition (3.25) in **(C2)** and the monotonicity condition which is deduced from the stochastic parabolicity property **(C1)**.

(ii) Suppose that Assumption **(A4(m,p))** holds for $p\in[2,\infty)$. Set $p=2\tilde{p}$ with $\tilde{p}\in[1,\infty)$; the Itô formula, the stochastic parabolicity condition **(C1)**, the growth conditions **(C2)**, the Burkholder–Davis–Gundy and Schwarz inequalities yield the existence of some constant C_p which also depends on the various constants in Assumptions **(A1)**–**(A4(m,p))** and conditions **(C1)**–**(C2)**, such that:

$$\mathbb{E}\Big(\sup_{s\in[0,t]}\|Z(s)\|_m^p\Big)\le C_p\Big[\mathbb{E}\|Z_0\|_m^p+\mathbb{E}\Big(\sup_{s\in[0,t]}\|Z(s)\|_m^{\tilde{p}}\Big|\sum_{r=0}^{d_1}\int_0^t\|F_r(s)\|_m dV_s^r\Big|^{\tilde{p}}\Big)$$

$$+\mathbb{E}\Big(\sup_{s\in[0,t]}\|Z(s)\|_m^{\tilde{p}}\Big|\sum_{l\ge1}\int_0^t\|G_l(s)\|_{m+1}d\langle M^l\rangle_s\Big|^{\tilde{p}}\Big)$$

$$+\mathbb{E}\Big(\sup_{s\in[0,t]}\|Z(s)\|_m^{\tilde{p}}\Big|\sum_{l\ge1}\int_0^t\big[\|Z(s)\|_m^2+\|G_l(s)\|_m^2\big]d\langle M^l\rangle_s\Big|^{\tilde{p}/2}\Big).$$

Using the Hölder and Young inequalities, (3.14) as well as Assumptions **(A1)** we deduce the existence of a constant $K > 0$ such that for any $\delta > 0$

$$\mathbb{E}\Big(\sup_{s\in[0,t]} \|Z(s)\|_m^p\Big) \leq 3\delta\mathbb{E}\Big(\sup_{s\in[0,t]} \|Z(s)\|_m^p\Big) + C(p)\Big[\mathbb{E}\|Z_0\|_m^p$$

$$+ K^{\tilde{p}}\delta^{-1}\mathbb{E}\Big|\int_0^t \sum_{r=0}^{d_1} \|F_r(s)\|_m^2 dV_s^r\Big|^{\tilde{p}} + K^{\tilde{p}}\delta^{-1}\mathbb{E}\Big|\int_0^t \sum_{l\geq 1} \|G_l(s)\|_{m+1}^2 d\langle M^l\rangle_s\Big|^{\tilde{p}}$$

$$+ K^{\tilde{p}-1}\delta^{-1}\mathbb{E}\int_0^t \|Z(s)\|_m^p dV_s + \delta^{-1}\Big|\mathbb{E}\int_0^t \sum_{l\geq 1} \|G_l(s)\|_m^2 d\langle M^l\rangle_s\Big|^{\tilde{p}}\Big].$$

Let $\delta = \frac{1}{6}$ and introduce the stopping time $\tau_N = \inf\{t \geq 0 : \|Z(t)\|_m \geq N\} \wedge T$. Replacing t by $t \wedge \tau_N$ in the above upper estimates, the Gronwall lemma and (3.14) prove that there exists a constant C such that $\mathbb{E}\Big(\sup_{s\in[0,t]\wedge\tau_N} \|Z(s)\|_m^{2p}\Big) \leq C\mathbb{E}\big(\|Z_0\|_m^p + K_m(T)^{p/2}\big)$ for every $N > 0$. As $N \to \infty$ the monotone convergence theorem concludes the proof of (3.17). This ends of the proof of Theorem 3.1. □

Remark 3.1. If $a_r^{j,k}(t,x) = 0$, for example for the Schrödinger equation, Assumption **(A2)** implies that $\sigma_l^j = 0$.

3.2.2 Further A Priori Estimates on the Difference

Theorem 3.1 is used to upper estimate moments of the difference of two processes solutions to equations of type (3.2). For $\varepsilon = 0, 1$, $r = 0, \ldots, d_1$, $l \geq 1$, $j,k = 1, \ldots, d$, let $a_{\varepsilon,r}^{j,k}(t,x), b_{\varepsilon,r}^{j,k}(t), a_{\varepsilon,r}^{j}(t,x)$, $a_{\varepsilon,r}(t,x), b_{\varepsilon,r}(t,x), \sigma_{\varepsilon,l}^{j}(t,x), \sigma_{\varepsilon,l}(t,x)$, $\tau_{\varepsilon,l}(t,x)$ be coefficients, $F_{\varepsilon,r}(t,x)$, $G_{\varepsilon,l}(t,x)$ be processes and let $Z_{\varepsilon,0}$ be random variables which satisfy the assumptions **(A1)**–**(A3(m))** and **(A4(m,p))** for some $m \geq 1$, $p \in [2,\infty)$, the same martingales $(M_t^l, t \in [0,T])$ and increasing processes $(V_t^r, t \in [0,T])$. Let $L_{\varepsilon,r}$ and $S_{\varepsilon,l}$ be defined as in (3.4) and (3.5), respectively. Extend these above coefficients, operators, processes and random variables to $\varepsilon \in [0,1]$ as follows: if f_0 and f_1 are given, for $\varepsilon \in [0,1]$, let $f_\varepsilon = \varepsilon f_1 + (1-\varepsilon) f_0$. Note that by convexity, all the previous assumptions are satisfied for any $\varepsilon \in [0,1]$. Given $\varepsilon \in [0,1]$, let Z_ε denote the solution to the evolution equation: $Z_\varepsilon(0,x) = Z_{\varepsilon,0}(x)$ and

$$dZ_\varepsilon(t,x) = \sum_{r=0}^{d_1} \big[L_{\varepsilon,r} Z_\varepsilon(t,x) + F_{\varepsilon,r}(t,x)\big]\, dV_t^r + \sum_{l\geq 1} \big[S_{\varepsilon,l} Z_\varepsilon(t,x) + G_{\varepsilon,l}(t,x)\big] dM_t^l. \tag{3.27}$$

Thus, Theorem 3.1 immediately yields the following.

Corollary 3.1. *With the notations above, the solution Z_ε to (3.27) with the initial condition $Z_{\varepsilon,0}$ exists and is unique with trajectories in $C([0,T];H^{m-1}) \cap L^\infty(0,T;H^m)$. Furthermore, the trajectories of Z_ε belong a.s. to $C_w([0,T];H^m)$ and there exists a constant $C_p > 0$ such that*

$$\sup_{\varepsilon\in[0,1]} \mathbb{E}\Big(\sup_{t\in[0,T]} \|Z_\varepsilon(t,\cdot)\|_m^p\Big) \le C_p \sup_{\varepsilon\in\{0,1\}} \mathbb{E}\Big(\|Z_{\varepsilon,0}\|_m^p + K_m(T)^{p/2}\Big) < \infty. \quad (3.28)$$

Following the arguments in [9], this enables us to estimate moments of $Z_1 - Z_0$ in terms of a process ζ_ε which is a formal derivative of Z_ε with respect to ε. Given operators or processes f_ε, $\varepsilon \in \{0,1\}$, set $f' = f_1 - f_0$.

Theorem 3.2. *Let $m \ge 3$, and $p \in [2,\infty)$. Then for any integer $\kappa = 0,\dots,m-2$*

$$\mathbb{E}\Big(\sup_{t\in[0,T]} \|Z_1(t) - Z_0(t)\|_\kappa^p\Big) \le \sup_{\varepsilon\in[0,1]} \mathbb{E}\Big(\sup_{t\in[0,T]} \|\zeta_\varepsilon(t)\|_\kappa^p\Big), \quad (3.29)$$

where ζ_ε is the unique solution to the following linear evolution equation:

$$d\zeta_\varepsilon(t) = \sum_{r=0}^{d_1} \big(L_{\varepsilon,r}\zeta_\varepsilon(t,x) + L_r' Z_\varepsilon(t,x) + F_r'(t,x)\big) dV_r(t)$$
$$+ \sum_{l\ge 1} \big(S_{\varepsilon,l}\zeta_\varepsilon(t,x) + S_l' Z_\varepsilon(t,x) + G_l'(t,x)\big) dM_t^l, \quad (3.30)$$

with the initial condition $Z_0' = Z_1 - Z_0$. Furthermore,

$$\sup_{\varepsilon\in[0,1]} \mathbb{E}\Big(\sup_{t\in[0,T]} \|\zeta_\varepsilon(t)\|_{m-2}^p\Big) < \infty. \quad (3.31)$$

Proof. Using (3.28) we deduce that the processes $\tilde{F}_r(t,x) = L_r' Z_\varepsilon(t,x) + F_r'(t,x)$ and $\tilde{S}_l(t,x) = S_l' Z_\varepsilon(t,x) + G_l'(t,x)$ satisfy the assumption (**A4(m-2,p)**) with $m-2 \ge 1$. Hence the existence and uniqueness of the process ζ_ε, solution to (3.30), as well as (3.31) can be deduced from Theorem 3.1.

We now prove (3.29) for $\kappa \in \{0,\dots,m-2\}$ and assume that the right-hand side is finite. Given $(f_\varepsilon, \varepsilon \in [0,1))$, for $h > 0$ and $\varepsilon \in [0,1]$ such that $\varepsilon + h \in [0,1]$, set $\delta_h f_\varepsilon = (f_{\varepsilon+h} - f_\varepsilon)/h$. We at first prove that (3.29) can be deduced from the following: for every $\varepsilon \in [0,1)$, as $h \to 0$ is such that $h + \varepsilon \in [0,1]$,

$$\mathbb{E}\Big(\sup_{t\in[0,T]} \|\delta_h Z_\varepsilon(t) - \zeta_\varepsilon(t)\|_0^p\Big) \to 0. \quad (3.32)$$

Indeed, assume that (3.32) holds and for $n > 0$ let $R_n = n^\kappa \Delta^\kappa (n\mathrm{Id} - \Delta)^{-\kappa}$ denote the *kapa*-fold composition of the resolvent of the Laplace operator Δ on the space $L^2 = H^0$. Then, by some classical estimates, there exists a constant

$C(\kappa)>0$ such that for any $\phi\in L^2$, $\|R_n h\|_\kappa \le C(\kappa)\|\phi\|_0$. Hence (3.32) yields that for every $n>0$, as $h\to 0$ with $\varepsilon+h\in[0,1]$, we have $\mathbb{E}\Big(\sup_{t\in[0,T]}\|\delta_h R_n Z_\varepsilon(t) - R_n\zeta_\varepsilon(t)\|_\kappa^p\Big)\to 0$. Furthermore, since for every integer $N\ge 1$, we have $Z_1 - Z_0 = \frac{1}{N}\sum_{k=0}^{N-1}\delta_{1/N}Z_{k/N} \le \sup_{\varepsilon\in[0,1]}\delta_{1/N}Z_\varepsilon$, we deduce that for every $n>0$ and $p\in[2,\infty)$:

$$\mathbb{E}\Big(\sup_{t\in[0,T]}\|R_n Z_0(t) - R_n Z_1(t)\|_\kappa^p\Big) \le \sup_{\varepsilon\in[0,1]}\mathbb{E}\Big(\sup_{t\in[0,T]}\|R_n\zeta_\varepsilon(t)\|_\kappa^p\Big).$$

Finally, if $\phi\in H^0$ is such that $\liminf_{n\to\infty}\|R_n\phi\|_\kappa = N_\kappa<\infty$, then $\phi\in H^\kappa$ and $\|\phi\|_\kappa\le N_\kappa$. Thus, by applying the Fatou Lemma and using estimate (3.31) we can conclude the proof of (3.29).

We will now prove the convergence (3.32). It is easy to see that the process $\eta_{\varepsilon,h}(t,\cdot) := \delta_h Z_\varepsilon(t,\cdot) - \zeta_\varepsilon(t,\cdot)$ has initial condition $\eta_{\varepsilon,h}(0)=0$ and is a solution of the evolution equation:

$$\begin{aligned} d\eta_{\varepsilon,h}(t) &= \sum_{r=0}^{d_1}\big[L_{\varepsilon,r}\eta_{\varepsilon,h}(t,\cdot) + L'_r\big(Z_{\varepsilon+h}(t,\cdot) - Z_\varepsilon(t,\cdot)\big)\big]dV_t^r \\ &\quad + \sum_{l\ge 1}\big[S_{\varepsilon,l}\eta_{\varepsilon,h}(t,\cdot) + S'_l\big(Z_{\varepsilon+h}(t,\cdot) - Z_\varepsilon(t,\cdot)\big)\big]dM_t^l. \end{aligned}$$

Hence, using once more Theorem 3.1, we deduce the existence of a constant $C_p>0$ independent of $\varepsilon\in[0,1)$ and $h>0$, such that $\varepsilon+h\in[0,1]$,

$$\mathbb{E}\Big(\sup_{t\in[0,T]}\|\delta_h Z_\varepsilon(t) - \zeta_\varepsilon(t)\|_0^p\Big) \le C_p\mathbb{E}\Big(\int_0^T \|Z_{\varepsilon+h}(t) - Z_\varepsilon(t)\|_2^2\,dV_t\Big)^{p/2}.$$

Using the interpolation inequality $\|\phi\|_2\le C\|\phi\|_0^{1/3}\|\phi\|_3^{2/3}$, see for instance Proposition 2.3 in [15], the Hölder inequality and the estimate (3.28) with $m=3$ from Corollary 3.1, we deduce that

$$\mathbb{E}\Big(\sup_{t\in[0,T]}\|\delta_h Z_\varepsilon(t) - \zeta_\varepsilon(t)\|_0^p\Big) \le C\Big[\mathbb{E}\Big(\sup_{t\in[0,T]}\|Z_{\varepsilon+h}(t) - Z_\varepsilon(t)\|_0^p\Big)\Big]^{1/3}.$$

Finally, the process $\Phi_{\varepsilon,h}(t,\cdot) = Z_{\varepsilon+h}(t,\cdot) - Z_\varepsilon(t,\cdot)$ is solution to the evolution equation

$$\begin{aligned} d\Phi_{\varepsilon,h}(t) &= \sum_{r=0}^{d_1}\big[L_{\varepsilon,r}\Phi_{\varepsilon,h}(t,\cdot) + hL'_r Z_{\varepsilon+h}(t,\cdot) + hF'_r(t,\cdot)\big]dV_t^r \\ &\quad + \sum_{l\ge 1}\big[S_{\varepsilon,l}\Phi_{\varepsilon,h}(t,\cdot) + hS'_l Z_{\varepsilon+h}(t,\cdot) + hG'_l(t,\cdot))\big]dM_t^l, \end{aligned}$$

with the initial condition $\Phi_{\varepsilon,h}(0) = h(Z_1 - Z_0)$. Thus, (3.28) and Theorem 3.1 prove the existence of a constant C, which does not depend on $\varepsilon \in [0,1)$ and $h > 0$ with $\varepsilon + h \in [0,1]$, and such that

$$\mathbb{E}\Big(\sup_{t\in[0,T]} \|Z_{\varepsilon+h}(t) - Z_\varepsilon(t)\|_0^p\Big) \le C h^{p/3}.$$

This concludes the proof of (3.32) and hence that of the Theorem 3.2. □

3.3 Speed of Convergence

3.3.1 Convergence for Time-Independent Coefficients

For $\varepsilon = 0, 1$, $r = 0, \dots, d_1$, let $(V^r_{\varepsilon,t}, t \in [0,T])$ be increasing processes which satisfy Assumptions **(A1)**, **(A2)**, **(A3(m+3))** and **(A4(m+3,p))** for some integer $m \ge 1$, some $p \in [2, +\infty)$ separately for the increasing processes $(V^r_{\varepsilon,t}, t \in [0,T])$, the same increasing process $(V_t, t \in [0,T])$ and the initial conditions $Z_{\varepsilon,0}$, $\varepsilon = 0, 1$. For $\varepsilon = 0, 1$, let Z_ε denote the solution to the evolution equation

$$dZ_\varepsilon(t,x) = \sum_{0\le r\le d_1} \big[L_r Z_\varepsilon(t,x) + F_r(x)\big]\, dV^r_{\varepsilon,t} + \sum_{l\ge 1} \big[S_l Z_\varepsilon(t,x) + G_l(x)\big] dM^l_t, \tag{3.33}$$

with the initial conditions $Z_0(0,\cdot) = Z_{0,0}$ and $Z_1(0,\cdot) = Z_{1,0}$, respectively. Let

$$A := \sup_{\omega\in\Omega} \sup_{t\in[0,T]} \max_{r=0,1,\dots,d_1} |V^r_{1,t} - V^r_{0,t}|.$$

Then the H^m norm of the difference $Z_1 - Z_0$ can be estimated in terms of A as follows when the coefficients of L_r and F_r are time-independent. Indeed, unlike the statements in [10], but as it is clear from the proof, the diffusion coefficients σ_l and G_l can depend on time.

Theorem 3.3. *Let L_r and F_r be time-independent, $\mathscr{F}_0$-measurable, V^r_ε, $\varepsilon = 0, 1$, M_l be as above and let Assumptions* **(A1)**, **(A2)**, **(A3(m+3))** *and* **(A4(m+3,p))** *be satisfied for some $m \ge 0$ and some $p \in [2, +\infty)$. Suppose furthermore that*

$$\mathbb{E}\Big(\Big|\sum_{r=0}^{d_1} \|F_r\|^2_{m+1}\Big|^{p/2} + \sup_{s\in[0,T]} \Big|\sum_{l\ge 1} \|G_r(s)\|^2_{m+2}\Big|^{p/2}\Big) < \infty. \tag{3.34}$$

Then there exists a constant $C > 0$, which only depends on d and the constants in the above assumptions, such that the solutions Z_0 and Z_1 to (3.33) satisfy the following inequality:

$$\mathbb{E}\Big(\sup_{t\in[0,T]} \|Z_1(t)-Z_0(t)\|_m^p\Big) \le C\,\Big(\mathbb{E}(\|Z_{1,0}-Z_{0,0}\|_m^p)+A^p\Big).$$

The proof of Theorem 3.3 will require several steps. Some of them do not depend on the fact that the coefficients are time-independent; we are keeping general coefficients whenever this is possible. The first step is to use Theorem 3.2 and hence to define a process Z_ε for any $\varepsilon \in [0,1]$; it does not depend on the fact that the coefficients are time-independent and extends to the setting of the previous section. For $\varepsilon \in [0,1]$, $r=0,\ldots,d_1$, $t\in[0,T]$ and $x\in\mathbb{R}^d$, let

$$V^r_{\varepsilon,t} = \varepsilon V^r_{1,t} + (1-\varepsilon)V^r_{0,t}, \quad \rho^r_{\varepsilon,t} = dV^r_{\varepsilon,t}/dV_t$$

and for $j,k=1,\ldots,d$, set $a^{j,k}_{\varepsilon,r}(t,x) = \rho^r_{\varepsilon,t} a^{j,k}_r(t,x)$, $b^{j,k}_{\varepsilon,r}(t) = \rho^r_{\varepsilon,t} b^{j,k}_r(t)$, $a^j_{\varepsilon,r}(t,x) = \rho^r_{\varepsilon,t} a^j_r(t,x)$, $a_{\varepsilon,r}(t,x) = \rho^r_{\varepsilon,t} a_r(t,x)$, $b_{\varepsilon,r}(t,x) = \rho^r_{\varepsilon,t} b_r(t,x)$, $L_{\varepsilon,r} = \rho^r_{\varepsilon,t} L_r$, $F_{\varepsilon,r}(t,x) = \rho^r_{\varepsilon,t} F_r(t,x)$. Then for $\varepsilon\in[0,1]$, the solution $Z_\varepsilon(t,\cdot)$ to Eq. (3.2) with the increasing processes $V^r_{\varepsilon,t}$ can be rewritten as (3.27) with the initial data $Z_\varepsilon(0) = \varepsilon Z_{1,0} + (1-\varepsilon)Z_{0,0}$ and the operators (resp. processes) $S_{\varepsilon,l} = S_l$ (resp. $G_{\varepsilon,l} = G_l$). Furthermore, we have

$$\sum_{r=0}^{d_1}\sum_{j,k=1}^{d} \lambda^j\lambda^k\big(a^{j,k}_{\varepsilon,r}(t,x) + i b^{j,k}_{\varepsilon}(t)\big)dV^r_t =$$

$$\sum_{0\le r\le d_1}\sum_{j,k=1}^{d} \lambda^j\lambda^k\big(a^{j,k}_{r}(t,x) + i b^{j,k}_{r}(t)\big)dV^r_{\varepsilon,t}.$$

Hence the conditions **(A1), (A2)**, **(A3(m+3))** and **(A4(m+3,p))** are satisfied. Therefore, using Theorem 3.2, one deduces that the proof of Theorem 3.3 reduces to check that

$$\sup_{\varepsilon\in[0,1]} \mathbb{E}\Big(\sup_{t\in[0,1]} \|\zeta_\varepsilon(t)\|_m^p\Big) \le C\big(\mathbb{E}\|Z_{1,0}-Z_{0,0}\|_m^p + A^p\big), \tag{3.35}$$

where if one lets $A^r_t = V^r_{1,t} - V^r_{0,t}$, the process ζ_ε is the unique solution to (3.30) which here can be written as follows: for $t\in[0,T]$ one has

$$\begin{aligned} d\zeta_\varepsilon(t) = {} & \sum_{r=0}^{d_1}\big[L_r\zeta_\varepsilon(t,x)dV^r_{\varepsilon,t} + \big(L_r Z_\varepsilon(t,x) + F_r(t,x)\big)dA^r_t\big] \\ & + \sum_{l\ge 1} S_l\zeta_\varepsilon(t,x)dM_l(t), \end{aligned} \tag{3.36}$$

and the initial condition is $\zeta_\varepsilon(0) = Z_{1,0} - Z_{0,0}$. To ease notations, given a multi-index α, $j,k\in\{1,\ldots,d\}$ and Z smooth enough, set

$$Z_\alpha = D^\alpha Z, \quad Z_{\alpha,j} = D^\alpha D_j Z \quad \text{and } Z_{\alpha,j,k} = D^\alpha D_j D_k Z,$$

so that for $Z, \zeta \in H^m$, $(Z,\zeta)_m = \sum_{|\alpha \le m}(Z_\alpha, \zeta_\alpha)_0$. Let

$$\mathscr{A} = \Big\{ \sum_\alpha \sum_\beta a^{\alpha,\beta} Z_\alpha Z_\beta \; ; \; a^{\alpha,\beta} \text{ uniformly bounded and complex-valued, } Z \in H^{m+3} \Big\}$$

and for $\Phi, \Psi \in \mathscr{A}$ set $\Phi \sim \Psi$ if there exists $Z \in H^m$ such that $\int_{\mathbb{R}^d} (\Phi - \Psi)(x)dx = \int_{\mathbb{R}^d} \Gamma(x)dx$, where Γ is a function defined by

$$\Gamma(x) = \sum_{|\alpha| \le m} Z_\alpha(x) \overline{P^\alpha Z(x)} \quad \text{with } P^\alpha Z = \sum_{|\beta| \le m} \gamma^{\alpha,\beta} Z_\beta, \tag{3.37}$$

for some complex-valued functions $\gamma^{\alpha,\beta}$ such that $|\gamma^{\alpha,\beta}|$ are estimated from above by the constants appearing in the assumptions **(A1), (A2)**, **(A3(m+3))**, and **(A4(m+3,p))**. Note that if Γ is as above, then for some constant $C_m(\Gamma)$ we have

$$\int_{\mathbb{R}^d} |\Gamma(x)| \, dx \le C_m(\Gamma) \|Z\|_m^2. \tag{3.38}$$

For $\varepsilon > 0$, $j, k = 1, \ldots, d$, $l \ge 1$ and $t \in [0,T]$, set $q_t^l = d\langle M^l \rangle_t / dV_t$ and let

$$\tilde{a}_\varepsilon^{j,k} := \tilde{a}_\varepsilon^{j,k}(t,\cdot) = \sum_{r=0}^{d_1} a_{\varepsilon,r}^{j,k}(t,\cdot) - \frac{1}{2} \sum_{l \ge 1} \sigma_l^j(t,\cdot) \sigma_l^k(t,\cdot)\, q_t^l.$$

For $m \ge 0$ and $z \in H^{m+1}$, set

$$[Z]_m^2 := [Z]_m^2(t) = \sum_{j,k=1}^{d} \big(\tilde{a}_\varepsilon^{j,k}(t) D_j Z \, , \, D_k Z\big)_m + C_m \|Z\|_m^2, \tag{3.39}$$

with $C_0 = 0$ and $C_m > 0$ to be chosen later on, so that the right-hand side of (3.39) is non-negative. Given $Z, \zeta \in H^{m+1}$, set $[Z,\zeta]_m = \frac{1}{4}\big([Z+\zeta]_m^2 + [Z-\zeta]_m^2\big)$. We at first prove that $[.]_m$ defines a non-negative quadratic form on H^{m+1} for some large enough constant C_m. Once more, this result does not require that the coefficients be time-independent.

Lemma 3.1. *Suppose that the conditions* **(A1), (A2)** *and* **(A3(m+1))** *are satisfied. Then there exists a large enough constant C_m such that* (3.39) *defines a non-negative quadratic form on H^{m+1}.*

Proof. Assumption **(A2)** for $\varepsilon \in \{0,1\}$ implies that (3.39) holds for $m = 0$ and $C_0 = 0$. Let $m \ge 1$ and α be a multi-index such that $1 \le |\alpha| \le m$. The Leibnitz formula implies the existence of constants $C(\alpha,\beta,\gamma)$ such that for $Z \in H^{m+1}$,

$$\Big(\sum_{j,k=1}^{d}\big(\tilde a_\varepsilon^{j,k}(t)D_jZ\big)_\alpha\,,\,Z_{\alpha,k}\Big)_0=\sum_{j,k=1}^{d}\big(\tilde a_\varepsilon^{j,k}(t)Z_{\alpha,j}\,,\,Z_{\alpha,k}\big)_0$$
$$+\sum_{\beta+\gamma=\alpha,|\beta|\geq 1}C(\alpha,\beta,\gamma)I_\varepsilon^{\alpha,\beta,\gamma}(t),$$

where $I_\varepsilon^{\alpha,\beta,\gamma}(t):=\sum_{j,k=1}^d\big(D^\beta\tilde a_\varepsilon^{j,k}(t)Z_{\gamma,j}\,,\,Z_{\alpha,k}\big)_0$. Furthermore given $m\geq 1$, multi-indices α with $|\alpha|\leq m$ and $Z\in H^{m+1}$, using Assumption **(A2)** we deduce that $\sum_{1\leq j,k\leq d}\big(\tilde a_\varepsilon^{j,k}(t)Z_{\alpha,j}\,,\,Z_{\alpha,k}\big)_0\geq 0$. Thus, the proof reduces to check that

$$I_\varepsilon^{\alpha,\beta,\gamma}(t)\sim 0. \tag{3.40}$$

Indeed, then the upper estimate (3.38) proves (3.39), which concludes the proof of the lemma. Integration by parts implies $I_\varepsilon^{\alpha,\beta,\gamma}(t)=-\sum_{j,k=1}^d\Big(D_k\big(D^\beta\tilde a_\varepsilon^{j,k}(t)Z_{\gamma,j}\big),\,Z_\alpha\Big)_0$. Since $|\beta|\leq m$, by Assumption **(A3(m+1))** we know that $D_kD^\beta\tilde a_\varepsilon^{j,k}(t)$ is bounded and hence $I_\varepsilon^{\alpha,\beta,\gamma}(t)\sim-\sum_{j,k=1}^d\big(D^\beta\tilde a_\varepsilon^{j,k}(t)Z_{\gamma,j,k}\,,\,Z_\alpha\big)_0$. If $|\beta|\geq 2$, then $|\gamma|\leq m-2$; hence by **(A3(m))** we deduce that $I_\varepsilon^{\alpha,\beta,\gamma}(t)\sim 0$. If $|\beta|=1$, then $\tilde a_\varepsilon^{j,k}(t)=\tilde a_\varepsilon^{k,j}(t)$, so that

$$I_\varepsilon^{\alpha,\beta,\gamma}(t)=\sum_{j,k=1}^d\big(D^\beta\tilde a_\varepsilon^{j,k}(t)Z_{\gamma,j}\,,\,D^\beta Z_{\gamma,k}\big)_0$$
$$\sim\frac12\sum_{j,k=1}^d\int_{\mathbb{R}^d}D^\beta\tilde a_\varepsilon^{j,k}(t,x)D^\beta\big(X_{\gamma,j}(.)X_{\gamma,k}(.)+Y_{\gamma,j}(.)Y_{\gamma,k}(.)\big)(x)dx.$$

Thus, integrating by parts and using **(A3(2))** and the inequality $|\gamma|+1\leq m$, we deduce that $I_\varepsilon^{\alpha,\beta,\gamma}(t)\sim-\frac12\sum_{j,k=1}^d\big(D^\beta D^\beta\tilde a_\varepsilon^{j,k}(t)Z_{\gamma,j},Z_{\gamma,k}\big)_0\sim 0$. This concludes the proof. □

The following lemma gathers some technical results which again hold for time-dependent coefficients.

Lemma 3.2. *Suppose that the assumptions of Theorem 3.3 hold. There exists a constant C such that for* $\zeta\in H^{m+1}$ *one has* dV_t *a.e.*

$$p(\zeta):=2\sum_{0\leq r\leq d_1}\rho_{\varepsilon,t}^r\langle\zeta,L_r\zeta\rangle_m+\sum_{l\geq 1}q_t^l\|S_l\zeta\|_m^2+2[\zeta]_m^2\leq C\|\zeta\|_m^2. \tag{3.41}$$

For any $\tilde r=0,1,\ldots,d_1$, $Z\in H^{m+3}$ *and* $\zeta\in H^{m+1}$ *let*

$$q_{\tilde r}(\zeta,Z)=\sum_{0\leq r\leq d_1}\rho_{\varepsilon,t}^r\big[\langle L_r\zeta,L_{\tilde r}Z\rangle_m+\langle\zeta,L_{\tilde r}L_rZ\rangle_m\big]+\sum_{l\geq 1}q_t^l(S_l\zeta,L_{\tilde r}S_lZ)_m.$$

Then there exists a constant C such that for any $Z \in H^{m+3}$ and $\zeta \in H^{m+1}$, one has dV_t a.e.

$$|q_{\tilde{r}}(\zeta, Z)| \leq C \|Z\|_{m+3}\big(\|\zeta\|_m + [\zeta]_m\big). \tag{3.42}$$

Proof. Suppose at first that $\zeta \in H^{m+2}$; since the upper estimates (3.41) and (3.42) only involve the H^{m+1}-norm of ζ, they will follow by approximation. Then we have

$$\sum_{0\leq r\leq d_1} 2\rho^r_{\varepsilon,t} \, (\zeta, L_r\zeta)_m + \sum_{l\geq 1} q^l_t \, \|S_l\zeta\|^2_m = \sum_{|\alpha|\leq m} Q^\alpha_t(\zeta, \zeta),$$

where

$$Q^\alpha_t(\zeta, \zeta) = 2 \sum_{0\leq r\leq d_1} \rho^r_{\varepsilon,t} \, \big(\zeta_\alpha \,, \, (L_r\zeta)_\alpha\big)_0 + \sum_{l\geq 1} q^l_t \, \|(S_l\zeta)_\alpha\|^2_0.$$

Integration by parts and assumption **(A3(m))** imply that for $|\alpha| \leq m$, one has

$$2\big(\zeta_\alpha \,, \, (a^j_{\varepsilon,r}\zeta_j)_\alpha\big)_0 \sim 2\int_{\mathbb{R}^d} a^j_{\varepsilon,r}(t,x)\big(X_\alpha(x)X_{\alpha,j}(x) + Y_\alpha(x)Y_{\alpha,j}(x)\big)dx$$

$$= \int_{\mathbb{R}^d} a^j_{\varepsilon,r}(t,x)\big(X^2_\alpha + Y^2_\alpha\big)_{,j}(x)\,dx \sim -\big(a^j_{\varepsilon,r}(t)_j \, \zeta_\alpha \,, \, \zeta_\alpha\big)_0 \sim 0,$$

$$\big(\zeta_\alpha \,, \, ((a_{\varepsilon,r} + i b_{\varepsilon,r})\zeta)_\alpha\big)_0 \sim 0,$$

$$2\Big((\sigma^j_l \zeta_j)_\alpha \,, \, \big((\sigma_l + i\tau_l)\zeta\big)_\alpha\Big)_0 \sim 2\Big(\sigma^j_l \, \zeta_{\alpha,j} \,, \, (\sigma_l + i\tau_l)\zeta_\alpha\Big)_0$$

$$\sim -\int_{\mathbb{R}^d} \big(\sigma^j_l \sigma_l)\big)_{,j}(x) \, |\zeta_\alpha(x)|^2 \, dx \sim 0,$$

$$\big\|\big((\sigma_k + i\tau_k)\zeta\big)_\alpha\big\|^2_0 \sim 0.$$

Finally, we have $\Big(\zeta_\alpha \,, \, \sum_{j,k}\big(D_k\big(i b^{j,k}_r(t) D_j\zeta\big)_\alpha\Big)_0 = 0$. Set $L^0_r\zeta = \sum_{j,k=1}^d D_k\big(a^{j,k}_r D_j\zeta\big)$ and $S^0_l\zeta = \sum_{j=1}^d \big(\sigma^j_l + i\tau^j_l\big) D_j\zeta$. Then we have

$$Q^\alpha_t(\zeta, \zeta) \sim 2\sum_{r=0}^{d_1} \rho^r_{\varepsilon,t}\big(\zeta_\alpha \,, \, (L^0_r\zeta)_\alpha\big)_0 + \sum_{l\geq 1} q^l_t \, \|(S^0_l\zeta)_\alpha\|^2_0. \tag{3.43}$$

If $m = 0$, integration by parts proves that the right-hand side of (3.43) is equal to $-2[\zeta]^2_0$ (with $C_0 = 0$). Let $m \geq 1$ and α be a multi-index such that $m \geq |\alpha| \geq 1$; set $\Gamma(\alpha) = \{(\beta, \gamma) : \alpha = \beta + \gamma, |\beta| = 1\}$. For $\phi, \psi \in H^m$, let $C(\beta, \gamma)$ be coefficients such that

$$D^\alpha(\phi\psi) = \phi D^\alpha \psi + \sum_{(\beta,\gamma)\in\Gamma(\alpha)} C(\beta,\gamma) D^\beta\phi\, D^\gamma\psi + \sum_{\beta+\gamma=\alpha, |\beta|\geq 2} C(\beta,\gamma) D^\beta\phi\, D^\gamma\psi.$$

This yields

$$\begin{aligned}\sum_{l\geq 1} q_t^l \|(S_l^0\zeta)_\alpha\|_0^2 &\sim \sum_{l\geq 1} q_t^l \sum_{j,k=1}^d \Big\{\big(\sigma_l^k\,\zeta_{\alpha,k}\,,\,\sigma_l^j\,\zeta_{\alpha,j}\big)_0 \\ &\quad + 2\sum_{(\beta,\gamma)\in\Gamma(\alpha)} C(\beta,\gamma)\big(D^\beta\sigma_l^k\zeta_{\gamma,k}\,,\,\sigma_l^j\,\zeta_{\alpha,j}\big)_0\Big\}. \\ &\sim \sum_{l\geq 1} q_t^l \sum_{j,k=1}^d \Big\{\Big(\sigma_l^k\sigma_l^j\,\zeta_{\alpha,k}\,,\,\zeta_{\alpha,j}\Big)_0 + 2C(\alpha,\beta)\Big(D^\beta\sigma_l^k\sigma_l^j\,\zeta_{\gamma,k}\,,\,\zeta_{\alpha,j}\Big)_0.\end{aligned}$$

Since for $(\beta,\gamma)\in\Gamma(\alpha)$ we have $|\gamma|+1\leq|\alpha|\leq m$ while $|\beta|+1=2$, integrating by parts and using **(A3(m))** we have for fixed l,

$$2q_t^l\sum_{j,k}\Big(D^\beta\sigma_l^k\sigma_l^j\,\zeta_{\gamma,k}\,,\,\zeta_{\alpha,j}\Big)_0 = -q_t^l\sum_{j,k}\big(D^\beta(\sigma_l^k\sigma_l^j)\zeta_{\gamma,j,k}\,,\,\zeta_\alpha\big)_0.$$

Furthermore, integration by parts and **(A3(m))** yield

$$\begin{aligned}2\rho_{\varepsilon,t}^r\big(\zeta_\alpha\,,\,(L^0\zeta)_\alpha\big)_0 &\sim -2\sum_{j,k}\Big\{\big(a_{\varepsilon,r}^{j,k}\zeta_{\alpha,j}\,,\,\zeta_{\alpha,k}\big)_0 \\ &\quad - \sum_{(\beta,\gamma)\in\Gamma(\alpha)} C(\beta,\gamma)\big(D^\beta(a_{\varepsilon,r}^{j,k})\zeta_{\gamma,j}\,,\,\zeta_{\alpha,k}\big)_0\Big\}.\end{aligned}$$

Therefore, the definition of $\tilde{a}_{\varepsilon,r}^{j,k}$, (3.43) and (3.40) yield

$$\begin{aligned}Q_t^\alpha(\zeta,\zeta) &\sim -2\sum_{j,k}\big(\tilde{a}_\varepsilon^{j,k}\,\zeta_{\alpha,j}\,,\,\zeta_{\alpha,k}\big)_0 - 2\sum_{j,k}\sum_{(\beta,\gamma)\in\Gamma(\alpha)} C(\beta,\gamma)\big(D^\beta(\tilde{a}_\varepsilon^{j,k})\zeta_{\gamma,j}\,,\,\zeta_{\alpha,k}\big)_0 \\ &\sim -2\sum_{j,k}\Big(\big(\zeta_j\tilde{a}_\varepsilon^{j,k}\big)_\alpha\,,\,\zeta_{\alpha,k}\Big)_0.\end{aligned}$$

Hence for $\zeta\in H^{m+1}$,

$$p(\zeta) = \sum_{|\alpha|\leq m}\int_{\mathbb{R}^d} Q_t^\alpha(\zeta,\zeta)dx + 2[\zeta]_m^2 = 2\sum_{|\alpha|\leq m}(\zeta_\alpha, P^\alpha\zeta)_0, \tag{3.44}$$

for some operator P^α which satisfies (3.37). Hence (3.38) concludes the proof of (3.41). Polarizing (3.44), we deduce that for $\tilde{Z},\zeta\in H^{m+1}$,

$$\sum_{r=0}^{d_1} \rho^r_{\varepsilon,t}\big[\langle \tilde{Z}, L_r\zeta\rangle_m + \langle L_r\tilde{Z}, \zeta\rangle_m\big] + \sum_{l\geq 1} q^l_t (S_l\tilde{Z}, S_l\zeta)_m + 2[\tilde{Z}, \zeta]_m$$
$$= \sum_{|\alpha|\leq m} \big[(\tilde{Z}_\alpha, P^\alpha\zeta)_0 + (\zeta_\alpha, P^\alpha\tilde{Z})_0\big].$$

Let $\tilde{r} \in \{0, 1, \dots, d_1\}$ and for $Z \in H^{m+3}$, $\zeta \in H^{m+1}$, set $\tilde{Z} = L_{\tilde{r}}Z$; then if one sets

$$q_{\tilde{r}}(\zeta, Z) := \sum_{r=0}^{d_1} \rho^r_{\varepsilon,t}\big[\langle L_r\zeta, L_{\tilde{r}}Z\rangle_m + \langle \zeta, L_{\tilde{r}}L_rZ\rangle_m\big] + \sum_{l\geq 1} q^l_t\, \big(S_l\zeta, L_{\tilde{r}}S_lZ\big)_m,$$

one deduces that

$$q_{\tilde{r}}(\zeta, Z) + \sum_r \rho^r_{\varepsilon,t}\big(\zeta, [L_rL_{\tilde{r}} - L_{\tilde{r}}L_r]Z\big)_m + 2\sum_l q^l_t\big(S_l\zeta, [S_lL_{\tilde{r}} - L_{\tilde{r}}S_l]Z\big)_m + 2[\zeta, L_{\tilde{r}}Z]_m$$
$$= \sum_{|\alpha|\leq m} \big[(D^\alpha L_{\tilde{r}}Z, P^\alpha\zeta)_0 + (D^\alpha\zeta, P^\alpha L_{\tilde{r}}Z)_0\big].$$

The operators $L_rL_{\tilde{r}} - L_{\tilde{r}}L_r$ and $S_lL_{\tilde{r}} - L_{\tilde{r}}S_l$ are of order 3 and 2, respectively. Hence integration by parts and the Cauchy–Schwarz inequality imply that

$$|q_{\tilde{r}}(\zeta, Z)| \leq C\,\|\zeta\|_m\|Z\|_{m+3} + C[\zeta]_m[Z]_m.$$

Finally, (3.39) and Assumption **(A3(m))** imply that for $Z \in H^{m+1}$,

$$[Z]^2_m \leq C\,\|Z\|^2_{m+1} + C_m\|Z\|^2_m \leq C\,\|Z\|^2_{m+1}.$$

This concludes the proof of (3.42). □

The following lemma is based on some time integration by parts and requires the coefficients of L_r and F_r to be time independent.

Lemma 3.3. *Let the assumptions of Theorem 3.3 be satisfied and Z_ε (resp. ζ_ε) denote the processes defined by (3.33) (resp. (3.36)). For $r = 0, \dots, d_1$ and $t \in [0, T]$, let $A^r_t = V^r_{1,t} - V^r_{0,t}$ and set*

$$J_{\varepsilon,t} := \sum_{r=0}^{d_1} \int_0^t \big(\zeta_\varepsilon(s), L_rZ_\varepsilon(s) + F_r\big)_m dA^r_s. \tag{3.45}$$

Then there exists a constant C such that for any stopping time $\tau \leq T$,

$$\mathbb{E}\Big[\sup_{t\in[0,\tau]}\Big(J_{\varepsilon,t}-\int_0^t[\zeta_\varepsilon(s)]_m^2 dV_s\Big)_+^{p/2}\Big]$$
$$\leq \frac{1}{4p}\mathbb{E}\Big(\sup_{t\in[0,\tau]}\|\zeta_\varepsilon(s)\|_m^2\Big)+C\Big(A^p+\mathbb{E}\int_0^\tau\|\zeta_\varepsilon(s)\|_m^p\,dV_s\Big). \tag{3.46}$$

Proof. The main problem is to upper estimate $J_{\varepsilon,t}$ in terms of A and not in terms of the total variation of the measures dA_t^r. This requires some integration by parts; Eqs. (3.33) and (3.36) imply

$$J_{\varepsilon,t}=\sum_{r=0}^{d_1}\big(\zeta_\varepsilon(t)\,,\,L_rZ_\varepsilon(t)+F_r\big)_m A_t^r-\sum_{1\leq k\leq 4}J_{\varepsilon,t}^k, \tag{3.47}$$

where for $t\in[0,T]$ we set

$$J_{\varepsilon,t}^1=\sum_r A_s^r\Big[\sum_{\tilde r}\langle L_{\tilde r}\zeta_\varepsilon(s),L_rZ_\varepsilon(s)+F_r\rangle_m+\langle\zeta_\varepsilon(s),L_r[L_{\tilde r}Z_\varepsilon(s)+F_{\tilde r}]\rangle_m\Big]dV_{\varepsilon,s}^r,$$

$$J_{\varepsilon,t}^2=\int_0^t\sum_r A_s^r\sum_{l\geq 1}\big(S_l(s)\zeta_\varepsilon(s),L_r[S_l(s)Z_\varepsilon(s)+G_l(s)]\big)_m d\langle M^l\rangle_s,$$

$$J_{\varepsilon,t}^3=\int_0^t\sum_r A_s^r\sum_{l\geq 1}\Big[\big(S_l(s)\zeta_\varepsilon(s),L_rZ_\varepsilon(s)+F_r\big)_m$$
$$+\big(\zeta_\varepsilon(s),L_r[S_l(s)Z_\varepsilon(s)+G_l(s)]\big)_m\Big]dM_s^l,$$

$$J_{\varepsilon,t}^4=\int_0^t\sum_r A_s^r\Big[\sum_{\tilde r}\big(L_{\tilde r}Z_\varepsilon(s)+F_{\tilde r},L_rZ_\varepsilon(s)+F_r\big)_m\Big]dA_s^{\tilde r}.$$

Note that

$$J_{\varepsilon,t}^4=\frac{1}{2}\int_0^t\sum_{r,\tilde r}\big(L_{\tilde r}Z_\varepsilon(s)+F_{\tilde r},L_rZ_\varepsilon(s)+F_r\big)_m d(A_s^rA_s^{\tilde r}).$$

Using (3.42), integration by parts, Assumption **(A3(m))**, the Cauchy–Schwarz and Young inequalities, we deduce that

$$J_{\varepsilon,t}^1+J_{\varepsilon,t}^2\leq CA\int_0^t\Big[\|Z_\varepsilon(s)\|_{m+3}\{[\zeta_\varepsilon(s)]_m+\|\zeta_\varepsilon(s)\|_m\}+\sum_r\|\zeta_\varepsilon(s)\|_m\,\|F_r\|_{m+2}\Big]dV_s$$
$$+CA\int_0^t\sum_l\|\zeta_\varepsilon(s)\|_m\,\|G_l(s)\|_{m+3}\,d\langle M^l\rangle_s$$
$$\leq\int_0^t\big([\zeta_\varepsilon(s)]_m^2+\|\zeta_\varepsilon(s)\|_m^2\big)\,dV_s$$

$$+ CA^2 \int_0^t \Big[\Big(\|Z_\varepsilon(s)\|_{m+3}^2 + \sum_r \|F_r\|_{m+2}^2\Big) dV_s + \sum_{l\geq 1} \|G_l(s)\|_{m+3}^2\, d\langle M^l\rangle_s\Big]$$

$$\leq \int_0^t \big([\zeta_\varepsilon(s)]_m^2 + \|\zeta_\varepsilon(s)\|_m^2\big)\, dV_s + CA^2\Big(\int_0^t \|Z_\varepsilon(s)\|_{m+3}^2\, dV_s + K_{m+2}(t)\Big),$$

where the last inequality is deduced from Assumption **(A4(m+2,2))**. The Cauchy–Schwarz inequality, integration by parts and Assumption **(A3(m+1))** imply that for fixed $r = 0, \ldots, d_1$ and $l \geq 1$,

$$\begin{aligned}&|\big(S_l(s)\zeta_\varepsilon(s), L_r Z_\varepsilon(s) + F_r\big)_m| + |\big(\zeta_\varepsilon(s), L_r[S_l(s)Z_\varepsilon(s) + G_l(s)]\big)_m|\\ &\quad\leq C\|\zeta_\varepsilon(s)\|_m\big[\|Z_\varepsilon(s)\|_{m+3} + \|F_r\|_{m+1} + \|G_l(s)\|_{m+2}\big].\end{aligned}$$

Therefore, the Burkholder–Davis–Gundy inequality and Assumption **(A1)** imply that for any stopping time $\tau \leq T$, we have

$$\begin{aligned}&\mathbb{E}\Big(\sup_{t\in[0,\tau]} |J_{\varepsilon,t}^3|^{p/2}\Big)\\
&\leq CA^{p/2}\mathbb{E}\Big(\int_0^\tau \|\zeta_\varepsilon(s)\|_m^2\Big[\|Z_\varepsilon(s)\|_{m+3}^2 + \sum_{0\leq r\leq d_1} \|F_r\|_{m+1}^2\\
&\qquad + \sup_{s\in[0,T]} \sum_{l\geq 1} \|G_l(s)\|_{m+2}^2\Big] d\langle M^l\rangle_s\Big)^{p/4}\\
&\leq CA^{p/2}\mathbb{E}\Big[\Big(\sup_{s\in[0,\tau]} \|Z_\varepsilon(s)\|_{m+3}^{p/2} + \Big|\sum_{r=0}^{d_1} \|F_r\|_{m+1}^2\Big|^{p/4} + \sup_{s\in[0,T]}\Big|\sum_{l\geq 1} \|G_l(s)\|_{m+2}^2\Big|^{p/4}\Big)\\
&\qquad \times \Big(\int_0^\tau \|\zeta_\varepsilon(s)\|_m^2 dV_s\Big)^{p/4}\Big]\\
&\leq CA^p\, \mathbb{E}\Big[\Big(\sup_{s\in[0,T]} \|Z_\varepsilon(s)\|_{m+3}^p\Big) + \Big|\tilde{K}\sum_{r=0}^{d_1} \|F_r\|_{m+1}^2\Big|^{p/2} + \sup_{s\in[0,T]}\Big|\sum_{l\geq 1} \|G_l(s)\|_{m+2}^2\Big|^{p/2}\Big]\\
&\qquad + \frac{1}{8p}\mathbb{E}\Big(\sup_{s\in[0,\tau]} \|\zeta_\varepsilon(s)\|_m^p\Big) + C\mathbb{E}\int_0^\tau \|\zeta(s)\|_m^p dV_s.\end{aligned}$$

Using the condition (3.34) and Theorem 3.1 with $m + 3$, we deduce that

$$\mathbb{E}\Big(\sup_{s\in[0,\tau]} |J_{\varepsilon,t}^3|^{p/2}\Big) \leq \frac{1}{8p}\mathbb{E}\Big(\sup_{t\in[0,\tau]} \|\zeta_\varepsilon(t)\|_m^p\Big) + C\mathbb{E}\int_0^\tau \|\zeta_\varepsilon(s)\|_m^p dV_s + CA^p.$$

Therefore,

$$\mathbb{E}\Big\{\sup_{t\in[0,\tau]}\Big(J_{\varepsilon,t}-\int_0^t[\zeta_\varepsilon(s)]_m^p(s)dV_s\Big)_+^{p/2}\Big\}\le 1/(8p)\mathbb{E}\Big(\sup_{t\in[0,\tau]}\|\zeta_\varepsilon(t)\|_m^2\Big)$$
$$+C\mathbb{E}\int_0^\tau\|\zeta_\varepsilon(s)\|_m^p dV_s+CA^p+C\mathbb{E}\Big(\sup_{t\in[0,\tau]}|J^4_{\varepsilon,t}|^{p/2}\Big).$$

Integrating by parts we obtain

$$2J^4_{\varepsilon,t}=\sum_{r,\tilde r}\big(L_{\tilde r}Z_\varepsilon(t)+F_{\tilde r}\,,\,L_rZ_\varepsilon(t)+F_r\big)_m A^r_t A^{\tilde r}_t-\sum_{j=1}^3 J^{4,j}_{\varepsilon,t}, \tag{3.48}$$

where

$$J^{4,1}_{\varepsilon,t}=2\sum_{r,\tilde r}\sum_{\bar r}\int_0^t A^r_sA^{\tilde r}_s\big\langle L_{\bar r}[L_{\tilde r}Z_\varepsilon(s)+F_{\tilde r}]\,,\,L_rZ_\varepsilon(s)+F_r\big\rangle_m dV^{\bar r}_{\varepsilon,s},$$

$$J^{4,2}_{\varepsilon,t}=2\sum_{r,\tilde r}\sum_{l\ge1}\int_0^t A^r_sA^{\tilde r}_s\big(L_{\tilde r}[S_l(s)Z_\varepsilon(s)+G_l(s)]\,,\,L_rZ_\varepsilon(s)+F_r\big)_m dM^l_s,$$

$$J^{4,3}_{\varepsilon,t}=\sum_{r,\tilde r}\sum_{l\ge1}\int_0^t A^r_sA^{\tilde r}_s\big(L_{\tilde r}[S_l(s)Z_\varepsilon(s)+G_l(s)],L_r[S_l(s)Z_\varepsilon(s)+G_l(s)]\big)_m d\langle M^l\rangle_s.$$

Integration by part, Assumption **(A3(m+2))**, the Cauchy–Schwarz and Young inequalities yield

$$|\langle L_{\bar r}[L_{\tilde r}Z_\varepsilon(s)+F_{\tilde r}]\,,\,L_rZ_\varepsilon(s)+F_r\rangle_m|\le C\big[\|Z_\varepsilon(s)\|^2_{m+3}+\|F_{\tilde r}\|^2_{m+2}+\|F_r\|^2_m\big],$$
$$|\big(L_{\tilde r}[S_l(s)Z_\varepsilon(s)+G_l(s)]\,,\,L_rZ_\varepsilon(s)+F_r\big)_m|\le C\big[\|Z_\varepsilon(s)\|^2_{m+3}+\|G_l(s)\|^2_{m+2}+\|F_r\|^2_m\big].$$

Hence, using Theorem 3.1, (3.34), Assumptions **(A1)** and **(A4(m+2))** we deduce

$$\mathbb{E}\Big(\sup_{t\in[0,\tau]}|J^{4,1}_{\varepsilon,t}+J^{4,3}_{\varepsilon,t}|^{p/2}\Big)\le CA^p.$$

Finally, the Burkholder–Davis–Gundy inequality implies that

$$\mathbb{E}\Big(\sup_{t\in[0,\tau]}|J^{4,2}_{\varepsilon,t}|^{p/2}\Big)$$
$$\le CA^p\mathbb{E}\Big|\int_0^\tau|\big(L_{\tilde r}[S_l(s)Z_\varepsilon(s)+G_l(s)]\,,\,L_rZ_\varepsilon(s)+F_r\big)_m|^2d\langle M^l\rangle_s\Big|^{p/4}\le CA^p.$$

Hence, $\mathbb{E}\big(\sup_{t\in[0,\tau]}|J^4_{\varepsilon,t}|^{p/2}\big)\le CA^p$, which concludes the proof. □

Using Lemmas 3.1–3.3, we now prove Theorem 3.3 for time-independent coefficients.

Proof of Theorem 3.3 Apply the operator D^α to both sides of (3.36) and use the Itô formula for $\| D^\alpha \zeta_\varepsilon(t)\|_0^2$. This yields

$$d\,\|\zeta_\varepsilon(t)\|_m^2 = 2\sum_{|\alpha|\le m}\sum_r \Big[\big\langle \zeta_\varepsilon(t), L_r\zeta_\varepsilon(t)\big\rangle_m \rho^r_{\varepsilon,t}\,dV_t + \big(\zeta_\varepsilon(t), L_r Z_\varepsilon(t) + F_r\big)_m dA^r_t\Big]$$
$$+\sum_{|\alpha|\le m}\sum_{l\ge 1}\Big[\|S_l(t)\zeta_\varepsilon(t)\|_m^2 q^l_t\,dV_t + 2\big(\zeta_\varepsilon(t), S_l(t)\zeta_\varepsilon(t)\big)_m dM^l_t\Big],$$

where $\langle Z,\zeta\rangle_m$ denotes the duality between H^{m-1} and H^{m+1} which extends the scalar product in H^m. Using (3.41) we deduce that

$$d\,\|\zeta_\varepsilon(t)\|_m^2 \le -2[\zeta_\varepsilon(t)]_m^2\,dV_t + C\,\|\zeta_\varepsilon(t)\|_m^2\,dV_t + 2dJ_{\varepsilon,t} + 2\sum_{l\ge 1}\big(\zeta_\varepsilon(t), S_l(t)\zeta_\varepsilon(t)\big)_m dM^l_t,$$

where $J_{\varepsilon,t}$ is defined by (3.45). Using (3.26) we deduce that $\big|\big(\zeta_\varepsilon(t), S_l(t)\zeta_\varepsilon(t)\big)_m\big| \le C\,\|\zeta_\varepsilon(t)\|_m^2$. Thus Lemma 3.3, the Burkholder–Davis–Gundy inequality and Assumption **(A1)** yield for any stopping time $\tau \le T$

$$\mathbb{E}\Big(\sup_{t\in[0,\tau]}\|\zeta_\varepsilon(t)\|_m^p\Big) \le C\mathbb{E}\|Z_{1,0} - Z_{0,0}\|_m^p + p\mathbb{E}\Big(\sup_{t\in[0,\tau]} J_{\varepsilon,t} - \int_0^t [\zeta_\varepsilon(s)]_m^2\,dV_s\Big)_+^{p/2}$$
$$+ C_p\mathbb{E}\Big|\int_0^\tau \|\zeta_\varepsilon(s)\|_m^4\,dV_s\Big|^{p/4} + C_p\mathbb{E}\Big|\int_0^\tau \|\zeta_\varepsilon(s)\|_m^2\,dV_s\Big|^{p/2}$$
$$\le C\mathbb{E}\|Z_{1,0} - Z_{0,0}\|_m^p + \frac{1}{4}\mathbb{E}\Big(\sup_{t\in[0,\tau]}\|\zeta_\varepsilon(t)\|_m^p\Big) + C\Big(A^p + \mathbb{E}\int_0^\tau \|\zeta_\varepsilon(s)\|_m^p\,dV_s\Big)$$
$$+ C\mathbb{E}\Big[\sup_{t\in[0,\tau]}\|\zeta_\varepsilon(t)\|_m^{p/2}\Big(\int_0^\tau \|\zeta_\varepsilon(s)\|_m^2\,dV_s\Big)^{p/4}\Big] + C_p\mathbb{E}\int_0^\tau \|\zeta_\varepsilon(s)\|_m^p\,dV_s$$
$$\le C\mathbb{E}\|Z_{1,0} - Z_{0,0}\|_m^p + CA^p + \frac{1}{2}\mathbb{E}\Big(\sup_{t\in[0,\tau]}\|\zeta_\varepsilon(t)\|_m^p\Big) + C\mathbb{E}\Big(\int_0^\tau \|\zeta_\varepsilon(s)\|_m^p\,dV_s\Big),$$

where the last upper estimate follows from the Young inequality. Let $\tau_N = \inf\{t : \|\zeta_\varepsilon(t)\|_m^p \ge N\} \wedge T$; then the Gronwall Lemma implies that

$$\mathbb{E}\Big(\sup_{t\in[0,\tau_N]}\|\zeta_\varepsilon(t)\|_m^p\Big) \le C\big(\mathbb{E}\|Z_{1,0} - Z_{0,0}\|_m^p + A^p\big).$$

Letting $N \to \infty$ concludes the proof. □

3.3.2 Case of the Time-Dependent Coefficients

In this section, we prove a convergence result similar to that in Theorem 3.3 when the coefficients of the operators depend on time. Integration by parts in Lemma 3.3 will give extra terms, which require more assumptions to be dealt with.

Assumption (A5(m)) There exists an integer number d_2, a $(\mathscr{F}_t)$-continuous martingale $N_t = (N_t^1, \ldots, N_t^{d_2})$ and for each $\gamma = 0, \ldots, d_2$ a bounded predicable process $h_\gamma : \Omega \times (0,T] \times \mathbb{R}^d \to \mathbb{R}^N$ for some N depending on d and d_1 such that

$$h_\gamma(t,x) := (a_{\gamma,r}^{j,k}(t,x), b_{\gamma,r}^{j,k}(t), a_{\gamma,r}^{j}(t,x), a_{\gamma,r}(t,x), b_{\gamma,r}(t,x), F_{\gamma,r}(t,x);$$
$$1 \le j,k \le d,\ 0 \le r \le d_1,\ 1 \le \gamma \le d_2),$$

for some symmetric non-negative matrices $(a_{\gamma,r}^{j,k}(t,x), j,k = 1, \ldots, d)$ and $(b_{\gamma,r}^{j,k}(t), j,k = 1, \ldots, d)$. Furthermore, we suppose that for every $\omega \in \Omega$ and $t \in [0,T]$, the maps $h_\gamma(t,\cdot)$ are of class $\mathscr{C}^{m+1}$ such that for some constant K we have $|D^\alpha h_\gamma(t,\cdot)| \le K$ for any multi-index α with $|\alpha| \le m+1$ and such that for $t \in [0,T]$,

$$\sum_{\gamma=1}^{d_2} d\langle N^\gamma\rangle_t \le dV_t,$$

$$h(t,x) = h(0,x) + \int_0^t h_0(s,x)dV_s + \sum_{\gamma=1}^{d_2} \int_0^t h_\gamma(s,x)dN_s^\gamma.$$

For $\gamma = 0, \ldots, d_2$, $r = 0, \ldots, d_1$, let $L_{\gamma,r}$ be the time-dependent differential operator defined by

$$L_{\gamma,r} Z(t,x) = \sum_{j,k=1}^{d} D_k\Big(\big[a_{\gamma,r}^{j,k}(t,x) + i b_{\gamma,r}^{j,k}(t)\big] D_j Z(t,x)\Big) + \sum_{j=1}^{d} a_{\gamma,r}^{j}(t,x) D_j Z(t,x)$$
$$+ \big[a_{\gamma,r}(t,x) + i b_{\gamma,r}(t,x)\big] Z(t,x).$$

For $r = 0, \ldots, d_1$, let

$$L_r Z(t,x) = L_r(0) Z(0,x) + \int_0^t L_{0,r} Z(s,x) dV_s + \sum_{\gamma=1}^{d_2} L_{\gamma,r} Z(s,x) dN_s^\gamma,$$

and $F_r(t,x) = F_r(0,x) + \int_0^t F_{0,r}(s,x)dV_s + \sum_{\gamma=1}^{d_2} F_{\gamma,r}(s,x)dN_s^\gamma$. We then have the following abstract convergence result which extends Theorem 3.3.

Theorem 3.4. *Suppose that Assumptions* **(A(1))**, **(A(2))**, **(A3(m+3))**, **(A4(m+3,p))** *and* **(A5(m))** *are satisfied and that*

$$\mathbb{E}\left(\sup_{t\in[0,T]}\Big|\sum_{r=0}^{d_1}\|F_r(t)\|_{m+1}^2\Big|^{p/2}+\sup_{t\in[0,T]}\Big|\sum_{l\geq 1}\|G_{m+2}(t)\|_{m+2}^2\Big|^{p/2}\right)<\infty. \quad (3.49)$$

Then there exists some constant $C>0$ such that

$$\mathbb{E}\Big(\sup_{t\in[0,T]}\|Z_1(t)-Z_0(t)\|_m^p\Big)\leq C\Big(\mathbb{E}\big(\|Z_1(0)-Z_0(0)\|_m^p\big)+A^p\Big). \quad (3.50)$$

Proof. Since Lemmas 3.1 and 3.2 did not depend on the fact that the coefficients are time-independent, only Lemma 3.3 has to be extended. For $t\in[0,T]$, let

$$J_{\varepsilon,t}=\sum_{r=0}^{d_1}\int_0^t\big(\zeta_\varepsilon(s)\,,\,L_r(s)Z_\varepsilon(s)+F_r(s)\big)_m dA_s^r.$$

Since $A_0^r=0$ for $r=0,\dots,d_1$, the integration by parts formula (3.47) has to be replaced by

$$J_{\varepsilon,t}=\sum_{r=0}^{d_1}\big(\zeta_\varepsilon(t)\,,\,L_r(t)Z_\varepsilon(t)+F_r(t)\big)_m A_t^r-\sum_{k=1}^{7}J_{\varepsilon,t}^k,\ t\in[0,T],$$

where the additional terms on the right-hand side are defined for $t\in[0,T]$ as follows:

$$J_{\varepsilon,t}^5=\sum_r A_s^r\big(\zeta_\varepsilon(s),L_{0,r}Z_\varepsilon(s)+F_{0,r}\big)_m dV_s,$$

$$J_{\varepsilon,t}^6=\int_0^t\sum_r A_s^r\sum_{\gamma=1}^{d_2}\big(\zeta_\varepsilon(s),L_{\gamma,r}Z_\varepsilon(s)+F_{\gamma,r}(s)\big)_m dN_s^\gamma,$$

$$J_{\varepsilon,t}^7=\int_0^t\sum_r A_s^r\sum_{l\geq 1}\sum_{\gamma=1}^{d_2}\big(S_l(s)\zeta_\varepsilon(s),L_{\gamma,r}Z^\varepsilon(s)+F_{\gamma,r}(s)\big)_m d\langle M^l,N^\gamma\rangle_s.$$

Arguments similar to those used in the proof of Lemma 3.3, using integration by parts and the regularity assumptions of the coefficients, prove that for $k=5,6$ there exists a constant $C>0$ such that for any stopping time $tau\leq T$ we have

$$\begin{aligned}\mathbb{E}\Big(\sup_{t\in[0,\tau]}|J_{\varepsilon,t}^k|^{p/2}\Big)&\leq CA^{p/2}\mathbb{E}\Big(\sup_{t\in[0,\tau]}\|\zeta_\varepsilon(t)\|_m^{p/2}\sup_{t\in[0,\tau]}\big(\|Z_\varepsilon(t)\|_{m+2}+C\big)^{p/2}\Big)\\&\leq\frac{1}{24p}\mathbb{E}\Big(\sup_{t\in[0,\tau]}\|\zeta_\varepsilon(t)\|_m^p\Big)+CA^p,\end{aligned}$$

where the last inequality follows from the Young inequality. Furthermore, the Burkholder–Davis–Gundy inequality and the upper estimates of the quadratic variations of the martingales N^γ yield for every $\gamma = 1, \dots, d_2$, $r = 0, \dots, d_1$ and $\tau \in [0, T]$,

$$\mathbb{E}\Big(\sup_{t\in[0,\tau]} |J^7_{\varepsilon,t}|^{p/2}\Big) \le C\mathbb{E}\Big(\int_0^\tau \big|A^r_s\big(\zeta_\varepsilon(s), L_{\gamma,r}Z_\varepsilon(s) + F_{\gamma,r}(s)\big)_m\big|^2 dV_s\Big)^{p/4}$$

$$\le CA^{p/2}\mathbb{E}\Big(\sup_{t\in[0,\tau]} \|\zeta_\varepsilon(t)\|_m^{p/2} \sup_{t\in[0,\tau]} \big(\|Z_\varepsilon(t)\|_{m+2} + C\big)^{p/2}\Big).$$

Hence, the proof will completed by extending the upper estimate (3.48) as follows:

$$2J^4_{\varepsilon,t} = \sum_{r,\tilde r} \big(L_{\tilde r} Z_\varepsilon(t) + F_{\tilde r}(t)\,,\, L_r Z_\varepsilon(t) + F_r(t)\big)_m A^r_t A^{\tilde r}_t - \sum_{j=1}^7 J^{4,j}_{\varepsilon,t}\,,$$

where for $j = 4, \dots, 7$ we have

$$J^{4,4}_{\varepsilon,t} = 2\sum_{r,\tilde r}\int_0^t A^r_s A^{\tilde r}_s \big(L_{\tilde r,0}(s)Z_\varepsilon(s) + F_{\tilde r,0}(s)\,,\, L_r(s)Z_\varepsilon(s) + F_r(s)\big)_m dV_s,$$

$$J^{4,5}_{\varepsilon,t} = \sum_{r,\tilde r}\sum_{\gamma,\tilde\gamma}\int_0^t A^r_s A^{\tilde r}_s \big(L_{\tilde\gamma,\tilde r}(s)Z_\varepsilon(s) + F_{\tilde\gamma,\tilde r}(s)\,,\, L_{\gamma,r}Z_\varepsilon(s) + F_{\gamma,r}(s)\big)_m d\langle N^{\tilde\gamma}, N^\gamma\rangle_s,$$

$$J^{4,6}_{\varepsilon,t} = 2\sum_{r,\tilde r}\sum_{\gamma}\sum_{l\ge1}\int_0^t A^r_s A^{\tilde r}_s \big(L_{\gamma,\tilde r}(s)Z_\varepsilon(s) + F_{\gamma,\tilde r}(s)\,,$$

$$L_r(s)[S_l(s)Z_\varepsilon(s) + G_l(s)]\big)_m d\langle N^\gamma, M^l\rangle_s,$$

$$J^{4,7}_{\varepsilon,t} = 2\sum_{r,\tilde r}\sum_{\gamma}\int_0^t A^r_s A^{\tilde r}_s \big(L_{\gamma,\tilde r}(s)Z_\varepsilon(s) + F_{\gamma,\tilde r}(s)\,,\, L_r(s)Z_\varepsilon(s) + F_r(s)\big)_m dN^\gamma_s\,.$$

We obtain upper estimates of the terms $\mathbb{E}\big(\sup_{t\in[0,\tau]} |J^{4,k}_{\varepsilon,t}|^{p/2}\big)$ for $k = 4, \dots, 7$ by arguments similar to that used for $k = 1, \dots, 3$, which implies $\mathbb{E}\big(\sup_{t\in[0,\tau]} |J^4_{\varepsilon,t}|^{p/2}\big) \le CA^p$. This concludes the proof. □

3.4 Speed of Convergence for the Splitting Method

The aim of this section is to show how the abstract convergence results obtained in Sect. 3.3 yield the convergence of a splitting method and extends the corresponding results from [9]. The proof, which is very similar to that in [9], is briefly sketched for the reader's convenience.

Assumption (A) *For $r = 0, \dots, d_1$, let L_r be defined by* (3.4) *and for $l \geq 1$ let S_l be defined by* (3.5). *Suppose that the Assumptions* **(A2)** *and* **(A3(m+3))** *are satisfied, and that for every $\omega \in \Omega$ and r, l, $F_r(t) = F_r(t, \cdot)$ is a weakly continuous H^{m+3}-valued function and $G_l(t) = G_l(t, \cdot)$ is a weakly continuous H^{m+4}-valued function. Suppose furthermore that the initial condition $Z_0 \in L^2(\Omega; H^{m+3})$ is $\mathscr{F}_0$-measurable, that F_r and G_l are predictable and that for some constant K one has*

$$\mathbb{E}\Big(\sup_{t\in[0,T]} \sum_{r=0}^{d_1} \|F_r(t,\cdot)\|_{m+3}^p + \sup_{t\in[0,T]} \sum_{l\geq 1} \|G(t,\cdot)\|_{m+4}^p + \|Z_0\|_{m+3}^p \Big) \leq K.$$

Let $V^0 = (V_t^0, t \in [0,T])$ be a predictable, continuous increasing process such that $V_0^0 = 0$ and that there exists a constant K such that $V_T^0 + \sum_{l\geq 1} \langle M^l\rangle_T \leq K$. Finally suppose that the following stochastic parabolicity condition holds. For every $(t,x) \in [0,T] \times \mathbb{R}^d$, every $\omega \in \Omega$ and every $\lambda \in \mathbb{R}^d$,

$$\sum_{j,k=1}^{d} \lambda_j \lambda_k \Big[2a_0^{j,k}(t,x) dV_t^0 + \sum_{l\geq 1} \sigma_l^j(t,x)\sigma_l^k(t,x) d\langle M\rangle_t \Big] \geq 0$$

in the sense of measures on $[0,T]$.

Let Z be the process solution to the evolution equation

$$\begin{aligned} dZ(t,x) =& \big(L_0 Z(t,x) + F_0(t,x)\big) dV_t^0 + \sum_{r=1}^{d_1} \big(L_r Z(t,x) + F_r(t,x)\big) dt \\ &+ \sum_{l\geq 1} \big(S_l Z(t,x) + G_l(t,x)\big) dM_t^l \end{aligned} \tag{3.51}$$

with the initial condition $Z(0,\cdot) = Z_0$. Then Theorem 3.1 proves the existence and uniqueness of the solution to (3.51), and that

$$\mathbb{E}\Big(\sup_{t\in[0,T]} \|Z(t)\|_{m+3}^p \Big) \leq C$$

for some constant C which depends only on d, d_1, K, m, p and T.

For every integer $n \geq 1$ let $\mathscr{T}_n = \{t_i := iT/n, i = 0, 1, \dots, n\}$ denote a grid on the interval $[0,T]$ with constant mesh $\delta = T/n$. For $n \geq 1$, let $Z^{(n)}$ denote the approximation of Z defined for $t \in \mathscr{T}_n$ using the following splitting method: $Z^{(n)}n(0) = 0$ and for $i = 0, 1, \dots, n-1$, let

$$Z^{(n)}(t_{i+1}) := P_\delta^{(d_1)} \cdots P_\delta^{(2)} P_\delta^{(1)} Q_{t_i, t_{i+1}} Z^{(n)}(t_i), \tag{3.52}$$

where for $r = 1, \dots, d_1$ and $t \in [0, T]$, $P_t^{(r)}\psi$ denotes the solution ζ_r of the evolution equation

$$d\zeta_r(t,x) = \big(L_r\zeta_r(t,x) + F_r(t,x)\big)dt \quad \text{and } \zeta_r(0,x) = \psi(x),$$

and for $s \in [0,t] \le T$, $Q_{s,t}\psi$ denotes the solution η of the evolution equation defined on $[s,T]$ by the "initial" condition $\eta(s,x) = \psi(x)$ and for $t \in [s,T]$ by

$$d\eta(t,x) = \big(L_0\eta(t,x) + F_0(t,x)\big)dV_t^0 + \sum_{l\ge 1}\big(S_l\eta(t,x) + G_l(t,x)\big)dM_t^l.$$

The following theorem gives the speed of convergence of this approximation.

Theorem 3.5. *Let $a_r^{j,k}, b_r^{j,k}, a_r^j, a_r, b_r, \sigma_l^j, \sigma_l, \tau_l, F_r, G_l$ satisfy the Assumption* **(A)**. *Suppose that $a_r^{j,k}, b_r^{j,k}, a_r^j, a_r, b_r$ and F_r are time-independent. Then there exists a constant $C > 0$ such that*

$$\mathbb{E}\Big(\sum_{t\in\mathcal{T}_n} \|Z^{(n)}(t) - Z(t)\|_m^p\Big) \le C n^{-p} \quad \textit{for every } n \ge 1.$$

Proof. Let $d' = d_1 + 1$ and let us introduce the following time change:

$$\kappa(t) = \begin{cases} 0 & \text{for } t \le 0, \\ t - k\delta d_1 & \text{for } t \in [kd'\delta, (kd'+1)\delta),\ k = 0,1,\dots,n-1, \\ (k+1)\delta & \text{for } t \in [(kd'+1)\delta, (k+1)d'\delta),\ k = 0,1,\dots,n-1. \end{cases}$$

Let, for every $t \in [0,T]$,

$$\begin{aligned} &\tilde{M}^l(t) = M^l_{\kappa(t)}, \quad \tilde{\mathcal{F}}_t = \mathcal{F}_{\kappa(t)}, \qquad \tilde{V}^0_{t,0} = \tilde{V}^0_{t,1} = V^0_{\kappa(t)}, \\ &\tilde{V}^r_{t,0} = \kappa(t), \qquad \tilde{V}^r_{t,1} = \kappa(t - r\delta) \text{ for } r = 1,2,\dots,d_1. \end{aligned}$$

For $\varepsilon = 0,1$, consider the evolution equations with the same initial condition $Z_0(0,x) = Z_1(0,x) = Z_0(x)$ and

$$dZ_\varepsilon(t) = \sum_{r=0}^{d_1}\big(L_r Z_\varepsilon(t) + F_r\big)d\tilde{V}^r_{t,\varepsilon} + \big(S_l Z_\varepsilon(t) + G_l\big)d\tilde{M}^l_t. \tag{3.53}$$

One easily checks that the Assumptions **(A1), (A2), (A3(m+3)), and (A4(m+3,p))** are satisfied with the martingales $\tilde{M}_l$ and the increasing processes $\tilde{V}^r_{\varepsilon,t}$ for $\varepsilon = 0,1$ and $r = 0,1,\dots,d_1$. Therefore, Theorem 3.1 implies that for $\varepsilon = 0,1$, Eq. (3.53) has a unique solution. Furthermore, since condition (3.34) holds, Theorem 3.3 proves the existence of a constant C such that

$$\mathbb{E}\Big(\sup_{t\in[0,d'T]}\|Z_1(t)-Z_0(t)\|_m^p\Big)\leq C\sup_{t\in[0,d'T]}\max_{1\leq r\leq d_1}|\kappa(t+r\delta)-\kappa(t)|^p=CT^pn^{-p}.$$

Since by construction, we have $Z_0(d't)=Z(t)$ and $Z_1(d't)=Z^{(n)}(t)$ for $t\in\mathcal{T}_n$, this concludes the proof. □

Note that the above theorem yields a splitting method for the following linear Schrödinger equation on $\mathbb{R}^d$:

$$dZ(t,x)=\Big(i\Delta Z(t,x)+\sum_{j=1}^{d}a^j(x)D_jZ(t,x)+F(x)\Big)dt$$
$$+\sum_{l\geq 1}\big[\big(\sigma_l(x)+i\tau_l(x)\big)Z(t,x)+G_l(x)\big]dM_t^l,$$

where a^j, F (resp. σ_l, τ_l and G_l) belong to H^{m+3} (resp. H^{m+4}). Indeed, this model is obtained with $a^{j,k}=\sigma_l^j=0$ and $b^{j,k}=1$ for $j,k=1,\ldots,d$ and $l\geq 1$.

Finally, Theorem 3.4 yields the following theorem for the splitting method in the case of time-dependent coefficients. The proof, similar to that of Theorem 3.5, will be omitted; see also [9], Theorem 5.2.

Theorem 3.6. *Let* $a_r^{j,k}, b_r^{j,k}, a_r^j, \sigma_l^j, \sigma_l, \tau_l, F, G_l$ *satisfy Assumptions* (**A**) *and* (**A5(m)**). *For every integer* $n\geq 1$ *let* $Z^{(n)}$ *be defined by* (3.52) *when the operators* L_r, S_l, *the processes* F_r *and* G_l *depend on time in a predictable way. Then there exists a constant* $C>0$ *such that for every* $n\geq 1$, *we have*

$$\mathbb{E}\Big(\sum_{t\in\mathcal{T}_n}\|Z^{(n)}(t)-Z(t)\|_m^p\Big)\leq Cn^{-p}.$$

References

1. Blanes, S., Moan, P.C.: Splitting methods for the time-dependent Schrödinger equation. Phys. Lett. A **265**(1–2), 3542 (2000)
2. de Bouard, A., Debussche, A.: A semi-discrete scheme for the nonlinear Schrödinger equation. Num. Math. **96**, 733–770 (2004)
3. de Bouard, A., Debussche, A.: Weak and strong order of convergence for a semidiscrete scheme for the stochastic nonlinear Schrödinger equation. Appl. Math. Optim. **54**, 369–399 (2006)
4. Debussche, A., Printems, J.: Numerical simulations for the stochastic Korteweg-de-Vries equation. Physica D **134**, 200–226 (1999)
5. Debussche, A., Printems, J.: Weak order for the discretization of the stochastic heat equation. Math. Comp. **78**(266), 845–863 (2009)
6. Dujardin, G., Faou, E.: Qualitative behavior of splitting methods for the linear Schrödinger equation in molecular dynamics. CANUM 2006—CongrÃšs National d'Analyse Numérique, pp. 234–239. ESAIM Proc., vol. 22. EDP Sci., Les Ulis (2008)
7. Gyöngy, I.: Lattice approximations for stochastic quasi-linear parabolic partial differential equations driven by space-time white noise I. Potential Anal. **9**, 1–25 (1998)

8. Gyöngy, I.: Lattice approximations for stochastic quasi-linear parabolic partial differential equations driven by space-time white noise II. Potential Anal. **11**, 1–37 (1999)
9. Gyöngy, I., Krylov, N.V.: On the splitting-up method and stochastic partial differential equations. Ann. Probab. **31–32**, 564–591 (2003)
10. Gyöngy, I., Millet, A.: On discretization schemes for stochastic evolution equations. Potential Anal. **23**, 99–134 (2005)
11. Gyöngy, I., Millet, A.: Rate of convergence of space time approximations for stochastic evolution equations. Potential Anal. **30**(1), 29–64 (2009)
12. Hausenblas, E.: Approximation for semilinear stochastic evolution equations. Potential Anal. **18**(2), 141–186 (2003)
13. Krylov, N.V., Rosovskii, B.L.: Characteristics of degenerating second-order parabolic Itô equations. Journal of Mathematical Sciences **32**(4), 336–348 (1986)
14. Krylov, N.V., Rosovskii, B.L.: Stochastic evolution equations. J. Soviet Math. **16**, 1233–1277 (1981)
15. Lions, J.L., Magenes, E.: Non-homogeneous boundary value problems and applications. Vol. I. Grundlehren der Mathematischen Wissenschaften [Fundamental Principles of Mathematical Sciences], vol. 181. Springer, New York (1972)
16. Millet, A., Morien, P.-L.: On implicit and explicit discretization schemes for parabolic SPDEs in any dimension. Stoch. Process. Appl. **115**(7), 1073–1106 (2005)
17. Neuhauser, C., Thalhammer, M.: On the convergence of splitting methods for linear evolutionary Schrödinger equations involving an unbounded potential. BIT **49**(1), 199–215 (2009)
18. Pardoux, E.: Équations aux dérivées partielles stochastiques nonlinéares monotones. Étude de solutions fortes de type Itô, Thèse Doct. Sci. Math. Univ. Paris Sud. (1975)
19. Printems, J.: On the discretization in time of parabolic stochastic differential equations. Math. Model Numer. Anal. **35**(6), 1055–1078 (2001)

Chapter 4
A Modelization of Public–Private Partnerships with Failure Time

Caroline Hillairet and Monique Pontier

Abstract A commissioning of public works (Maitrise d'ouvrage publique, namely MOP in French) is a system where the community has commissioned equipment (hospital, prison, etc.) for its own needs and to bear the cost, partly by self and partly by a loan from a bank. On another hand, public–private partnership (PPP) means that community agrees on a period (15–25 years) with the contractor and is billed rent. More or less it means "leasing" purchase, covering three parts: depreciation of equipment, maintenance costs, and financial costs. This new formula is based on an "ordonnance" of June 17, 2004, amended by the law of 28 July 2008 (see legifrance.gouv.fr), justified by the emergency of requested equipment construction or its complexity. Our aim is to study the advantages and disadvantages of the new PPP formula. Here is a particular case of a risk-neutral consortium. We discuss the advantages of outsourcing ("externalization") in terms of model parameters and prove that externality is interesting only in case of large enough noise when we exclude the risk of bankruptcy. Indeed, this risk does not seem covered under current legislation. Finally, we study what could happen in case of failure penalties to be paid by the private consortium. In such a case, externality could be interesting in some context as high noise, high reference cost, short maturity, and high enough penalty.

C. Hillairet
Laboratoire du CMAP Ecole Polytechnique, Route de Saclay, 91128 Palaiseau Cedex, France
e-mail: hillaire@cmapx.polytechnique.fr

M. Pontier (✉)
Institut Mathématique de Toulouse, Laboratoire de Statistique et Probabilités, Université Paul Sabatier, 31062 Toulouse Cedex 9, France
e-mail: pontier@math.univ-toulouse.fr

L. Decreusefond and J. Najim (eds.), *Stochastic Analysis and Related Topics*, Springer Proceedings in Mathematics & Statistics 22, DOI 10.1007/978-3-642-29982-7_4,

4.1 Introduction

In the classic formula of a public project, a commissioning of public works (Maitrise d'ouvrage publique, namely MOP in French) the community realizes equipments (hospital, prison, etc.) for its own needs and to bear the cost, partly by self and partly by a loan from a bank.

In the formula "partnership agreement" (or public–private partnership, namely PPP) community agrees on a period (15–25 years) with the contractor and is charged a rent. Somehow, it is a lease purchase, covering three parts: depreciation of equipment, maintenance costs, and financial costs. This formula is based on an "ordonnance" of June 17, 2004, amended by the law of 28 July 2008 (see legifrance.gouv.fr). The justification for this device is mainly based on the urgency of requested equipment construction or its complexity. Here are studied the advantages and disadvantages of the PPP contract system. We have not included problems of taxation (as the "ordonnance" and the law do). However, it should be noted that part of value added tax (VAT) is recoverable in case of MOP, whereas in the framework of a PPP, not only it is not, but it is also added at the VAT payable on the loan. This particularity could have an influence, but this problem is not addressed in this paper. Here in particular, in the case of a risk-neutral consortium, we discuss the benefits of outsourcing in terms of model parameters. We show that when including the risk of bankruptcy, the externality can be interesting when a penalty is imposed on the consortium in case of bankruptcy and in a certain context: for instance when uncertainty is high enough, or the reference cost is important, or short maturity, or sufficient penalty. In fact, this corresponds to a risk transfer from public to private.

Section 4.2 sets the problem, following Iossa et al. model [1] and introduces the various parameters of the problem. In Sect. 4.3 we solve an optimization problem simultaneously for the consortium and the public community. Then we study in Sect. 4.4 the effects of introducing a bankruptcy time whose risk does not seem covered under the legislation above-named; this changes the model. If no penalty is required, the result of the optimization yields to choose a minimal externality in contrast to the result in case of absence of bankruptcy (Sect. 4.3). Finally, in Sect. 4.5, the consortium is obliged to pay penalties in case of bankruptcy: it is the only case discussed here where in a particular configuration of the game settings, outsourcing can be interesting for both parties. Section 4.6 gathers these results.

4.2 The Problem Setting

We follow here the framework of [1] by adding a stochastic view point. To modelize the randomness of the model, we introduce a filtered probability space $(\Omega, \mathbf{F} = (\mathcal{F}_t)_{t\in[0,T]}, \mathbf{P})$.

The *operational cost* $(C_s)_{s\in[0,T]}$ of the infrastructure maintenance is a nonnegative $\mathbf{F}$-adapted process. $(C_s)_{s\in[0,T]}$ is a rate (its unit is euros per unit time) and can be written as

$$C_s = \theta_0 - e_s - \delta a + \varepsilon_s, \quad s \in [0, T], \tag{4.1}$$

where

- θ_0 is the benchmark cost of the maintenance.
- e_s is the effort on the maintenance done at date s to reduce the cost, it is a nonnegative $\mathbf{F}$-adapted process.
- a is the effort on the construction to improve the infrastructure quality, it is a parameter in $\mathbb{R}^+$.
- $(\varepsilon_s)_{s\in[0,T]}$ is a centered bounded $\mathbf{F}$-adapted process that modelizes the random operational risk of the activity. We will assume that $\varepsilon \in [-m, M]$, $dt \otimes d\mathbf{P}\,a.s.$
- δ is the externality, it is a parameter in $\mathbb{R}^+$.

The externality represents the impact of the infrastructure on the maintenance cost. We assume that improving the infrastructure quality reduces the maintenance operational cost, thus the externality δ is nonnegative. The maintenance cost is payable by the consortium until the maturity T or until a possible default of the consortium. We will assume the natural condition that the costs are nonnegative a.s. This condition leads to some constraint detailed in Sect. 4.3.2, using the expression of the optimal efforts.

The community pays to the consortium a rent $t(c)$ which is a function of the cost c: this rent permits both to pay the consortium for its work and to cover the maintenance costs that are in its charge. We assume that the community chooses a linear expression for the rent:

$$t(c) = \alpha - \beta c, \text{ with } \beta \geq -1, \text{ and } \alpha \text{ such that a.s. } t(C_s) \geq C_s \ \forall s \in [0, T].$$

$t(c) - c$ is a decreasing function of the costs C_s, and thus an increasing function of the efforts e_s. The larger β is, the greater is the incitement to the consortium to make effort on the infrastructure, but at the cost of a greater risk premium α.

Remark 4.1. ε_s being in the interval $[-m, M]$ $\forall s \in [0, T]$ $(m > 0,\ M > 0)$ the condition $t(C_s) \geq C_s$ $\forall s \in [0, T]$ is satisfied as soon as $\alpha \geq (\beta + 1)(\theta_0 + M)$.

The consortium aims to maximize its terminal utility, discounted at the rate $r \geq 0$, its optimization problem can be formulated as follows:

$$\max_{(a,e)\in[0,+\infty[\times E} \left(\mathbf{E}\left(\int_0^T e^{-rs}\,(U(t(C_s) - C_s, s) - \phi(e_s))\,ds \right) - \psi(a) \right) \tag{4.2}$$

with $E = \{(e_s)_{s\in[0,T]}\ \mathbf{F}$ adapted such that $\forall s \in [0, T]\ e_s \geq 0\,a.s.\}$. The functions ϕ and ψ represent the effort cost, and following [1] we will choose $\phi(a) = \frac{a^2}{2}$ and $\psi(e) = \frac{e^2}{2}$. U is a utility function that modelizes the consortium risk aversion.

Definition 4.1. A function $U : (t, c) \to U(t, c)$ is called utility function if

(i) $U : [0, T] \times]0, +\infty[\to \mathbf{R}$ is continuous.
(ii) $\forall t \in [0, T]$, $U(t, \cdot)$ is strictly increasing and strictly concave.
(iii) The derivatives $\frac{\partial}{\partial t}U$, $\frac{\partial}{\partial c}U$ exist and are continuous on $[0, T] \times]0, +\infty[$.

This optimization problem (4.2) will be reformulated in Sect. 4.4 in the case of a possible default of the consortium at a random time τ.

On the other hand, the community aims to maximize the social welfare defined as the social value of the project minus the rent paid at the consortium. The community optimization problem can be formulated as follows:

$$\max_{(\alpha,\beta)\in\mathscr{A}} SW : (\alpha, \beta) \mapsto \left(\mathbf{E}\left[B_0 + \int_0^T e^{-rs} b(e_s) ds - \left(\int_0^T e^{-rs} t(C_s) ds - C_0\right)\right]\right) \tag{4.3}$$

with $B_0 \geq 0$, $b : \mathbf{R}^+ \to \mathbf{R}^+$ C^1 and increasing (e.g., $b(x) = bx$, $b > 0$), b represents the community utility.

$\mathscr{A} = \{(\alpha, \beta), \alpha \geq 0, \beta \geq -1$ such that $t(C_s) \geq C_s \ \forall s \in [0, T]\}$ in order that the consortium is refund of the maintenance costs; C_0 is the initial cost payable by the consortium, B_0 is the initial social value of the project.

4.3 Solution of the Problem Without Default

4.3.1 Maximization of the Consortium Utility

Proposition 4.1. *The parameters of the rent* (α, β) *being fixed, there exists a unique solution* $(\hat{a}, \widehat{e_s})$ *at the optimization problem (4.1), given by*

$$\begin{cases} \widehat{e_s} = (\beta + 1)U'(\alpha - (\beta + 1)(\theta_0 - e_s - \delta a + \varepsilon_s)) \\ \widehat{a} = \delta \mathbf{E}(\int_0^T e^{-rs} e_s ds). \end{cases}$$

Furthermore,

$$\widehat{a} \mapsto \widehat{e_s}(a)$$

is decreasing: the more effort the consortium makes for the construction, the less effort it has to do for the maintenance.

Proof. We have to optimize the function

$$(a, e) \mapsto= \mathbf{E}\left(\int_0^T e^{-rs}\left(U(\alpha - (\beta + 1)(\theta_0 - e_s - \delta a + \varepsilon_s)) - \frac{1}{2}e_s^2\right) ds\right) - \frac{1}{2}a^2,$$

which is concave in a and in e_s and thus which is maximum when its gradient is zero. This leads to the couple (a, e_s) solution of system in Proposition 4.1. We claim that for all a, there exists a unique nonnegative solution $e_s(a)$. Indeed, since U is strictly concave and increasing, we have

$$U'(\alpha - (\beta + 1)(\theta_0 - e_s - \delta a + \varepsilon_s)) > 0,$$

and $g : x \mapsto (\beta+1)U'(\alpha - (\beta + 1)(\theta_0 - x - \delta a + \varepsilon_s))$ is decreasing. Thus $e_s(a)$ is the abscissa of the intersection of the bisector and the function g graph, and it is solution of the implicit equation

$$F(x, a) = x - (\beta + 1)U'(\alpha - (\beta + 1)(\theta_0 - x - \delta a + \varepsilon_s)) = 0.$$

The relation between e_s and a is reflected by the derivative

$$\frac{de}{da} = -\frac{\partial_a F}{\partial_x F} = \frac{(\beta + 1)^2 \delta U''(\alpha - (\beta + 1)(\theta_0 - x - \delta a + \varepsilon_s))}{1 - (\beta + 1)^2 U''(\alpha - (\beta + 1)(\theta_0 - x - \delta a + \varepsilon_s))}.$$

Since $U'' < 0$, $\frac{de}{da} < 0$, $a \mapsto e_s(a)$ is decreasing. We do the same for the function $h : a \mapsto \delta \mathbf{E}[\int_0^T e^{-rs} e_s(a) ds]$ and we conclude by the existence of an unique optimal $\hat{a}$ solution of equation $a = h(a)$. □

Notation. We introduce the following notation, useful for the rest of the paper:

$$A_t := \int_0^t e^{-rs} ds.$$

Example 4.1 (linear utility). $U(x) = \eta + \gamma x$. In this case, the consortium is risk neutral. The rent rule being fixed, the optimal efforts for the consortium are given by

$$\begin{cases} \widehat{e_s} = \gamma(\beta + 1) \quad \forall s \in [0, T] \\ \widehat{a} = \delta \widehat{e_s} (\int_0^T e^{-rs} ds) = \delta \gamma (\beta + 1) A_T. \end{cases}$$

Example 4.2 (quadratic utility). $U(x) = x - \frac{\gamma}{2} x^2$ with $\gamma > 0$ such that $t(C_s) < \frac{1}{\gamma}$ $\forall s \in [0, T]$.

In this case, the risk aversion of the consortium $\frac{\gamma}{1-\gamma x}$ is an increasing function of his wealth. The rent rule being fixed, the optimal efforts for the consortium are given by

$$\begin{cases} \widehat{a} = \delta \frac{(\beta+1)(1-\gamma(\alpha-(\beta+1)\theta_0))^+ A_T}{1+\gamma(\beta+1)^2(1+\delta^2 A_T)} \\ \widehat{e_s} = \frac{(\beta+1)}{1+\gamma(\beta+1)^2} (1 - \gamma(\alpha - (\beta + 1)\theta_0 + (\beta + 1)\delta \widehat{a} - (\beta + 1)\varepsilon_s)) \quad \forall s \in [0, T] \end{cases}$$

The noise (ε_s) being centered, $\mathbf{E}(e_s)$ does not depend on s. Furthermore, ε_s taking values in $[-m, M]$ $\forall s \in [0, T]$ $(m > 0, M > 0)$, the condition $t(C_s) < \frac{1}{\gamma}$ $\forall s \in [0, T]$ is satisfied as soon as $\alpha - (\beta + 1)(\theta_0 - m) + \frac{(\beta+1)^2}{1+\gamma(\beta+1)^2}((1 + \delta^2)(1 - \gamma(\alpha - (\beta + 1)\theta_0))^+ + \delta(\beta + 1)M) < \frac{1}{\gamma}$.

Remark 4.2. Comparison between these two examples: in the case of a quadratic utility function, $\mathbf{E}(e_s) \leq (\beta + 1)$ and is a decreasing function of the risk aversion, whereas for a linear utility with slop $\gamma > 1$, $\mathbf{E}(e_s) = e_s > (\beta + 1)$. Thus the more risk averse the consortium is, the less effort it will do both for the construction and for the maintenance of the infrastructure.

Finding an explicit solution of the community optimization problem being tedious in a general setting, we will from now on focus on the framework of linear utility functions both for the community and the consortium

$$U(x) = \gamma x,\ \gamma > 0\ ;\ b(x) = b.x, b > 0.$$

This leads to the following constraints on the externality.

4.3.2 *Constraints on the Externality*

It seems natural to assume the cost being nonnegative a.s. This leads to some constraint on the parameters that we will explicit, using the expression of the optimal efforts in the framework of a linear utility. Moreover, in practice, the community cannot outsource more than a given level δ_{max}. We fix δ_{max} such that $C_s \geq 0$ almost surely:

$$C_s \geq 0 \Longleftrightarrow \theta_0 \geq e_s + \delta a - \varepsilon_s.$$

Using the expressions of a and e (see Example 4.1):

$$C_s \geq 0 \Longleftrightarrow \theta_0 - m \geq \gamma(\beta + 1)(1 + \delta^2 A_T). \tag{4.4}$$

This is a constraint linking δ and β. Note that this induces $\theta_0 \geq m$ since $\beta \geq -1$.

4.3.3 *Maximization of the Community Social Welfare*

Our aim is to find explicit solutions in order to quantify the advantages of outsourcing, with the linear utilities $U(x) = \gamma x,\ \gamma > 0$ et $b(x) = b.x, b > 0$. The rent rule being fixed, the consortium optimal efforts are given by

$$\begin{cases} \widehat{e_s} = \gamma(\beta + 1) & \forall s \in [0, T] \\ \widehat{a} = \delta\gamma(\beta + 1)A_T. \end{cases}$$

Proposition 4.2. *We recall the social welfare (4.3):*

$$SW(\alpha,\beta) = \left(\mathbf{E}\left[B_0 + \int_0^T e^{-rs} b(e_s)ds - \left(\int_0^T e^{-rs} t(C_s)ds - C_0\right)\right]\right)$$

with $C_s = \theta_0 - \widehat{e_s} - \delta\widehat{a} + \varepsilon_s$ *and* $t(C_s) = \alpha - \beta C_s$. *We assume that*

$$0 \le \delta^2 \le \delta^2_{max} = \frac{\theta_0 - 2m - \gamma(b+1)}{\gamma A_T}. \tag{4.5}$$

Then the community optimal policy is given by

$$\widehat{\beta} = \frac{\gamma b + \theta_0 - \gamma(1+\delta^2 A_T)}{2\gamma(1+\delta^2 A_T)} \tag{4.6}$$

$$\widehat{\alpha} = \frac{\gamma b + \theta_0 + \gamma(1+\delta^2 A_T)}{2\gamma(1+\delta^2 A_T)}(M - b\gamma - \gamma(1+\delta^2 A_T)). \tag{4.7}$$

Remark here that Assumption (4.5) implies that the benchmark cost θ_0 is bounded from below, otherwise negative costs can occur.

Proof. Since $\widehat{e_s} = \gamma(\beta+1)$ is constant:

$$\begin{aligned}\frac{SW(\alpha,\beta) - B_0 - C_0}{A_T} &= \mathbf{E}[be_s - \alpha + \beta C_s] = be_s - \alpha + \beta(\theta_0 - e_s(1+\delta^2 A_T)) \\ &= (b - \beta(1+\delta^2 A_T))e_s - \alpha + \beta\theta_0 \\ &= (b - \beta(1+\delta^2 A_T))\gamma(\beta+1) - \alpha + \beta\theta_0.\end{aligned}$$

SW is a polynomial function of degree 2 in β:

$$\frac{SW(\alpha,\beta) - B_0 - C_0}{A_T} = -\beta^2\gamma(1+\delta^2 A_T) + \beta(\gamma b + \theta_0 - \gamma(1+\delta^2 A_T)) - \alpha + b\gamma$$

The dominating coefficient is negative, thus there exists a unique maximum achieved for

$$\widehat{\beta} = \frac{\gamma b + \theta_0 - \gamma(1+\delta^2 A_T)}{2\gamma(1+\delta^2(A_T))} \tag{4.8}$$

that can be also written as

$$\gamma(1+\delta^2 A_T) = \frac{\gamma b + \theta_0}{2\widehat{\beta}+1}, \tag{4.9}$$

as soon as the constraint (4.4) is satisfied, that is as soon as

$$\theta_0 - m - \frac{1}{2}(b\gamma + \theta_0 + \gamma(1+\delta^2 A_T)) \ge 0,$$

which is indeed satisfied since the externality δ is bounded from above by $\delta_{max}^2 = \frac{\theta_0-2m-\gamma(b+1)}{\gamma A_T}$. The choice of α must satisfy the constraint $t(C_s) \geq C_s$, that is

$$\widehat{\alpha} \geq (\widehat{\beta}+1)(\theta_0+\varepsilon_s-\widehat{e_s}-\delta\widehat{a})\ ds \otimes d\mathbf{P}\, a.s.$$

The constraint must be satisfied in the linear case where $\widehat{e_s} = \gamma(\beta+1)$, $\widehat{a} = \delta\gamma(\beta+1)(A_T)$, and $\varepsilon_s \leq M$. We choose α saturating this constraint, $\widehat{\alpha}$ is given in terms of $\widehat{\beta}$ and using relation (4.9)

$$\begin{aligned}\widehat{\alpha} &= (\widehat{\beta}+1)(\theta_0+M-\gamma(\widehat{\beta}+1)(1+\delta^2 A_T))\\ &= (\widehat{\beta}+1)\left(\theta_0+M-(\widehat{\beta}+1)\frac{\gamma b+\theta_0}{2\widehat{\beta}+1}\right)\\ \widehat{\alpha} &= \frac{\widehat{\beta}+1}{2\widehat{\beta}+1}[\hat{\beta}(\theta_0+2M-b\gamma)+M-b\gamma].\end{aligned}$$

Since $\widehat{\alpha} \geq (1+\widehat{\beta})C_s$ and since the cost is nonnegative via (4.4), we necessarily have $\widehat{\alpha} \geq 0$. □

Remark that (4.9) implies that $\widehat{\beta}$ is a decreasing function of the externality δ, which satisfies

$$\frac{b\gamma+m}{\theta_0-2m-b\gamma} \leq \widehat{\beta} \leq \frac{b\gamma+\theta_0-\gamma}{2\gamma}.$$

The upper bound corresponds to $\delta=0$ (no outsourcing), the lower bound corresponds to δ maximum (4.5). Thus the study of the impact of the externality δ on the social welfare can be done through the study of the function $\beta \mapsto SW(\alpha(\beta),\beta)$ where we replace δ by its function of β using (4.9).

Proposition 4.3. *We assume (4.5). If $b\gamma - \frac{\gamma^2}{b\gamma+\theta_0} \leq M$—that is if the noise level is high enough—the social welfare is optimal for the maximal externality $\delta = \sqrt{\frac{\theta_0-2m-\gamma(b+1)}{\gamma A_T}}$. Otherwise, if the noise level is lower, the social welfare is optimal for $\delta = \delta_{max}$ or for $\delta = 0$ (depending on whether $SW(\beta_{max}) < SW(\beta_{min})$ or not).*

In conclusion, if we exclude the default risk, the externality is attractive only if the noise level is high enough.

Proof. We want to optimize the following function:

$$\beta \mapsto \frac{SW(\alpha,\beta)-B_0-C_0}{A_T} = \beta^2\frac{\gamma b+\theta_0}{2\beta+1} - \alpha + b\gamma,$$

that is, replacing α by its optimal value function of β

$$\beta \mapsto \frac{1}{2\beta+1}[\beta^2 2(b\gamma - M) - \beta(3M + \theta_0 - 4b\gamma) - M + 2b\gamma].$$

We recall the constraint on β

$$\beta_{min} = \frac{\gamma b + m}{\theta_0 - 2m - b\gamma} \leq \beta \leq \beta_{max} = \frac{\gamma b + \theta_0 - \gamma}{2\gamma}.$$

More precisely, we study on the interval $[\beta_{min}, \beta_{max}]$ the function

$$f(\beta) = \frac{1}{2\beta+1}[\beta^2 2(b\gamma - M) - \beta(3M + \theta_0 - 4b\gamma) - M + 2b\gamma]. \tag{4.10}$$

Differentiating with respect to β, we get

$$\frac{2\beta+1}{4(b\gamma - M)} f'(\beta) = \beta^2 + \beta - \frac{\theta_0 + M}{4(b\gamma - M)}$$

The discriminant of this polynomial function of degree 2 is $\Delta = 1 + \frac{\theta_0 + M}{b\gamma - M} = \frac{\theta_0 + b\gamma}{b\gamma - M}$. If $\Delta > 0$, the positive root is $\beta_r = \frac{-1+\sqrt{\frac{\theta_0+b\gamma}{b\gamma-M}}}{2}$. Remark that

$$\beta_r < \beta_{max} \Longleftrightarrow M < b\gamma - \frac{\gamma^2}{b\gamma + \theta_0}.$$

This can only happen if $b\gamma - \frac{\gamma^2}{b\gamma+\theta_0} > 0$, that is by solving the second degree inequation $(b\gamma)^2 + b\gamma\theta_0 - \gamma^2 > 0$ if $b\gamma > \frac{\theta_0}{2}(\sqrt{1 + \frac{4\gamma^2}{\theta_0^2}} - 1)$.

- First case: $b\gamma \leq M$ (i.e., high level of noise)
  f (and thus SW) is a strictly decreasing function of β (and thus strictly increasing in δ). The social welfare is optimal for $\delta = \delta_{max}$ (maximal externality) for a high level of noise.
- Second case: $b\gamma - \frac{\gamma^2}{b\gamma+\theta_0} \leq M \leq b\gamma$ (i.e., medium level of noise)
  Since $\beta_r > \beta_{max}$, f (and thus SW) is again a strictly decreasing function of β (and thus strictly increasing in δ). The social welfare is optimal for $\delta = \delta_{max}$.
- Third case: $b\gamma - \frac{\gamma^2}{b\gamma+\theta_0} \geq M$ (i.e., low level of noise and $b\gamma$ large enough). The optimal externality depends on whether or not β_r is greater than β_{min}:
  - If $\beta_r > \beta_{min}$, then SW is a strictly decreasing function of β on $[\beta_{min}, \beta_r]$ and strictly increasing on $[\beta_r, \beta_{max}]$. Thus, the social welfare is optimal for $\delta = \delta_{max}$ or for $\delta = 0$ (whether $SW(\beta_{max}) < SW(\beta_{min})$ or not).

- If $\beta_r \leq \beta_{min}$, then SW is a strictly increasing function of β (and a strictly decreasing function of δ). Thus, the social welfare is optimal when there is no outsourcing ($\delta = 0$).

To summarize this third case, the derivative $f'(\beta)$ being successively negative then positive, the optimum is achevied at one of the interval bound and is equal to $SW(\beta_{max}) \vee SW(\beta_{min})$. □

In conclusion, when we exclude the default risk of the consortium, outsourcing becomes attractive only if the noise level is high enough. This corresponds to a risk transfer from the community to the consortium. We conclude this section with a toy numerical example in order to quantify numerically this benchmark noise level under which outsourcing is not attractive.

4.3.4 Numerical Example

In this example we take $\theta_0 = 100$ euros per unit time. The noise represents the randomness of the cost around this value, that is $\Theta_0 = \theta_0 + \epsilon$ is a random variable with values in $[\theta_0 - M, \theta_0 + M]$ (here we take $m = M$). In the case of a linear utility $U(x) = \gamma x$, we let $\gamma = 25$ euros per unit time and $b = 1$.

$$\theta_0 = 100 \; ; \; \gamma = 25 \; ; \; b = 1. \tag{4.11}$$

Proposition 4.4. *For $\theta_0 = 100$; $\gamma = 25$; $b = 1$, outsourcing is attractive if and only if noise level M is greater than $\frac{50}{3}$ (i.e., around 16.7% of the benchmark cost θ_0).*

Proof. First, the level $b\gamma - \frac{\gamma^2}{b\gamma+\theta_0}$ given in Proposition 4.3 is equal to 20 in this example, thus if $M \geq 20$ the maximal externality is optimal. Remark that in this case

$$\beta_{max} = 2, \; \forall M \; ; \; f(\beta_{max}) = 50 - 3M,$$

where f (which has the same behavior as SW) was defined in (4.10):

$$f(\beta) = \frac{1}{2\beta + 1}[\beta^2 2(b\gamma - M) - \beta(3M + \theta_0 - 4b\gamma) - M + 2b\gamma].$$

Now we study the case where $M < 20$. Proposition 4.3 says that the optimum depends on the position of $f(\beta_{min})$ with respect to this value $50 - 3M$. We have

$$\beta_{min} = \frac{25 + M}{75 - 2M}.$$

We set $M = 5\mu$, then we obtain

$$f(\beta_{min}) = \frac{2500 - 20 \times 50\mu + 95\mu^2}{5(15 - 2\mu)},$$

to be compared to $f(\beta_{max}) = 5(10 - 3\mu)$. Then, $f(\beta_{min}) < f(\beta_{max})$ if and only if $M < \frac{50}{3} \sim 16.7$, leading β optimal equal to β_{max}, and $\widehat{\delta} = 0$. □

4.4 Introduction of a Default Time, Without Penalty

We extend here the previous model in a more dynamic point of view and by introducing a default time. We still consider linear utilities ($U(x) = \gamma x$ and $b(x) = bx$) and we consider the operational cost as a semimartingale

$$dC_s = (\theta_0 - e_s - \delta a)ds + \sigma dW_s.$$

The community chooses then the following expression for the rent:

$$dt(C_s) = \alpha ds - \beta dC_s.$$

We define the default time τ as the first time as the consortium cannot refund its debt anymore. In a first step, we assume that no penalty is imposed to the consortium in case of default (the case of penalty will be studied in Sect. 4.5).

4.4.1 *Utility Maximization for the Consortium*

The consortium must refund the debt at a rate D ($dD_s = De^{-rs}ds$) that is deducted from its profit. Its aim is to optimize (with $U(x) = \gamma x$ being its utility function)

$$(e, a, \tau) \mapsto \mathbf{E}\left(\int_0^{\tau\wedge T} e^{-rs}\left(\gamma[dt(C_s) - dC_s - dD_s] - \frac{1}{2}e_s^2\right)\right) - \frac{1}{2}a^2$$

$$= \mathbf{E}\left(\int_0^{\tau\wedge T} e^{-rs}\left[\gamma(\alpha - D) - \gamma(\beta+1)(\theta_0 - e_s - \delta a) - \frac{1}{2}e_s^2\right]ds\right) - \frac{1}{2}a^2$$

since $\mathbf{E}\left(\int_0^{\tau\wedge T} e^{-rs} dW_s\right) = 0$.

Proposition 4.5. *We assume that the initial effort does not depend on the default time (which is unknown at date 0). Then the optimal policy of a risk neutral consortium is given by*

$$\begin{cases} \widehat{e_s} = \gamma(\beta+1)\mathbf{1}_{[0,\tau\wedge T]}(s) \\ \widehat{a} = \delta\gamma(\beta+1)A_T. \end{cases}$$

Proof. Since the default time τ is unknown at the initial date, we do the optimization only in (e, a) with the fact that the effort e is done on the interval $[0, \tau]$, thus

$$\begin{cases} \widehat{e_s} = \gamma(\beta + 1)\mathbf{1}_{[0,\tau\wedge T]}(s) \\ \widehat{a} = \delta\gamma(\beta + 1)\mathbf{E}(\int_0^{\tau\wedge T} e^{-rs}ds) \end{cases}$$

But rather than taking an initial effort a depending on τ, it is more relevant to take the optimal initial effort as in the case with no default (thus we may overevaluate it):

$$\widehat{a} = \delta\gamma(\beta+1)\mathbf{E}\left(\int_0^T e^{-rs}ds\right) = \delta\gamma(\beta+1)A_T. \qquad \square$$

4.4.2 Definition of the Default Time and of the Community Optimization Problem

We introduce the initial fund financing the project: $DA_T = \int_0^T e^{-rs}Dds$. The consortium must refund its debt, $dt \otimes d\mathbf{P}$ a.s.:

$$DA_T + t(C_t) - C_t - D_t \geq 0,$$

that is

$$Y_t = DA_T + \int_0^t e^{-rs}(\alpha - D - (\beta + 1)(\theta_0 - \gamma(\beta + 1)(1 + \delta^2 A_T))ds$$
$$- \int_0^t e^{-rs}(\beta + 1)\sigma dW_s \geq 0.$$

Thus the default occurs when this constraint is not satisfied anymore.

Definition 4.2. The default time τ is defined as

$$\tau = \inf\{t : Y_t < 0\}.$$

If $r = 0$ (then $A_t = t$):

$$\tau = \inf\{t : \int_0^t e^{-rs}(\beta + 1)\sigma dW_s > DA_T$$
$$+ (\alpha - D - (\beta + 1)(\theta_0 - \gamma(\beta + 1)(1 + \delta^2 A_T))A_t\}.$$

If $r > 0$:

$$\tau = \inf\{t : \int_0^t e^{-rs} dW_s > A_r - B_r e^{-rt}\},$$

where

$$A_r = \frac{rDA_T + (\alpha - D - (\beta+1)(\theta_0 - \gamma(\beta+1)(1+\delta^2 A_T))}{r(\beta+1)\sigma},$$

$$B_r = \frac{\alpha - D - (\beta+1)(\theta_0 - \gamma(\beta+1)(1+\delta^2 A_T))}{r(\beta+1)\sigma}.$$

We remark here that this default time τ is increasing in α: the greater the community rent is, the longer the consortium avoids the default (and this $\forall r$). Considering that it is optimal for the community to postpone the default as longer as possible, we will choose the maximum α satisfying the constraints detailed in the following.

We adapt the definition of SW because in case of default, the community should take over from the consortium to refund the debt

$$SW(\alpha,\beta) - B_0 - C_0 = \mathbf{E}\left(\int_0^T e^{-rs} be_s ds - \int_0^\tau e^{-rs} dt(C_s) - \int_\tau^T e^{-rs} D ds\right)$$

$$= \mathbf{E}\left(\int_0^{\tau\wedge T} e^{-rs}[b\gamma(\beta+1) - \alpha + \beta(\theta_0 - \gamma(\beta+1)(1+\delta^2 A_T))]ds\right.$$
$$\left. - D\int_\tau^T e^{-rs} ds\right)$$
$$= [D + b\gamma(\beta+1) - \alpha + \beta(\theta_0 - \gamma(\beta+1)(1+\delta^2 A_T))]\mathbf{E}[A_{\tau\wedge T}] - DA_T.$$

Introducing

$$H := D + b\gamma(\beta+1) + \beta(\theta_0 - \gamma(\beta+1)(1+\delta^2 A_T)),$$

$$SW(\alpha,\beta) - B_0 - C_0 + DA_T = (H - \alpha)\mathbf{E}[A_{\tau\wedge T}],$$

which is the product of a decreasing and an increasing function in α. If $\alpha \geq H$, then $\alpha \mapsto SW(\alpha,\beta) - B_0 - C_0 + DA_T$ is decreasing, thus an optimal α must be less than H and we get

$$SW(\alpha,\beta) - B_0 - C_0 + DA_T = (H - \alpha)\mathbf{E}[A_{\tau\wedge T}] \geq 0.$$

Therefore, the optimum exists in the interval $[0, H]$ and we will study the following function of the parameters α, β, δ:

$$E[A_{\tau\wedge T}] = \mathbf{E}[A_\tau \mathbf{I}_{\tau<T}] + A_T \mathbf{P}(\tau > T).$$

4.4.3 Solution in the Case $r = 0$

If $r = 0$,

$$\tau = \inf\{t : W_t > A - Bt\},$$

where

$$A = \frac{DA_T}{(\beta+1)\sigma},$$
$$B = \frac{D + (\beta+1)(\theta_0 - \gamma(\beta+1)(1+\delta^2 T)) - \alpha}{(\beta+1)\sigma}.$$

We define

$$K := D + (\beta+1)(\theta_0 - \gamma(\beta+1)(1+\delta^2 A_T)). \tag{4.12}$$

4.4.3.1 The Law of the Default Time, Case $r = 0$

Using 3.2.3. page 148 [2], the law of τ is given by

Proposition 4.6. *If $r = 0$, the default time is defined by*

$$\tau = \inf\left\{t : W_t > \frac{DA_T}{(\beta+1)\sigma} + \frac{\alpha - K}{(\beta+1)\sigma}t\right\}$$

where K is defined in (4.12). Then the density of τ on $\mathbb{R}^+$ is

$$t \mapsto \frac{DA_T}{(\beta+1)\sigma\sqrt{2\pi t^3}} \exp\left[-\frac{1}{2t}\left(\frac{DA_T - (\alpha - K)t}{(\beta+1)\sigma}\right)^2\right].$$

If $r = 0$, $\mathbf{P}\{\tau < \infty\} = \exp(A(K-\alpha) - |A(K-\alpha)|)$ (cf. [3] page 197). Thus, $\mathbf{E}[\tau]$ is finite if and only if $\alpha < K$. In order to postpone the default, we take $\alpha \geq K$.

Corollary 4.1. *If $r = 0$, we choose $\alpha = K$, and the default time is defined as $\tau = \inf\{t : W_t > \frac{DA_T}{(\beta+1)\sigma}\}$, the density of τ on $\mathbb{R}^+$ is*

$$t \mapsto \frac{DA_T}{(\beta+1)\sigma\sqrt{2\pi t^3}} \exp\left[-\frac{1}{2t}\left(\frac{DA_T}{(\beta+1)\sigma}\right)^2\right].$$

4.4.3.2 Constraints on the Parameters, Case $r = 0$

A reasonable constraint is to take a nonnegative instantaneous operational cost (in expectation):

$$\mathbf{E}[C_s ds] = (\theta_0 - \gamma(\beta + 1)(1 + \delta^2 T))ds \geq 0.$$

that is

$$\theta_0 \geq \gamma(\beta + 1)(1 + \delta^2 T).$$

We have previously justified choosing $\alpha \geq K$ to move the default back, such that $\mathbf{E}[\tau] = +\infty$.

Proposition 4.7. *The expected instantaneous cost being nonnegative, and $\alpha \geq K$ (such that $E[\tau] = +\infty$) induce the following constraints:*

$$\theta_0 - b\gamma(\beta + 1) \leq \gamma(\beta + 1)(1 + \delta^2 T) \leq \theta_0. \tag{4.13}$$

Furthermore $0 \leq K \leq H$ and this proves the existence of an optimal α in the interval $[K, H]$.

Proof. The expected instantaneous cost being nonnegative is equivalent to

$$\mathbf{E}[C_s ds] = (\theta_0 - \gamma(\beta + 1)(1 + \delta^2 T))ds \geq 0,$$

thus we get the right-hand side inequality

$$\theta_0 \geq \gamma(\beta + 1)(1 + \delta^2 T).$$

This implies

$$K = D + (\beta + 1)(\theta_0 - \gamma(\beta + 1)(1 + \delta^2 T)) \geq D \geq 0.$$

The left-hand size inequality follows from

$$H - K = (\beta + 1)b\gamma - \theta_0 + \gamma(\beta + 1)(1 + \delta^2 T) \geq 0.$$

□

We now choose $\alpha = K$, which maximizes the first factor $SW - B_0 - C_0 + DA_T$. We remark that in this case, in expectation, the instantaneous rent is positive:

$$E[t(C_s) - C_s]ds = \alpha - (\beta + 1)(\theta_0 - \gamma(\beta + 1)(1 + \delta^2 T)) = D > 0.$$

4.4.3.3 Study of the Social Welfare, Function of β, δ, Case $r = 0$

If $r = 0$, the law of τ is explicit, furthermore (Corollary 4.1) *we choose* $\alpha = K = D + (\beta + 1)(\theta_0 - \gamma(\beta + 1)(1 + \delta^2 T))$ and thus the factor of the expectation is

$$H - K = (\beta + 1)b\gamma - \theta_0 + \gamma(\beta + 1)(1 + \delta^2 T).$$

Corollary 4.2. *If $r = 0$, we choose $\alpha = K < H$, and the function*

$$SW(\alpha, \beta) - B_0 - C_0 + DA_T = (H - K)\mathbf{E}[\tau \wedge T]$$
$$= [(\beta + 1)b\gamma - \theta_0 + \gamma(\beta + 1)(1 + \delta^2 T)]$$
$$\times \left[\int_0^\infty (t \wedge T)\frac{DA_T}{(\beta + 1)\sigma\sqrt{2\pi t^3}} \exp\left[-\frac{1}{2t}\left(\frac{DA_T}{(\beta + 1)\sigma}\right)^2\right] dt\right].$$

With the choice $\alpha = K$, the default time is the hitting time of $\frac{DA_T}{(\beta+1)\sigma}$ by a Brownian motion, thus the density of τ is $\frac{DA_T}{(\beta+1)\sigma\sqrt{2\pi t^3}} \exp\left[-\frac{1}{2t}\left(\frac{DA_T}{(\beta+1)\sigma}\right)^2\right]$. Assuming that β, δ satisfy (4.13), Corollary 4.2 gives the function we want to optimize

$$(\beta, \delta) \mapsto [(\beta + 1)b\gamma - \theta_0 + \gamma(\beta + 1)(1 + \delta^2 T)] \int_{\mathbb{R}+} t \wedge T \frac{DA_T}{\sigma(\beta + 1)\sqrt{2\pi t^3}}$$
$$\times \exp\left[-\frac{1}{2t}\left(\frac{DA_T}{(\beta + 1)\sigma}\right)^2\right] dt.$$

Proposition 4.8. *Let $r = 0$. We assume that the default time is postponed as longer as possible and that, the consortium optimal policy $(\widehat{e_s}, \widehat{a})$ being established, the PPP contract requires nonnegative (in expectation) operational cost and rent. Then the optimal rent rule and the optimal externality are*

$$\alpha^* = D,$$
$$-1 < \beta^* = \frac{\theta_0}{\gamma} - 1, \qquad (4.14)$$
$$\delta^* = 0.$$

Outsourcing is not optimal in this case.

Proof. The function

$$\delta \mapsto [(\beta + 1)b\gamma - \theta_0 + \gamma(\beta + 1)(1 + \delta^2 T)] \int_{\mathbb{R}+} t \wedge T \frac{DA_T}{\sigma(\beta + 1)\sqrt{2\pi t^3}}$$
$$\times \exp\left[-\frac{1}{2t}\left(\frac{DA_T}{(\beta + 1)\sigma}\right)^2\right] dt$$

is increasing in δ and using (4.13), $(1 + \delta^2 T)^* = \frac{\theta_0}{\gamma(\beta+1)}$. This optimum is greater than 1, thus we get the constraint for β:

$$\gamma(\beta + 1) \le \theta_0.$$

Replacing $1 + \delta^2 T$ by its optimal value, we want to optimize the function

$$\beta \mapsto b\gamma \int_{\mathbb{R}+} t \wedge T \frac{DA_T}{\sigma\sqrt{2\pi t^3}} \exp\left[-\frac{1}{2t}\left(\frac{DA_T}{(\beta+1)\sigma}\right)^2\right] dt.$$

This function is increasing, thus $\beta^* = \frac{\theta_0}{\gamma} - 1$, and using the expression of $(1+\delta^2 T)^*$ we get that $\delta^* = 0$. Finally,

$$\alpha^* = D + (\beta^* + 1)(\theta_0 - \gamma(\beta^* + 1)) = D.$$

□

The interpretation is the following: if there is no penalty in case of a default, the community optimal policy is to outsource the less possible (and MOP are better and more secure than PPP). Furthermore, we remark that at the optimum, $\mathbf{E}(C_s) = 0$, and the rent is $t(C_s) - C_s = D - \frac{\theta_0}{\gamma} C_s$, $\mathbf{E}(t(C_s) - C_s) = D$. Thus, the rent coincides, in expectation, to the refund of the consortium debt.

4.4.4 Solution of the Problem in the Case r > 0

If $r > 0$:

$$\tau = \inf\left\{t : \int_0^t e^{-rs} dW_s > A_r - B_r e^{-rt}\right\},$$

where

$$A_r = \frac{rDA_T + (\alpha - D - (\beta+1)(\theta_0 - \gamma(\beta+1)(1+\delta^2 A_T))}{r(\beta+1)\sigma},$$

$$B_r = \frac{\alpha - D - (\beta+1)(\theta_0 - \gamma(\beta+1)(1+\delta^2 A_T))}{r(\beta+1)\sigma}.$$

4.4.4.1 Constraints on the Parameters

By continuity, we have almost surely

$$\int_0^\tau e^{-rs}(\beta+1)\sigma dW_s = DA_T + (\alpha - D - (\beta+1)(\theta_0 - \gamma(\beta+1)(1+\delta^2 A_T))A_\tau$$

that implies in the case $r > 0$

$$\begin{aligned} 0 &= \mathbf{E}\left[\int_0^\tau e^{-rs}(\beta+1) dW_s\right] \\ &= DA_T + (\alpha - D - (\beta+1)(\theta_0 - \gamma(\beta+1)(1+\delta^2 A_T))\mathbf{E}[A_\tau]. \end{aligned}$$

Thus

$$\mathbf{E}[A_\tau] = \frac{DA_T}{D - \alpha + (\beta+1)(\theta_0 - \gamma(\beta+1)(1+\delta^2 A_T))}, \tag{4.15}$$

this implies the constraint on the parameters (since $0 \leq A_\tau \leq 1/r$):

$$0 \leq \mathbf{E}[A_\tau] = \frac{DA_T}{D - \alpha + (\beta+1)(\theta_0 - \gamma(\beta+1)(1+\delta^2 A_T))} \leq 1/r. \tag{4.16}$$

We recall

$$K = D + (\beta+1)(\theta_0 - \gamma(\beta+1)(1+\delta^2 A_T)),$$

thus we have the condition on α:

$$rDA_T \leq D - \alpha + (\beta+1)(\theta_0 - \gamma(\beta+1)(1+\delta^2 A_T)) = K - \alpha, \ \alpha \leq K - rDA_T. \tag{4.17}$$

Furthermore, the instantaneous cost being nonnegative (in expectation):

$$\mathbf{E}[C_s ds] = (\theta_0 - \gamma(\beta+1)(1+\delta^2 T))ds \geq 0.$$

implies that

$$\theta_0 \geq \gamma(\beta+1)(1+\delta^2 T).$$

4.4.4.2 Study of the Social Welfare, Function of β, δ; Case $r \neq 0$

Proposition 4.9. *Let $r > 0$. We assume that the default time is postponed as longer as possible and that, the consortium optimal policy $(\widehat{e_s}, \widehat{a})$ being established, the PPP contract requires nonnegative (in expectation) operational cost and rent. Then the optimal rent rule and the optimal externality are*

$$\begin{aligned} \widehat{\alpha} &= De^{-rT}, \\ -1 < \widehat{\beta} &= \frac{\theta_0}{\gamma} - 1, \\ \widehat{\delta} &= 0. \end{aligned} \tag{4.18}$$

Proof. We summarize the constraints: the cost rate is nonnegative (in expectation):

$$\theta_0 \geq \gamma(\beta+1)(1+\delta^2 A_T)$$

as for the rent:

$$\alpha \geq (\beta+1)(\theta_0 - \gamma(\beta+1)(1+\delta^2 A_T)) = (K - D).$$

The optimal parameters must satisfy

$$\alpha \leq H \wedge (K - rDA_T).$$

As in the case $r = 0$, it seems to be relevant to choose α such that to postpone the default as longer as possible, that is $\alpha = K - rDA_T$ (that satisfies the constraint $\alpha \geq K - D$ since $rA_T \leq 1$). We get

$$H - \alpha = (\beta + 1)b\gamma - \theta_0 + \gamma(\beta + 1)(1 + \delta^2 A_T) + rDA_T$$

that is increasing in δ and must be nonnegative. This implies a constraint linking β and δ:

$$b\gamma(\beta + 1) - \theta_0 + \gamma(\beta + 1)(1 + \delta^2 A_T) + rDA_T \geq 0.$$

Furthermore, with this choice of α, we get

$$A_r = 0, \ B_r = \frac{-DA_T}{(\beta + 1)\sigma}.$$

Thus $\tau = \inf\{t / \int_0^t e^{-rs} dW_s > \frac{DA_T}{(\beta+1)\sigma} e^{-rt}\}$ does not dependent on δ. For continuity reason,

$$\int_0^\tau e^{-rs} dW_s = \frac{DA_T}{(\beta + 1)\sigma} e^{-r\tau}$$

thus $\mathbf{E}[e^{-r\tau}] = 0$, that is $\tau = +\infty$ a.s. and $\tau \wedge T = T$, $A_{\tau \wedge T} = A_T$. Therefore, for this choice of α,

$$SW(\alpha, \beta, \delta) = [(\beta + 1)b\gamma - \theta_0 + \gamma(\beta + 1)(1 + \delta^2 A_T) + rDA_T]A_T.$$

SW is increasing in δ and the optimal δ is given by

$$\widehat{(1 + \delta^2 A_T)} = \frac{\theta_0}{\gamma(\beta + 1)}$$

with the constraint $\frac{\theta_0}{\gamma(\beta+1)} \geq 1$, that is $\beta \leq \frac{\theta_0}{\gamma} - 1$. Finally,

$$\widehat{\alpha} = De^{-rT}$$

and we easily check that $H - \alpha = rDA_T + (\beta + 1)b\gamma$ is positive. The last step is to find the optimum $\beta + 1$ for the function

$$\beta \mapsto f(\beta + 1) = (rDA_T + (\beta + 1)b\gamma)A_T.$$

This function is increasing, the optimal β is given such as in the case $r = 0$:

$$\widehat{\beta} = \frac{\theta_0}{\gamma} - 1$$

and $\hat{\delta} = 0$. The community optimal policy is the same as in the case $r = 0$. □

In conclusion, whatever the interest rate is, outsourcing is NEVER optimal if we consider the possibility that the consortium defaults and if no penalty is administered in case of default. Given the maturity of PPP contract, it is obvious that we have to take into account the possibility of default. We will now focus on finding some case where outsourcing is attractive if a penalty is administered in case of default.

4.5 Penalty in Case of Default with $r = 0$

Here we add in Sect. 4.4 model a penalty $\rho V(T-t)^+$ that the consortium should pay in case of default, whereas the community receives the compensation $\rho' V(T-t)^+$. We assume the natural constraint $\rho V \leq D$ and we denote $\rho = \rho' + \varepsilon$ where εV is used to pay the liquidation cost. We summarize the constraints

$$\rho V \leq D \; ; \; \rho = \rho' + \varepsilon, \; \varepsilon > 0. \tag{4.19}$$

In such a case, we only consider the case $r = 0$, since in that case the law of τ has a very sophisticated expression not so easy to manage [4]. We consider the rent $dt(C_s) = \alpha ds - \beta dC_s$, thus the consortium optimal policy remains the following.

Proposition 4.10. *Considering the rent dynamic $dt(C_s) = \alpha ds - \beta dC_s$ and the operational cost dynamic $dC_s = (\theta_0 - e_s - \delta a)ds + \sigma dW_s$, the consortium optimal policy is*

$$\hat{e}_t = \gamma(\beta+1)\mathbf{1}_{[0,\tau]}(t), \; \hat{a} = \gamma(\beta+1)\delta T.$$

The default time is now defined as

$$\tau = \inf\{t : (\beta+1)\sigma W_t > DT + (\alpha - K)t - \rho V(T-t)^+\} \wedge T\},$$

where K is defined in (4.12). Thus $\tau = \tilde{\tau} \wedge T$ with

$$\tilde{\tau} := \inf\{t : (\beta+1)\sigma W_t > DT + (\alpha - K + \rho V)t - \rho VT\}.$$

4.5.1 *Constraints on the Parameters*

As in the previous Sect. 4.4, we choose to postpone the default as longer as possible, for both the consortium and the community interest:

$$\alpha \geq K - \rho V = D + (\beta+1)(\theta_0 - \gamma(\beta+1)(1+\delta^2 T)) - \rho V.$$

Using the fact that the operational cost and the rent are nonnegative, we precise the constraints on the parameters.

Proposition 4.11. *We assume that the operational cost and the rent are nonnegative (in expectation) and we choose the bigger externality satisfying this assumption. Then the optimal parameters α, β, δ satisfy the following constraints:*

$$\gamma(\beta + 1)(1 + \delta^2 T) = \theta_0, \tag{4.20}$$

$$\gamma(\beta + 1) \leq \theta_0, \tag{4.21}$$

$$0 \leq D - \rho V \leq \alpha < D + b\gamma(\beta + 1) - \rho' V. \tag{4.22}$$

This last interval is not empty since $\rho > \rho'$ and $\gamma(\beta + 1) \geq 0$.

Corollary 4.3. *With the choice of a maximal externality, the consortium optimal effort can be written with respect to (θ_0, δ, T):*

$$\hat{e}_t = \frac{\theta_0}{1 + \delta^2 T} \mathbf{1}_{[0,\tau]}(t), \ \hat{a} = \frac{\theta_0}{1 + \delta^2 T} \delta T.$$

In this case $dC_s = \sigma dW_s$ on $[0, \tau]$.

Proof. The expectation of the instantaneous cost being nonnegative

$$\mathbf{E}[C_s ds] = (\theta_0 - \gamma(\beta + 1)(1 + \delta^2 T))ds \geq 0$$

that is $\theta_0 \geq \gamma(\beta + 1)(1 + \delta^2 T)$. Our goal here is to find situations where outsourcing is attractive, thus we "a priori" choose δ maximum

$$\widehat{1 + \delta^2 T} = \frac{\theta_0}{\gamma(\beta + 1)}.$$

This leads to the following constraint on β (since $1 + \delta^2 T \geq 1$):

$$\beta + 1 \leq \frac{\theta_0}{\gamma}.$$

With this choice of externality, the decision of postponing the default as longer as possible leads to the constraint on α

$$\alpha \geq K - \rho V = D - \rho V. \tag{4.23}$$

Furthermore, the instantaneous rent is nonnegative (in expectation)

$$\mathbf{E}[t(C_s) - C_s] = \alpha - (\beta + 1)(\theta_0 - \gamma(\beta + 1)(1 + \delta^2 T)) \geq 0,$$

thus, with the choice of δ maximum, $\alpha \geq 0$. We compute the social welfare, with δ maximum and taking into account the compensation received in case of default:

$$SW(\alpha, \beta) - B_0 - C_0 = [D + b\gamma(\beta + 1) - \alpha]\mathbf{E}[\tau \wedge T] - DT + \rho' VE[(T - \tau)^+],$$

that is using $(T-\tau)^+ = T - \tilde{\tau} \wedge T$,

$$SW(\alpha,\beta) - B_0 - C_0 + (D-\rho' V)T = [D + b\gamma(\beta+1) - \alpha - \rho' V]\mathbf{E}[\tilde{\tau} \wedge T]. \quad (4.24)$$

This expression of SW requires the following constraint:

$$D + b\gamma(\beta+1) - \alpha - \rho' V > 0, \text{ i.e., } \alpha < D + b\gamma(\beta+1) - \rho' V.$$

Furthermore, the constraints on α ($\alpha \geq 0$ and (4.23)) lead to

$$(D-\rho V)^+ < \alpha < D + b\gamma(\beta+1) - \rho' V.$$

Using assumption (4.19), $(D-\rho V)^+ = D - \rho V$ and

$$0 \leq D - \rho V \leq \alpha < D + b\gamma(\beta+1) - \rho' V.$$

This interval is not empty since $\rho > \rho'$ and $b\gamma(\beta+1) \geq 0$. □

4.5.2 Maximization of the Social Welfare

To emphasize the dependency on β, we now denotes $\tilde{\tau}$ by

$$\tau_\beta := \inf\{t : (\beta+1)\sigma W_t > (D-\rho V)T + (\alpha - D + \rho V)t\}.$$

We remark that $\beta \mapsto \tau_\beta$ is decreasing. Using (4.24), we express the social welfare SW as a function of β.

Lemma 4.1. *Up to an additive constant, the social welfare is the sum of two functions of $\beta+1$:*

$$f(\beta+1) = b\gamma(\beta+1)\mathbf{E}[\tilde{\tau} \wedge T] = b\gamma \int_0^\infty t \wedge T \frac{(D-\rho V)T}{\sigma\sqrt{2\pi t^3}} \times \exp\left[-\frac{1}{2t}\left(\frac{(D-\rho V)T - (\alpha - D + \rho V)t}{(\beta+1)\sigma}\right)^2\right] dt$$

and

$$g(\beta+1) = [D - \alpha - \rho' V]\mathbf{E}[\tau_\beta \wedge T].$$

The following proposition gives the community optimal policy.

Proposition 4.12. *We assume (4.19). We assume that we postpone the default as longer as possible and that, the consortium optimal policy $(\widehat{e_s}, \widehat{a})$ being established, the PPP contract requires nonnegative (in expectation) operational cost and rent. Then the optimal rent rule and the optimal externality are*

(i) If $D - \alpha - \rho' V \leq 0$, $\widehat{\beta} = \frac{\theta_0}{\gamma}$ and the same conclusions as in the case with no penalty hold (4.14).
(ii) If $D - \alpha - \rho' V > 0$, we choose $\widehat{\alpha} = D - \rho V$ (this does not contradict (ii) since $\rho' < \rho$) and we denote $A = \gamma \frac{(D-\rho V)\sqrt{T}}{\theta_0 \sigma}$. Then the sign $(f+g)'(\frac{\theta_0}{\gamma})$ is the one of the following expressions:

$$(b\theta_0 + 2(\rho - \rho')V)A(1 - \Phi(A)) + b\theta_0 A^{-1}\left(\Phi(A) - \frac{1}{2} - A\phi(A)\right)$$
$$- [2(\rho - \rho')V]\phi(A).$$

For a "small" A, $(f+g)'(\frac{\theta_0}{\gamma}) < 0$, and there exists an optimal β strictly less than $\frac{\theta_0}{\gamma}$, thus the optimal externality $\hat{\delta}$ is strictly positive.

Proof. On the one hand, the function f is increasing from $f(0) = 0$ to

$$f(\infty) = b\gamma \int_{\mathbb{R}+} t \wedge T \frac{(D - \rho V)T}{\sigma\sqrt{2\pi t^3}} dt = \frac{4b\gamma(D - \rho V)T\sqrt{T}}{\sigma\sqrt{2\pi}}.$$

On the other hand, concerning the function g, two cases may occur:

(i) If $D - \alpha - \rho' V \leq 0$, g is also increasing, the optimal β is $\frac{\theta_0}{\gamma}$ and the same conclusions as in the case without penalty hold (4.14).
(ii) If $D - \alpha - \rho' V > 0$, g is decreasing and it is necessary to go into detail, using the constraints (4.21) and (4.22):

$$\gamma(\beta + 1) \leq \theta_0,$$
$$0 \leq D - \rho V \leq \alpha < D - \rho' V.$$

□

We will study the functions f and g in the interval $]0, \frac{\theta_0}{\gamma}]$. To do this, and in order to simplify the computations, we choose $\alpha = D - \rho V$ (thus τ_β is a.s. finite with an infinite expectation). We do the change of parameter:

$$\zeta = \frac{(D - \rho V)T}{(\beta + 1)\sigma}, \; \zeta \geq \frac{\gamma(D - \rho V)T}{\theta_0 \sigma}.$$

Thus

$$\tilde{f}(\zeta) = b\gamma \frac{(D-\rho V)T}{\sigma} \int_{\mathbb{R}+} t \wedge T \frac{1}{\sqrt{2\pi t^3}} \exp\left[-\frac{1}{2t}\zeta^2\right] dt, \qquad (4.25)$$

$$\tilde{g}(\zeta) = [(\rho-\rho')V]\zeta \int_{\mathbb{R}+} t \wedge T \frac{1}{\sqrt{2\pi t^3}} \exp\left[-\frac{1}{2t}(\zeta)^2\right] dt.$$

We will use the following technical lemma.

Lemma 4.2. *Let ϕ be the density of a standard centered Gaussian random variable and Φ its cumulative function. Then for all positive A:*

$$\int_0^A u^2\phi(u)du = -A\phi(A) + \Phi(A) - \frac{1}{2}; \quad \int_A^\infty u^{-2}\phi(u)du = A^{-1}\phi(A) - 1 + \Phi(A).$$

We now compute the derivative function of $\tilde{g}$.

Lemma 4.3.

$$\tilde{g}'(\zeta) = [(\rho-\rho')V]4\zeta\left[\frac{\sqrt{T}}{\zeta}\phi\left(\frac{\zeta}{\sqrt{T}}\right) - 1 + \Phi\left(\frac{\zeta}{\sqrt{T}}\right)\right].$$

Proof. Before computing the derivative, we do the following change of variable in $\tilde{g}$: $u^2 = \frac{\zeta^2}{t}$, $t = \frac{\zeta^2}{u^2}$, $dt = -2\frac{\zeta^2}{u^3}du$, and $\phi(u) = \frac{1}{\sqrt{2\pi}}\exp\left[-\frac{u^2}{2}\right]$:

$$\begin{aligned}
\tilde{g}(\zeta) &= [(\rho-\rho')V]\zeta \int_{\mathbb{R}+} \frac{\zeta^2}{u^2} \wedge T \frac{u^3}{\zeta^3}\frac{2\zeta^2}{u^3}\phi(u)du \\
&= [(\rho-\rho')V]2 \int_{\mathbb{R}+} \frac{\zeta^2}{u^2} \wedge T\phi(u)du \\
&= 2[(\rho-\rho')V]\left(\int_0^{\frac{\zeta}{\sqrt{T}}} T\phi(u)du + \int_{\frac{\zeta}{\sqrt{T}}}^\infty \frac{\zeta^2}{u^2}\phi(u)du\right).
\end{aligned}$$

The previous lemma leads to

$$\tilde{g}(\zeta) = 2[(\rho-\rho')V]\left(T\left(\Phi\left(\frac{\zeta}{\sqrt{T}}\right) - \frac{1}{2}\right) + \zeta^2\left(\frac{\sqrt{T}}{\zeta}\phi\left(\frac{\zeta}{\sqrt{T}}\right) - 1 + \Phi\left(\frac{\zeta}{\sqrt{T}}\right)\right)\right).$$

Up to the multiplicative constant $2(\rho-\rho')V$, the derivative of the first term is $\sqrt{T}\phi(\frac{\zeta}{\sqrt{T}})$ and the second term is

$$\sqrt{T}\zeta\phi\left(\frac{\zeta}{\sqrt{T}}\right) - \zeta^2 + \zeta^2\Phi\left(\frac{\zeta}{\sqrt{T}}\right)$$

whose derivative is ($\phi'(u) = -u\phi(u)$):

$$\sqrt{T}\phi\left(\frac{\zeta}{\sqrt{T}}\right) - \frac{\zeta^2}{\sqrt{T}}\phi\left(\frac{\zeta}{\sqrt{T}}\right) - 2\zeta\left(1 - \Phi\left(\frac{\zeta}{\sqrt{T}}\right)\right) + \frac{\zeta^2}{\sqrt{T}}\phi\left(\frac{\zeta}{\sqrt{T}}\right)$$

and reduction leads to the result. □

The derivative of $\tilde{f}$ is

Lemma 4.4.

$$\tilde{f}'(\zeta) = -2b\gamma\frac{(D-\rho V)T}{\sigma}\left(1 - \Phi\left(\frac{\zeta}{\sqrt{T}}\right) + T\zeta^{-2}\left[\Phi\left(\frac{\zeta}{\sqrt{T}}\right) - \frac{1}{2} - \frac{\zeta}{\sqrt{T}}\phi\left(\frac{\zeta}{\sqrt{T}}\right)\right]\right).$$

Proof. We deduce from (4.25):

$$\tilde{f}'(\zeta) = -b\gamma\frac{(D-\rho V)T}{\sigma}\zeta\int_{\mathbb{R}+} t \wedge T\frac{1}{t\sqrt{2\pi t^3}}\exp\left[-\frac{1}{2t}\zeta^2\right]dt.$$

Doing a change of variable in $\tilde{f}'$:

$$\tilde{f}'(\zeta) = -2b\gamma\frac{(D-\rho V)T}{\sigma}\zeta\int_{\mathbb{R}+}\left(\frac{\zeta^2}{u^2}\right) \wedge T\frac{1}{\frac{\zeta^2}{u^2}\sqrt{2\pi\left(\frac{\zeta^2}{u^2}\right)^3}}\exp\left[-\frac{u^2}{2}\right]\frac{\zeta^2}{u^3}du =$$

$$= -2b\gamma\frac{(D-\rho V)T}{\sigma}\int_{\mathbb{R}+}\left(\frac{\zeta^2}{u^2}\right) \wedge T\frac{1}{\sqrt{2\pi}}\frac{u^2}{\zeta^2}\exp\left[-\frac{u^2}{2}\right]du =$$

$$-2b\gamma\frac{(D-\rho V)T}{\sigma}\left(\int_0^{\frac{\zeta}{\sqrt{T}}} T\frac{u^2}{\zeta^2}\phi(u)du + \int_{\frac{\zeta}{\sqrt{T}}}^{\infty}\phi(u)du\right).$$

Lemma 4.2 yields

$$\tilde{f}'(\zeta) = -2b\gamma\frac{(D-\rho V)T}{\sigma}\left(T\zeta^{-2}\left[-\frac{\zeta}{\sqrt{T}}\phi\left(\frac{\zeta}{\sqrt{T}}\right) + \Phi\left(\frac{\zeta}{\sqrt{T}}\right) - \frac{1}{2}\right]\right.$$
$$\left. +1 - \Phi\left(\frac{\zeta}{\sqrt{T}}\right)\right).$$

□

Proof of Proposition 4.12, case (ii) : We are looking at the sign of $(f+g)'(\frac{\theta_0}{\gamma})$ which is the sign of $-(\tilde{f}+\tilde{g})'(\zeta)$ (in $\zeta = \frac{\gamma(D-\rho V)T}{\theta_0\sigma}$). Using the two lemmas,

$$-(\tilde{f}+\tilde{g})'(\zeta) =$$

$$2b\gamma\frac{(D-\rho V)T}{\sigma}\left(1-\Phi\left(\frac{\zeta}{\sqrt{T}}\right)+\frac{T}{\zeta^2}\left[\Phi\left(\frac{\zeta}{\sqrt{T}}\right)-\frac{1}{2}-\frac{\zeta}{\sqrt{T}}\phi\left(\frac{\zeta}{\sqrt{T}}\right)\right]\right)$$

$$-4\zeta(\rho-\rho')V\left[\frac{\sqrt{T}}{\zeta}\phi\left(\frac{\zeta}{\sqrt{T}}\right)-1+\Phi\left(\frac{\zeta}{\sqrt{T}}\right)\right].$$

Thus the sign of $(f+g)'$ is the one of

$$b\gamma\frac{(D-\rho V)\sqrt{T}}{\sigma}\left(1-\Phi\left(\frac{\zeta}{\sqrt{T}}\right)+T\zeta^{-2}\left[\Phi\left(\frac{\zeta}{\sqrt{T}}\right)-\frac{1}{2}-\frac{\zeta}{\sqrt{T}}\phi\left(\frac{\zeta}{\sqrt{T}}\right)\right]\right)$$

$$-2\frac{\zeta}{\sqrt{T}}[(\rho-\rho')V]\left[\frac{\sqrt{T}}{\zeta}\phi\left(\frac{\zeta}{\sqrt{T}}\right)-1+\Phi\left(\frac{\zeta}{\sqrt{T}}\right)\right].$$

For $\zeta=\gamma\frac{(D-\rho V)}{\theta_0\sigma}$, we set $A:=\gamma\frac{(D-\rho V)\sqrt{T}}{\theta_0\sigma}$, and the sign of $(f+g)'(\frac{\theta_0}{\gamma})$ is the one of

$$b\theta_0 A\left(1-\Phi(A)+A^{-2}\left[\Phi(A)-\frac{1}{2}-A\phi(A)\right]\right)-2A[(\rho-\rho')V][A^{-1}\phi(A)-1+\Phi(A)]$$

which is the expected expression of Proposition 4.12(ii)

$$(b\theta_0+2(\rho-\rho')V)A(1-\Phi(A))+b\theta_0A^{-1}\left(\Phi(A)-\frac{1}{2}-A\phi(A)\right)-[2(\rho-\rho')V]\phi(A).$$

The asymptotics around zero of the two first terms are

$$[2(\rho-\rho')V+b\theta_0]A[1-\Phi(A)]\sim[2(\rho-\rho')V+b\theta_0]\frac{A}{2},$$

$$b\theta_0A^{-1}\left(\left[\Phi(A)-\frac{1}{2}-A\phi(A)\right]\right)\sim b\theta_0\frac{5A^2}{6\sqrt{2\pi}},$$

and the third term is equal for $A=0$ to $-2(\rho-\rho')V\phi(0)=-\frac{2(\rho-\rho')V}{\sqrt{2\pi}}<0$. Thus, for A small enough, $(f+g)'(\frac{\theta_0}{\gamma})<0$. •

Remark 4.3. This condition "$A=\gamma\frac{(D-\rho V)\sqrt{T}}{\theta_0\sigma}$ small enough" is satisfied if

- The noise level is high (large σ)
- The benchmark cost θ_0 is high
- The maturity T is short
- $D-\rho V$ is small, that is the penalty ρ is large enough

In this section that modelizes the better the reality, we show that outsourcing is attractive for the community in case of high uncertainty or high noise, short maturity, high benchmark operational cost or a sufficiently high penalty in case of default.

4.6 Conclusion

Three models of PPP contracts have been studied in this paper:

- The first one assumes that there is no default risk and that the contract does not end before maturity.
- The second one introduces the default risk of the consortium, without any compensation for the community in case of an unreciprocated contract breaking-off.
- The third one also considers the default risk of the consortium, and the consortium has to pay penalty in case of default, the community receiving a part of this penalty as a compensation.

In the second model, whatever is the discount rate (positive or zero), the community optimal policy is to give up for outsourcing. In the first model, outsourcing is optimal if the noise level around the maintenance benchmark cost is higher than a threshold: this corresponds to a risk transfer from the community to the consortium. Remark that this threshold is an increasing function of the benchmark cost and of the coefficient of the consortium utility. Similarly, in the third model with penalty in case of default, outsourcing is optimal if the randomness is high enough, or if the contract maturity is short, if the benchmark cost or the penalty is high enough.

Acknowledgements We thank Jerôme Pouyet who introduced us the economic bases of this system of public–private partnerships. We also thank the colleagues who heard to us and asked some question which allowed us to improve our paper.

References

1. Iossa, E., Martimort, D., Pouyet, J.: Partenariats Public-Privé, quelques réflexions. Revue économique **59**(3) (2008)
2. Jeanblanc, M., Yor, M., Chesney, M.: Mathematical Methods for Financial Markets. Springer, Berlin (2009)
3. Karatzas, I., Shreve, S.: Brownian Motion and Stochastic Calculus. Springer, Berlin (1988)
4. Patie P.: On some first passage times problems motivated by financial applications. PhD Thesis, ETH Zürich (2004)

Chapter 5
On a Flow of Transformations of a Wiener Space

Joseph Najnudel, Daniel Stroock, and Marc Yor

Abstract In this chapter, we define, via Fourier transform, an ergodic flow of transformations of a Wiener space which preserves the law of the Ornstein–Uhlenbeck process and which interpolates the iterations of a transformation previously defined by Jeulin and Yor. Then, we give a more explicit expression for this flow, and we construct from it a continuous gaussian process indexed by $\mathbb{R}^2$, such that all its restriction obtained by fixing the first coordinate are Ornstein–Uhlenbeck processes.

5.1 Introduction

An abstract Wiener space is a triple $(H, E, \mathscr{W})$ consisting of a separable, real Hilbert space H, a separable real Banach space E in which H is continuously embedded as a dense subspace, and a Borel probability measure $\mathscr{W}$ on E with the property that, for each $x^* \in E^*$, the $\mathscr{W}$-distribution of the map $x \in E \longmapsto \langle x, x^* \rangle \in \mathbb{R}$, from E to $\mathbb{R}$, is a centered gaussian random distribution with variance $\|h_{x^*}\|_H^2$, where h_{x^*} is the element of H determined by $(h, h_{x^*})_H = \langle h, x^* \rangle$ for all $h \in H$. See Chap. 8 of [5] for more information on this topic.

Because $\{h_{x^*} : x^* \in E^*\}$ is dense in H and $\|h_{x^*}\|_H = \|\langle \cdot, x^* \rangle\|_{L^2(\mathscr{W})}$, there is a unique isometry, known as the Paley–Wiener map, $\mathscr{I} : H \longmapsto L^2(\mathscr{W})$ such that $\mathscr{I}(h) = \langle \cdot, x^* \rangle$ if $h = h_{x^*}$. In fact, for each $h \in H$, $\mathscr{I}(h)$ under $\mathscr{W}$ is a centered

J. Najnudel
Institut für Mathematik, Universität Zürich, Winterthurerstrasse 190, 8057 Zürich, Switzerland

D. Stroock
Massachusetts Institute of Technology, Department of Mathematics, Cambridge MA, USA

M. Yor (✉)
Laboratoire de Probabilités et Modèles Aléatoires, Université Paris VI, 75252 Paris, France

L. Decreusefond and J. Najim (eds.), *Stochastic Analysis and Related Topics*, Springer Proceedings in Mathematics & Statistics 22, DOI 10.1007/978-3-642-29982-7_5,

Gaussian variable with variance $\|h\|_H^2$. Because when $h = h_{x^*}$, $\mathscr{I}(h)$ provides an extension of $(\cdot, h)_H$ to E, for intuitive purposes one can think of $x \rightsquigarrow [\mathscr{I}(h)](x)$ as a giving meaning to the inner product $x \rightsquigarrow (x, h)_H$, although for general h this will be defined only up to a set of $\mathscr{W}$-measure 0.

An important property of abstract Wiener spaces is that they are invariant under orthogonal transformations on H. To be precise, given an orthogonal transformation $\mathscr{O}$ on H, there is a $\mathscr{W}$-almost surely unique $T_{\mathscr{O}} : E \longrightarrow E$ with the property that, for each $h \in H$, $\mathscr{I}(h) \circ T_{\mathscr{O}} = \mathscr{I}(\mathscr{O}^\top h)$ $\mathscr{W}$-almost surely. Notice that this is the relation which one would predict if one thinks of $[\mathscr{I}(h)](x)$ as the inner product of x with h. In general, $T_{\mathscr{O}}$ can be constructed by choosing $\{x_m^* : m \geq 1\} \subseteq E^*$, so the $\{h_{x_m^*} : m \geq 1\}$ is an orthonormal basis in H and then taking

$$T_{\mathscr{O}} x = \sum_{m=1}^{\infty} \langle x, x_m^* \rangle \mathscr{O} h_{x_m^*},$$

where the series converges in E for $\mathscr{W}$-almost every x as well as in $L^p(\mathscr{W}; E)$ for every $p \in [1, \infty)$. See Theorem 8.3.14 in [5] for details. In the case when $\mathscr{O}$ admits an extension as a continuous map on E into itself, $T_{\mathscr{O}}$ can be the taken equal to that extension. In any case, it is an easy matter to check that the measure $\mathscr{W}$ is preserved by $T_{\mathscr{O}}$. Less obvious is a theorem, originally formulated by I.M. Segal (cf. [4]), which says that $T_{\mathscr{O}}$ is ergodic if and only $\mathscr{O}$ admits no nontrivial, finite dimensional, invariant subspace. Equivalently, $T_{\mathscr{O}}$ is ergodic if and only if the complexification $\mathscr{O}_{\mathrm{c}}$ has a continuous spectrum as a unitary operator on the complexification H_{c} of H.

The classical Wiener space provides a rich source of examples to which the preceding applies. Namely, take $H = H_0^1$ to be the space of absolutely continuous $h \in \Theta$ whose derivative $\dot{h}$ is in $L^2([0, \infty))$, and set $\|h\|_{H_0^1} = \|\dot{h}\|_{L^2([0,\infty))}$. Then H_0^1 with norm $\|\cdot\|_{H_0^1}$ is a separable Hilbert space. Next, take $E = \Theta$, where Θ is the space of continuous paths $\theta : [0, \infty) \longrightarrow \mathbb{R}$ such that $\theta(0) = 0$ and

$$\frac{|\theta(t)|}{t^{\frac{1}{2}} \log(e + |\log t|)} \longrightarrow 0 \quad \text{as } t > 0 \text{ tends to } 0 \text{ or } \infty,$$

and set

$$\|\theta\|_\Theta = \sup_{t>0} \frac{|\theta(t)|}{t^{\frac{1}{2}} \log(e + |\log t|)}.$$

Then Θ with norm $\|\cdot\|_\Theta$ is a separable Banach space in which H_0^1 is continuously embedded as a dense subspace. Finally, the renowned theorem of Wiener combined with the Brownian law of the iterated logarithm says that there is a Borel probability measure $\mathscr{W}_{H_0^1}$ on Θ for which $(H_0^1, \Theta, \mathscr{W}_{H_0^1})$ is an abstract Wiener space. Indeed, it is the classical Wiener space on which the abstraction is modeled, and $\mathscr{W}_{H_0^1}$ is the distribution of an $\mathbb{R}$-valued Brownian motion.

One of the simplest examples of an orthogonal transformation on H_0^1 for which the associated transformation on Θ is ergodic is the Brownian scaling map S_α given by $S_\alpha\theta(t) = \alpha^{-\frac{1}{2}}\theta(\alpha t)$ for $\alpha > 0$. It is an easy matter to check that the restriction $\mathcal{O}_\alpha$ of S_α to H_0^1 is orthogonal, and so, since S_α is continuous on Θ, we can take $T_{\mathcal{O}_\alpha} = S_\alpha$. Furthermore, as long as $\alpha \neq 1$, an elementary computation shows that $\lim_{n\to\infty}\big(g, \mathcal{O}_\alpha^n h\big)_H = 0$, first for smooth $g, h \in H_0^1$ with compact support in $(0,\infty)$ and thence for all $g, h \in H_0^1$. Hence, when $\alpha \neq 1$, $\mathcal{O}_\alpha$ admits no nontrivial, finite dimensional subspace, and therefore S_α is ergodic; and so, by the Birkoff's Individual Ergodic Theorem, for $p \in [1,\infty)$ and $f \in L^p(\mathcal{W}_{H_0^1})$,

$$\lim_{n\to\infty} \frac{1}{n}\sum_{m=0}^{n-1} f \circ S_\alpha^n = \int f \, d\mathcal{W}_{H_0^1}$$

both $\mathcal{W}_{H_0^1}$-almost surely and in $L^p(\mathcal{W}_{H_0^1})$. Moreover, since $\{S_\alpha : \alpha \in (0,\infty)\}$ is a multiplicative semigroup in the sense that $S_{\alpha\beta} = S_\alpha \circ S_\beta$, one has the continuous parameter version

$$\lim_{a\to\infty} \frac{1}{\log a}\int_1^a (f \circ S_\alpha)\frac{d\alpha}{\alpha} = \int f \, d\mathcal{W}_{H_0^1}$$

of the preceding result.

A more challenging ergodic transformation of the classical Wiener space was studied by Jeulin and Yor (see [1, 3, 6]), and, in the framework of this chapter, it is obtained by considering the transformation $\mathcal{O}$ on H_0^1, defined by

$$[\mathcal{O}h](t) = h(t) - \int_0^t \frac{h(s)}{s}\, ds. \tag{5.1}$$

An elementary calculation shows that $\mathcal{O}$ is orthogonal. Moreover, $\mathcal{O}$ admits a continuous extension to Θ given by replacing $h \in H_0^1$ in (5.1) by $\theta \in \Theta$. That is

$$[T_{\mathcal{O}}\theta] = \theta(t) - \int_0^t \frac{\theta(s)}{s}\, ds \quad \text{for } \theta \in \Theta \text{ and } t \geq 0. \tag{5.2}$$

In addition, one can check that $\lim_{n\to\infty}\big(g, \mathcal{O}^n h\big)_{H_0^1} = 0$ for all $g, h \in H_0^1$, which proves that $T_{\mathcal{O}}$ is ergodic for $\mathcal{W}_{H_0^1}$.

In order to study the transformation $T_{\mathcal{O}}$ in greater detail, it will be convenient to reformulate it in terms of the Ornstein–Uhlenbeck process. That is, take H^U to be the space of absolutely continuous functions $h : \mathbb{R} \longrightarrow \mathbb{R}$ such that

$$\|h\|_{H^U} \equiv \sqrt{\int_{\mathbb{R}} \left(\tfrac{1}{4}h(t)^2 + \dot{h}(t)^2\right) dt} < \infty.$$

Then H^U becomes a separable Hilbert space with norm $\|\cdot\|_{H^U}$. Moreover, the map $F : H_0^1 \longrightarrow H^U$ given by

$$[F(g)](t) = e^{-\frac{t}{2}} g(e^t), \quad \text{for } g \in H_0^1 \text{ and } t \in \mathbb{R}, \tag{5.3}$$

is an isometric surjection which extends as an isometry from Θ onto Banach space $\mathscr{U}$ of continuous $\omega :\longrightarrow \mathbb{R}$ satisfying $\lim_{|t|\to\infty} \frac{|\omega(t)|}{\log|t|} = 0$ with norm $\|\omega\|_{\mathscr{U}} = \sup_{t\in\mathbb{R}} \big(\log(e + |t|)\big)^{-1} |\omega(t)|$. Thus, $(H^U, \mathscr{U}, \mathscr{W}_{H^U})$ is an abstract Wiener space, where $\mathscr{W}_{H^U} = F_* \mathscr{W}_{H_0^1}$ is the image of $\mathscr{W}_{H_0^1}$ under the map F. In fact, $\mathscr{W}_{H^U}$ is the distribution of a standard, reversible Ornstein–Uhlenbeck process.

Note that the scaling transformations for the classical Wiener space become translations in the Ornstein–Uhlenbeck setting. Namely, for each $\alpha > 0$, $F \circ S_\alpha = \tau_{\log\alpha} \circ F$, where τ_s denotes the time-translation map given by $[\tau_s\omega](t) = \omega(s+t)$. Thus, for $s \neq 0$, the results proved about the scaling maps say that τ_s is an ergodic transformation for $\mathscr{W}_{H^U}$. In particular, for $p \in [1,\infty)$ and $f \in L^p(\mathscr{W}_{H^U})$,

$$\lim_{n\to\infty} \frac{1}{n} \sum_{m=0}^{n-1} f \circ \tau_{ns} = \lim_{T\to\infty} \frac{1}{T} \int_0^T f \circ \tau_s \, ds = \int f \, d\mathscr{W}_{H^U}$$

both $\mathscr{W}_{H^U}$-almost surely and in $L^p(\mathscr{W}_{H^U})$.

The main goal of this chapter is to show that the reformulation of transformation $T_{\mathscr{O}}$ coming from the Jeulin–Yor transformation in terms of the Ornstein–Uhlenbeck process allows us to embed $T_{\mathscr{O}}$ in a continuous-time flow of transformations on the space $\mathscr{U}$, each of which is $\mathscr{W}_{H_0^1}$-measure preserving and all but one of which is ergodic. In Sect. 5.2, this flow is described via Fourier transforms. In Sect. 5.3, a direct and more explicit expression, involving hypergeometric functions and principal values, is computed. In Sect. 5.4, we study the two-parameter gaussian process which is induced by the flow introduced in Sect. 5.2. In particular, we compute its covariance and prove that it admits a version which is jointly continuous in its parameters.

5.2 Preliminary Description of the Flow

Let $\mathscr{O}$ and $T_{\mathscr{O}}$ be the transformations on H_0^1 and Θ given by (5.1) and (5.2), and recall the unitary map $F : H_0^1 \longrightarrow H^U$ in (5.3) and its continuous extension as an isometry from Θ onto $\mathscr{U}$. Clearly, the inverse of F is given by

$$F^{-1}(\omega)(t) = \sqrt{t}\,\omega(\log t) \quad \text{for } t > 0.$$

Because F is unitary and $\mathscr{O}$ is orthogonal on H_0^1, $-F \circ \mathscr{O} \circ F^{-1}$ is an orthogonal transformation on H^U, and because

$$S := -F \circ T_{\mathscr{O}} \circ F^{-1}$$

is continuous extension of $-F \circ \mathscr{O} \circ F^{-1}$ to $\mathscr{U}$, we can identify S as $T_{-F\circ\mathscr{O}\circ F^{-1}}$.

Another expression for action of S is

$$[S(\omega)](t) = -\omega(t) + \int_0^\infty e^{-\frac{s}{2}}\omega(t-s)\,ds \quad \text{for } t \in \mathbb{R}.$$

Equivalently,

$$S(\omega) = \omega * \mu,$$

where μ is the finite, signed measure μ given by

$$\mu := -\delta_0 + e^{-\frac{t}{2}}\mathbf{1}_{t\ge 0}dt.$$

To confirm that $\omega * \mu$ is well defined as a Lebesgue integral and that it maps $\mathscr{U}$ continuously into itself, note that, for any $\omega \in \mathscr{U}$ and $t \in \mathbb{R}$,

$$\int_0^\infty e^{-\frac{s}{2}}|\omega(t-s)|\,ds \le \|\omega\|_{\mathscr{U}} \int_0^\infty e^{-\frac{s}{2}} \log(e+|t|+s)\,ds$$
$$\le \|\omega\|_{\mathscr{U}} \log(e+|t|) \int_0^\infty e^{-\frac{s}{2}}(1+s)\,ds \le 9\|\omega\|_{\mathscr{U}} \log(e+|t|)$$

The Fourier transform $\widehat{\mu}$ of μ is given by

$$\widehat{\mu}(\lambda) = \int_{\mathbb{R}} e^{-i\lambda x}\,d\mu(x) = -1 + \int_0^\infty e^{-x(1/2+i\lambda)}dx = -1 + \frac{1}{1/2+i\lambda} = \frac{1-2i\lambda}{1+2i\lambda} = e^{-2i\,\mathrm{Arctg}(2\lambda)}.$$

Hence, for all $h \in H^U$ and $\lambda \in \mathbb{R}$,

$$\widehat{h * \mu}(\lambda) = e^{-2i\,\mathrm{Arctg}(2\lambda)}\widehat{h}(\lambda), \tag{5.4}$$

which, since

$$\|h\|_{H^U}^2 = \frac{1}{8\pi}\int_{\mathbb{R}} |\widehat{h}(\lambda)|^2(1+4\lambda^2)\,d\lambda,$$

provides another proof that $S \restriction H^U$ is isometric.

The preceding, and especially (5.4), suggests a natural way to embed $S \restriction H^U$ into a continuous group of orthogonal transformations. Namely, for $u \in \mathbb{R}$, let μ^{*u} to be the unique tempered distribution whose Fourier transform is given by

$$\widehat{\mu^{*u}}(\lambda) = e^{-2iu\,\mathrm{Arctg}(2\lambda)}, \tag{5.5}$$

and define $\mathscr{S}^u\varphi = \varphi * \mu^{*u}$ for φ in the Schwartz test function class $\mathscr{S}$ of smooth functions which, together with all their derivatives, are rapidly decreasing. Because

$$\widehat{\mathscr{S}^u\varphi}(\lambda) = e^{-2iu\,\mathrm{Arctg}(2\lambda)}\hat{\varphi}(\lambda),$$

it is obvious that $\mathscr{S}^u$ has a unique extension as an orthogonal transformation on H^U, which we will again denote by $\mathscr{S}^u$. Furthermore, it is clear that $\mathscr{S}^{u+v} = \mathscr{S}^u \circ \mathscr{S}^v$ for all $u, v \in \mathbb{R}$. Finally, for all $g, h \in H^U$, $u \in \mathbb{R}$,

$$(g, \mathscr{S}^u h)_{H^U} = \frac{1}{8\pi} \int_{\mathbb{R}} \overline{\widehat{g}(\lambda)}\, \widehat{h}(\lambda)\, e^{-2iu\,\mathrm{Arctg}(2\lambda)} (1 + 4\lambda^2)\, d\lambda$$
$$= \frac{1}{16\pi} \int_{-\pi/2}^{\pi/2} \overline{\widehat{g}\left(\frac{\tan(\tau)}{2}\right)}\, \widehat{h}\left(\frac{\tan(\tau)}{2}\right) \left(1 + \tan^2(\tau)\right)^2 e^{-2iu\tau}\, d\tau,$$

where

$$\frac{1}{16\pi} \int_{-\pi/2}^{\pi/2} \left|\widehat{g}\left(\frac{\tan(\tau)}{2}\right)\right| \left|\widehat{h}\left(\frac{\tan(\tau)}{2}\right)\right| \left(1 + \tan^2(\tau)\right)^2 d\tau$$
$$= \frac{1}{8\pi} \int_{\mathbb{R}} |\widehat{g}(\lambda)|\, |\widehat{h}(\lambda)|\, (1 + 4\lambda^2)\, d\lambda$$
$$\leq \frac{1}{8\pi} \left(\int_{\mathbb{R}} |\widehat{g}(\lambda)|^2 (1 + 4\lambda^2) d\lambda \right)^{1/2} \left(\int_{\mathbb{R}} |\widehat{h}(\lambda)|^2 (1 + 4\lambda^2) d\lambda \right)^{1/2}$$
$$= ||g||_{H^U} ||h||_{H^U} < \infty.$$

Hence, by Riemann–Lebesgue lemma, shows that $(g, \mathscr{S}^u h)_{H^U}$ tends to zero when $|u|$ goes to infinity.

Now define the associated transformations $S^u := T_{\mathscr{S}^u}$ on $\mathscr{U}$ for each $u \in \mathbb{R}$. By the general theory summarized in the introduction and the preceding discussion, we know that $\{S^u : u \in \mathbb{R}\}$ is a flow of $\mathscr{W}_{H^U}$-measure preserving transformations and that for each $u \neq 0$, S^u is ergodic.

5.3 A More Explicit Expression

So far we know very little about the transformations S^u for general $u \in \mathbb{R}$. By getting a handle on the tempered distributions μ^{*u}, in this section we will attempt to find out a little more.

We begin with the case when u is an integer $n \in \mathbb{Z}$. Recalling that $\mu = -\delta_0 + e^{-\frac{t}{2}} \mathbf{1}_{t \geq 0}\, dt$, one can use induction to check that, for $n \geq 0$,

$$\mu^{*n} = (-1)^n \left(\delta_0 + e^{-\frac{t}{2}} L_n'(t) \mathbf{1}_{t \geq 0} dt\right),$$

where L_n is the nth Laguerre polynomial. Indeed, the Laguerre polynomials satisfy the following relations: for all $n \geq 0$,

$$L_n(0) = 1$$

and for all $n \geq 0$, $t \in \mathbb{R}$,

$$L'_{n+1}(t) = L'_n(t) - L_n(t).$$

Similarly, starting from $\mu^{*-1} = -\delta_0 + e^{\frac{t}{2}}\mathbf{1}_{t\geq 0}\,dt$, one finds that

$$\mu^{*n} = (-1)^n\big(\delta_0 + e^{\frac{t}{2}}L'_n(-t)\mathbf{1}_{t\leq 0}dt\big)$$

for $n \leq 0$. In particular, μ^{*n} is a finite, signed measure for $n \in \mathbb{Z}$, and $S^n\omega$ can be identified as $\mu^{*n} * \omega$ for all $\omega \in \mathscr{U}$ and $n \in \mathbb{Z}$.

As the next result shows, when $u \notin \mathbb{Z}$, μ^{*u} is more singular tempered distribution than a finite, signed measure.

Proposition 5.1. *For each $u \notin \mathbb{Z}$, the distribution μ^{*u} is given by the following formula:*

$$\mu^{*u} = \cos(\pi u)\delta_0(x) + \frac{\sin(\pi u)}{\pi}pv(1/x) + \Phi_u(x), \tag{5.6}$$

where pv denotes the principal value, and $\Phi_u \in L^2(\mathbb{R})$ is the function for which $\Phi_u(x)$ equals

$$e^{-|x|/2}\left(-\frac{u\sin(\pi u)}{\pi}\sum_{k=0}^{\infty}\frac{(1-u\,\mathrm{sgn}(x))_k|x|^k}{k!(k+1)!}\left[\frac{\Gamma'}{\Gamma}(1+k-u\,\mathrm{sgn}(x)) - \frac{\Gamma'}{\Gamma}(1+k)\right.\right.$$
$$\left.\left. -\frac{\Gamma'}{\Gamma}(2+k) + \log(|x|)\right] + \frac{\sin(\pi u)}{\pi x}\right) - \frac{\sin \pi u}{\pi x},$$

Γ'/Γ being the logarithmic derivative of the Euler gamma function and $(\ )_k$ being the Pochhammer symbol.

Proof. Define the functions ψ_u and θ_u from $\mathbb{R}^* = \mathbb{R}\setminus\{0\}$ to $\mathbb{R}$ so that $\theta_u(x) = e^{-\frac{x}{2}}\psi_u(x)$ and $\psi_u(x)$ equals

$$-\frac{u\sin(\pi u)}{\pi}\sum_{k=0}^{\infty}\frac{(1-u\,\mathrm{sgn}(x))_k|x|^k}{k!(k+1)!}\left[\frac{\Gamma'}{\Gamma}(1+k-u\,\mathrm{sgn}(x)) - \frac{\Gamma'}{\Gamma}(1+k)\right.$$
$$\left. -\frac{\Gamma'}{\Gamma}(2+k) + \log(|x|)\right] + \frac{\sin(\pi u)}{\pi x}.$$

From Lebedev [2], p. 264, Eq. (9.10.6), with the parameters $\alpha = 1-u$ or $\alpha = 1+u$, $n = 1$, $z = x$ or $z = -x$, the function ψ_u satisfies, for all $x \in \mathbb{R}^*$, the differential equation:

$$x\psi''_u(x) + (2-|x|)\psi'_u(x) + (u - \mathrm{sgn}(x))\psi_u(x) = 0,$$

and grows at most polynomially at infinity. One then deduces that θ_u decreases as least exponentially at infinity and satisfies (for $x \neq 0$) the following equation:

$$x\theta''_u(x) + 2\theta'_u(x) + \left(u - \frac{x}{4}\right)\theta_u(x) = 0. \tag{5.7}$$

At the same time, by writing

$$e^{-|x|/2} = (e^{-|x|/2} - 1) + 1$$

and expanding $\theta_u(x)$ accordingly, we obtain:

$$\theta_u(x) = \frac{\sin(\pi u)}{\pi x} - \frac{u\sin(\pi u)}{\pi}\left[\frac{\Gamma'}{\Gamma}(1 - u\,\mathrm{sgn}(x)) - \frac{\Gamma'}{\Gamma}(1) - \frac{\Gamma'}{\Gamma}(2) + \log(|x|)\right]$$
$$- \frac{\sin(\pi u)}{2\pi}\,\mathrm{sgn}(x) + \eta_u(x),$$

for

$$\eta_u(x) = x\eta_u^{(1)}(x) + |x|\eta_u^{(2)}(x) + x\log(|x|)\eta_u^{(3)}(x) + |x|\log(|x|)\eta_u^{(4)}(x),$$

where $\eta_u^{(1)}, \eta_u^{(2)}, \eta_u^{(3)}, \eta_u^{(4)}$ are all smooth functions. The derivatives of the functions x, $|x|$, $x\log|x|$, $|x|\log|x|$ in the sense of the distributions are obtained by interpreting their ordinary derivatives as distributions. Similarly, the product by x of their second distributional derivatives are obtained by multiplying their ordinary second derivatives by x. Hence, both $\eta_u'(x)$ and $x\eta_u''(x)$ as distributions can be obtained by computing $\eta_u'(x)$ and $x\eta_u''(x)$ as functions on $\mathbb{R}^*$.

Now, let ν_u be the distribution given by the expression:

$$\nu_u(x) = \cos(\pi u)\delta_0(x) + \frac{\sin(\pi u)}{\pi} pv(1/x) + \left[\theta_u(x) - \frac{\sin(\pi u)}{\pi x}\right]. \tag{5.8}$$

Note that the term in brackets, in the definition of ν_u, is a locally integrable function, and that ν_u coincides with the function θ_u in the complement of the neighborhood of zero. Let us now prove that ν_u satisfies the analog of the Eq. (5.7), in the sense of the distributions. One has:

$$\nu_u(x) = \cos(\pi u)\delta_0(x) + \frac{\sin(\pi u)}{\pi} pv(1/x) - \frac{u\sin(\pi u)}{\pi}\left[\frac{\Gamma'}{\Gamma}(1 - u\,\mathrm{sgn}(x))\right.$$
$$\left. - \frac{\Gamma'}{\Gamma}(1) - \frac{\Gamma'}{\Gamma}(2) + \log(|x|)\right] - \frac{\sin(\pi u)}{2\pi}\,\mathrm{sgn}(x) + \eta_u(x).$$

Since

$$\frac{\Gamma'}{\Gamma}(1+u) - \frac{\Gamma'}{\Gamma}(1-u) = \frac{\frac{d}{du}(\Gamma(1+u)\Gamma(1-u))}{\Gamma(1+u)\Gamma(1-u)} = \frac{\frac{d}{du}(\pi u/\sin(\pi u))}{\pi u/\sin(\pi u)} = \frac{1}{u} - \pi\cot(\pi u),$$

one obtains, after straightforward computation,

$$\nu_u(x) = \cos(\pi)\delta_0(x) + \frac{\sin(\pi u)}{\pi} pv(1/x) - \frac{u\cos(\pi u)}{2}\,\mathrm{sgn}(x)$$
$$- \frac{u\sin(\pi u)}{\pi}\log(|x|) + c(u) + \eta_u(x),$$

where $c(u)$ does not depend on x. One deduces that

$$\nu_u(x) = \cos(\pi u)\delta_0(x) + \frac{\sin(\pi u)}{\pi} pv(1/x) + \chi_{u,1}(x),$$

where $\chi_{u,1}$ denotes a locally integrable function. Moreover,

$$\nu_u'(x) = \cos(\pi u)\delta_0'(x) - \frac{\sin(\pi u)}{\pi} fp(1/x^2)$$
$$- u\cos(\pi u)\delta_0(x) - \frac{u\sin(\pi u)}{\pi} pv(1/x) + \eta_u'(x),$$

where $fp(1/x^2)$ denotes the finite part of $1/x^2$, and then

$$x\nu_u'(x) = -\cos(\pi u)\delta_0(x) - \frac{\sin(\pi u)}{\pi} pv(1/x) - \frac{u\sin(\pi u)}{\pi} + x\eta_u'(x).$$

By differentiating again, one obtains:

$$\nu_u'(x) + x\nu_u''(x) = -\cos(\pi u)\delta_0'(x) + \frac{\sin(\pi u)}{\pi} fp(1/x^2) + \eta_u'(x) + x\eta_u''(x).$$

Therefore,

$$x\nu_u''(x) + 2\nu_u'(x) + \left(u - \frac{x}{4}\right)\nu_u(x) = \chi_{u,2}(x) + \left(-\cos(\pi u)\delta_0'(x) + \frac{\sin(\pi u)}{\pi} fp(1/x^2)\right)$$
$$+ \left(\cos(\pi u)\delta_0'(x) - \frac{\sin(\pi u)}{\pi} fp(1/x^2) - u\cos(\pi u)\delta_0(x) - \frac{u\sin(\pi u)}{\pi} pv(1/x)\right)$$
$$+ u\left(\cos(\pi u)\delta_0(x) + \frac{\sin(\pi u)}{\pi} pv(1/x)\right) = \chi_{u,2}(x),$$

where $\chi_{u,2}$ is a locally integrable function. Since θ_u satisfies (5.7), $\chi_{u,2}$ is identically zero. Hence, ν_u is a tempered distribution solving the differential equation:

$$x\nu_u''(x) + 2\nu_u'(x) + \left(u - \frac{x}{4}\right)\nu_u(x) = 0,$$

or equivalently,

$$\frac{x}{4}\nu_u(x) - \frac{d^2}{d^2x}(x\nu_u(x)) - u\nu_u(x) = 0.$$

Multiplying by $-4i$ and taking the Fourier transform (in the sense of the distributions), one deduces:

$$\widehat{\nu_u}'(\lambda)(1+4\lambda^2) = -4iu\widehat{\nu_u}(\lambda).$$

This linear equation admits a unique solution, up to a multiplicative factor c:

$$\widehat{\nu_u}(\lambda) = c\exp\left(\int_0^\lambda \frac{-4iu}{1+4t^2}dt\right) = c\exp(-2iu\,\mathrm{Arctg}(2\lambda)).$$

Hence, ν_u is proportional to μ^{*u}. In order to determine the constant c, let us observe that the distribution $\nu_{u,0}$ given by

$$\nu_{u,0}(x) = \nu_u(x) - c\cos(\pi u)\delta_0(x) - \frac{c\sin(\pi u)}{\pi}pv(1/x)$$

admits the Fourier transform:

$$\widehat{\nu_{u,0}}(\lambda) = c\,e^{-2iu\,\mathrm{Arctg}(2\lambda)} - c\,e^{-\pi i u\,\mathrm{sgn}(\lambda)}.$$

One deduces that $\widehat{\nu_{u,0}}$ is a function in L^2, which implies that $\nu_{u,0}$ is also a function in L^2, and then locally integrable. Since the last term in (5.8) is also a locally integrable function, one deduces that $c = 1$, and then

$$\mu^{*u} = \nu_u,$$

which proves Proposition 5.1. □

The reasonably explicit expression for μ^{*u} found in Proposition 5.1 yields a reaonably explicit expression for the action of $\mathscr{S}^u$. Indeed, only the term $pv(1/x)$ is a source of concern. However, convolution with respect of $pv(1/x)$ is, apart from a multiplicative constant, just the Hilbert transform, whose properties are well known. In particular, it is a translation invariant, bounded map on $L^2(\mathbb{R})$, and as such it is also a bounded map on H^U. Thus, we can unambiguously write $\mathscr{S}^u(h) = h * \mu^{*u}$ for all $h \in H^U$. On the other hand, the interpretation of $\omega * \mu^{*u}$ for $\omega \in \mathscr{U}$ needs some thought. No doubt, $\omega * \mu^{*u}$ is well defined as an element of $\mathscr{S}'$, the space tempered distributions, but it is not immediately obvious that it can be represented by an element of $\mathscr{U}$ or, if it can, that the element of $\mathscr{U}$ which represents it can be identified as $S^u\omega$. In fact, the best that we should expect is that such statements will be true of $\mathscr{W}_{H^U}$-almost every $\omega \in \mathscr{U}$. The following result justifies that expectation.

Proposition 5.2. *For $\mathscr{W}_{H^U}$-almost every $\omega \in \mathscr{U}$, the tempered distribution $\omega * \mu^{*u}$ is represented by an element of $\mathscr{U}$ which can be can be identified as $S^u\omega$.*

Proof. Recall that, for $\varphi \in \mathscr{S}$, $\varphi * \mu^{*-u}$ is the element of $\mathscr{S}$ whose Fourier transform is given by

$$\widehat{\varphi * \mu^{*-u}}(\lambda) = \widehat{\varphi}(\lambda)e^{2iu\,\mathrm{Arctg}(2\lambda)} \quad \text{for all } \lambda \in \mathbb{R}.$$

Also, if $T \in \mathscr{S}'$, then $T * \mu^{*u}$ is the tempered distribution whose action on $\varphi \in \mathscr{S}$ is given by

$$_{\mathscr{S}}\langle \varphi, T * \mu^{*u}\rangle_{\mathscr{S}'} = {}_{\mathscr{S}}\langle \varphi * \mu^{*-u}, T\rangle_{\mathscr{S}'}.$$

Now choose an orthonormal basis $\{h_n : n \geq 1\}$ for H^U all of whose members are elements of $\mathscr{S}$, and, for each $n \geq 1$, set $g_n = \frac{1}{4}h_n + h_n''$. Next, think of g_n as the element of $\mathscr{U}^*$ whose action on $\omega \in \mathscr{U}$ is given by

$$_{\mathscr{U}}\langle \omega, g_n\rangle_{\mathscr{U}^*} = {}_{\mathscr{S}}\langle g_n, \omega\rangle_{\mathscr{S}'}.$$

It is then an easy matter to check that, in the notation of the introduction, $h_n = h_{g_n}$. Hence, if B is the subset of $\omega \in \mathscr{U}$ for which

$$\omega = \lim_{n\to\infty} \sum_{m=1}^{n} {}_{\mathscr{S}}\langle g_n, \omega\rangle_{\mathscr{S}'} h_n \quad \text{and} \quad S^u\omega = \lim_{n\to\infty} \sum_{m=1}^{n} {}_{\mathscr{S}}\langle g_n, \omega\rangle_{\mathscr{S}'} h_n * \mu^{*u},$$

where the convergence is in $\mathscr{U}$, then $\mathscr{W}_{H^U}(B) = 1$.

Now let $\omega \in B$. Then, for each $\varphi \in \mathscr{S}$,

$$\begin{aligned} {}_{\mathscr{S}}\langle \varphi, \omega * \mu^{*u}\rangle_{\mathscr{S}'} &= {}_{\mathscr{S}}\langle \varphi * \mu^{*-u}, \omega\rangle_{\mathscr{S}'} = \lim_{n\to\infty} \sum_{m=1}^{n} {}_{\mathscr{S}}\langle g_n, \omega\rangle_{\mathscr{S}'}\, {}_{\mathscr{S}}\langle \varphi, h_n * \mu^{*u}\rangle_{\mathscr{S}'} \\ &= \lim_{n\to\infty} \sum_{m=1}^{n} {}_{\mathscr{S}}\langle g_n, \omega\rangle_{\mathscr{S}'}\, {}_{\mathscr{S}}\langle \varphi, \mathscr{S}^u h_n\rangle_{\mathscr{S}'} = {}_{\mathscr{S}}\langle \varphi, S^u\omega\rangle_{\mathscr{S}'}. \end{aligned}$$

Thus, for $\omega \in B$, $\omega * \mu^{*u} \in \mathscr{S}'$ is represented by $S^u\omega \in \mathscr{U}$. □

5.4 A Two Parameter Gaussian Process

By construction, $\{S^u\omega(t) : (u,t) \in \mathbb{R}^2\}$ is a gaussian family in $L^2(\mathscr{W}_{H^U})$. In this concluding section, we will show that this family admits a modification which is jointly continuous in (u,t).

Let $\varphi, \psi \in \mathscr{S}$ and $u, v \in \mathbb{R}^2$ be given. Then, by Proposition 5.2, for $\mathscr{W}_{H^U}$-almost every $\omega \in \mathscr{U}$,

$$\iint_{\mathbb{R}^2} \varphi(s)\psi(t)(S^u(\omega))(s)(S^v(\omega))(t)\, ds dt = {}_{\mathscr{S}}\langle \varphi, \omega * \mu^{*u}\rangle_{\mathscr{S}'}\, {}_{\mathscr{S}}\langle \psi, \omega * \mu^{*v}\rangle_{\mathscr{S}'},$$

where the integral in the left-hand side is absolutely convergent. Because $\mathbb{E}^{\mathscr{W}_{H^U}}\left[S^u\omega(t)^2\right]$ is finite and independent of $(u,t) \in \mathbb{R}^2$, by taking the expectation with respect to $\mathscr{W}_{H^U}$ and using (5.5), one can pass from this to

$$\iint_{\mathbb{R}^2} \varphi(s)\psi(t)\mathbb{E}_{\mathscr{W}_{HU}}\left[(S^u(\omega))(s)(S^v(\omega))(t)\right] dsdt$$

$$= \mathbb{E}_{\mathscr{W}_{HU}}\left[{}_{\mathscr{S}}\langle\varphi, \omega * \mu^{*u}\rangle_{\mathscr{S}'}\, {}_{\mathscr{S}}\langle\psi, \omega * \mu^{*v}\rangle_{\mathscr{S}'}\right]$$

$$= \frac{2}{\pi}\int_{-\infty}^{\infty} \frac{e^{2i(u-v)\,\mathrm{Arctg}(2\lambda)}}{1+4\lambda^2}\widehat{\varphi}(\lambda)\,\overline{\widehat{\psi}(\lambda)}d\lambda$$

$$= \frac{2}{\pi}\iiint_{\mathbb{R}^3} \frac{e^{i[(t-s)\lambda+2(u-v)\,\mathrm{Arctg}(2\lambda)]}}{1+4\lambda^2}\varphi(s)\psi(t)\,dsdtd\lambda.$$

Hence,

$$\mathbb{E}_{\mathscr{W}_{HU}}[S^u(\omega))(s)(S^v(\omega))(t)] = \frac{2}{\pi}\int_{-\infty}^{\infty} \frac{e^{i[(t-s)\lambda+2(u-v)\,\mathrm{Arctg}(2\lambda)]}}{1+4\lambda^2}d\lambda, \tag{5.9}$$

first for almost every and then, by continuity, for all $(s,t) \in \mathbb{R}^2$. In particular, we now know that the $\mathscr{W}_{HU}$-distribution of $\{S^u(\omega))(t) : (u,t) \in \mathbb{R}^2\}$ is stationary.

To show that there is a continuous version of this process, we will use Kolmogorov's continuity criterion, which, because it is stationary and gaussian, comes down to showing that

$$\left|1 - \mathbb{E}_{\mathscr{W}_{HU}}[(S^u(\omega))(s)(S^v(\omega))(t)]\right| \le C\left|(u,s)-(v,t)\right|^\alpha$$

for some $C < \infty$ and $\alpha > 0$. But

$$\left|1 - \mathbb{E}_{\mathscr{W}_{HU}}[(S^u(\omega))(s)(S^v(\omega))(t)]\right| \le \frac{2}{\pi}\int_{-\infty}^{\infty}\frac{d\lambda}{1+4\lambda^2}\left|e^{i[(t-s)\lambda+2(u-v)\,\mathrm{Arctg}(2\lambda)]} - 1\right|$$

$$\le \frac{2}{\pi}\int_{-\infty}^{\infty}\frac{d\lambda}{1+4\lambda^2}\left|e^{i(t-s)\lambda}-1\right| + \frac{2}{\pi}\int_{-\infty}^{\infty}\frac{d\lambda}{1+4\lambda^2}\left|e^{2i(u-v)\,\mathrm{Arctg}(2\lambda)}-1\right|$$

$$\le \frac{2}{\pi}\int_{-\infty}^{\infty}\frac{d\lambda}{1+4\lambda^2}(|t-s||\lambda| \wedge 2) + \frac{4}{\pi}\int_{-\infty}^{\infty}\frac{d\lambda}{1+4\lambda^2}|(u-v)\,\mathrm{Arctg}(2\lambda)|,$$

and, after simple estimation, this shows that

$$|1-\mathbb{E}[(S^u(\omega))(s)(S^v(\omega))(t)]| \le C\left[|u-v| + |t-s|\left(1+\log\left(1+\frac{1}{(t-s)^2}\right)\right)\right],$$

where $C < \infty$. Clearly, the desired conclusion follows.

Remark 5.1. A question about filtrations comes naturally when one considers the group of transformations $(S^u)_{u\in\mathbb{R}}$ on the space $\mathscr{U}$. Indeed, for all $t, u \in \mathbb{R}$, let $\mathscr{F}_t^u$ be the σ-algebra generated by the $\mathscr{W}_{HU}$-negligible subsets of $\mathscr{U}$ and the variables $(S^u(\omega))(s)$, for $s \in (-\infty, t]$ (these variables are well defined up to a negligible set). From the results of Jeulin and Yor, one quite easily deduces the following properties of the filtrations of the form $(\mathscr{F}_t^u)_{t\in\mathbb{R}}$ for $u \in \mathbb{R}$:

- For all $t, u \in \mathbb{R}$, $\mathscr{F}_t^u$ is generated by $\mathscr{F}_t^{u+1}$ and $(S^u(\omega))(t)$.
- For all $t, u \in \mathbb{R}$, $\mathscr{F}_t^{u+1}$ and $(S^u(\omega))(t)$ are independent under $\mathscr{W}_{H^U}$.
- For all $t, u \in \mathbb{R}$, the decreasing intersection of $\mathscr{F}_t^{u+n}$ for $n \in \mathbb{Z}$ is trivial (i.e., it satisfies the zero-one law).
- If $u \in \mathbb{R}$ is fixed, the σ-algebra generated by $\mathscr{F}_t^{u+n}$ for $t \in \mathbb{R}$ does not depend on $n \in \mathbb{Z}$.

All these statements concern the sequence of filtrations $(\mathscr{F}^{u+n})_{n\in\mathbb{Z}}$ for fixed $u \in \mathbb{R}$. A natural question arises: how can these results be extended to the continuous family of filtrations $(\mathscr{F}^u)_{u\in\mathbb{R}}$? Unfortunately, for the moment, we have no answer to this question (in particular the family does not seem to be decreasing with u).

References

1. Jeulin, T., Yor, M.: Filtration des ponts browniens et équations différentielles stochastiques linéaires, Séminaire de Probabilités, XXIV. Lecture Notes in Math., vol. 1426, pp. 227–265. Springer, Berlin (1990)
2. Lebedev, N.-N.: Special Functions and Their Applications (Translated from Russian). Dover, New York (1972)
3. Meyer, P.-A.: Sur une transformation du mouvement brownien due à Jeulin et Yor, Séminaire de Probabilités, XXVIII. Lecture Notes in Math., vol. 1583, pp. 98–101. Springer, Berlin (1994)
4. Stroock, D.: Some thoughts about Segal's ergodic theorem. Colloq. Math. **118**(1), 89–105 (2010)
5. Stroock, D.: Probability Theory, an Analytic View, 2nd edn. Cambridge University Press, London (2011)
6. Yor, M.: Some Aspects of Brownian Motion. Part I: Some Special Functionals. Lectures in Mathematics. Birkhaüser, ETH Zürich (1992)

Chapter 6
Measure Invariance on the Lie-Wiener Path Space

Nicolas Privault

Abstract In this chapter we extend some recent results on moment identities, Hermite polynomials, and measure invariance properties on the Wiener space, to the setting of path spaces over Lie groups. In particular we prove the measure invariance of transformations having a quasi-nilpotent covariant derivative via a Girsanov identity and an explicit formula for the expectation of Hermite polynomials in the Skorohod integral on path space.

Keywords Covariant derivatives • Lie groups • Malliavin calculus • Measure invariance • Path space • Quasi-nilpotence • Skorohod integral

Mathematics Subject Classification: 60H07, 58G32.

6.1 Introduction

The Wiener measure is known to be invariant under random isometries whose Malliavin gradient satisfies a quasi-nilpotence condition, cf. [12]. In particular, the Skorohod integral $\delta(Rh)$ is known to have a Gaussian law when $h \in H = L^2(\mathbb{R}_+, \mathbb{R}^d)$ and R is a random isometry of H such that DRh is a.s. a quasi-nilpotent operator. Such results can be proved using the Skorohod integral operator δ and its adjoint the Malliavin derivative D on the Wiener space, and have been recently recovered under simple conditions and with short proofs in [5] using moment identities and in [6] via an exact formula for the expectation of random

N. Privault (✉)
Division of Mathematical Sciences, School of Physical and Mathematical Sciences, Nanyang Technological University, SPMS-MAS-05-43, 21 Nanyang Link, Singapore 637371, Singapore
e-mail: nprivault@ntu.edu.sg

L. Decreusefond and J. Najim (eds.), *Stochastic Analysis and Related Topics*, Springer Proceedings in Mathematics & Statistics 22, DOI 10.1007/978-3-642-29982-7_6,

Hermite polynomials. Indeed it is well known that the Hermite polynomial defined by its generating function

$$e^{xt-t^2\mu^2/2} = \sum_{n=0}^{\infty} \frac{t^n}{n!} H_n(x,\mu), \qquad x,t \in \mathbb{R},$$

satisfies the identity

$$E[H_n(X,\sigma^2)] = 0, \tag{6.1}$$

when $X \simeq \mathcal{N}(0,\sigma^2)$ is a centered Gaussian random variable with variance $\sigma^2 \geq 0$, and that the generating function can be used to characterize the gaussianity of X. In [6], conditions on the process $(u_t)_{t\in\mathbb{R}_+}$ have been deduced for the expectation $E[H_n(\delta(u), \|u\|^2)]$, $n \geq 1$, to vanish. Such conditions cover the quasi-nilpotence condition of [12] and include the adaptedness of $(u_t)_{t\in\mathbb{R}_+}$, which recovers the above invariance result using the characteristic function of $\delta(u)$.

On the other hand, the Skorohod integral and Malliavin gradient can also be defined on the path space over a Lie group, cf. [1, 3, 10]. In this chapter we prove an extension of (6.1) to the path space case, by computing in Theorem 6.1 the expectation

$$E[H_n(\delta(u), \|u\|^2)], \qquad n \geq 1,$$

of the random Hermite polynomial $H_n(\delta(u), \|u\|^2)$, where $\delta(u)$ is the Skorohod integral of a possibly anticipating process $(u_t)_{t\in\mathbb{R}_+}$. This result also recovers the above conditions for the invariance of the path space measure, and extends the results of [6] and [5] to path spaces over Lie group.

In Corollary 6.4 below, we summarize our results in the derivation formula

$$\begin{aligned}\frac{\partial}{\partial\lambda} E\left[e^{\lambda\delta(u)-\frac{\lambda^2}{2}\|u\|^2}\right] &= -E\left[e^{\lambda\delta(u)-\lambda^2\langle u,u\rangle/2}\frac{\partial}{\partial\lambda}\log \det{}_2(I-\lambda\nabla u)\right] \\ &\quad -\lambda E\left[e^{\lambda\delta(u)-\lambda^2\langle u,u\rangle/2}\langle (I-\lambda\nabla u)^{-1}u, D\log\det{}_2(I-\lambda\nabla u)\rangle\right],\end{aligned} \tag{6.2}$$

for λ in a neighborhood of 0, in which D, ∇ respectively denote the Malliavin gradient and covariant derivative on path space, and $\det_2(I-\lambda\nabla u)$ denotes the Carleman–Fredholm determinant of $I-\lambda\nabla u$. When ∇u is quasi-nilpotent, we have $\det_2(I-\lambda\nabla u) = 1$, cf. Theorem 3.6.1 of [13], or [14], hence the derivative (6.2) vanishes, which yields

$$E\left[e^{\lambda\delta(u)-\frac{\lambda^2}{2}\|u\|^2}\right] = 1,$$

for λ in a neighborhood of 0, cf. Corollary 6.3. If in addition $\langle u,u\rangle$ is a.s. constant, this implies

$$E\left[e^{\lambda\delta(u)}\right] = e^{-\frac{\lambda^2}{2}\|u\|^2}, \qquad \lambda \in \mathbb{R},$$

showing that $\delta(u)$ is centered Gaussian with variance $\|u\|^2$.

This chapter is organized as follows. In Sect. 6.2 we review some notation on closable gradient and divergence operators, and associated commutation relations. In Sect. 6.3 we derive moment identities for the Skorohod integral on path spaces. In Sect. 6.4 we consider the expectation of Hermite polynomials, and in Sect. 6.5 we derive Girsanov identities on path space.

6.2 The Lie-Wiener Path Space

In this section we recall some notation on the Lie-Wiener path space, cf. [1, 3, 10, 11], and we prove some auxiliary results. Let G denote either $\mathbb{R}^d$ or a compact connected d-dimensional Lie group with associated Lie algebra $\mathscr{G}$ identified to $\mathbb{R}^d$ and equipped with an Ad-invariant scalar product on $\mathbb{R}^d \simeq \mathscr{G}$, also denoted by $(\cdot,\cdot)$. The commutator in $\mathscr{G}$ is denoted by $[\cdot,\cdot]$. Let $\mathrm{ad}\,(u)v = [u,v]$, $u,v \in \mathscr{G}$, with $\mathrm{Ad}\,e^u = e^{\mathrm{ad}\,u}$, $u \in \mathscr{G}$.

The Brownian motion $(\gamma(t))_{t\in\mathbb{R}_+}$ on G with paths in $\mathscr{C}_0(\mathbb{R}_+,\mathscr{G})$ is constructed from $(B_t)_{t\in\mathbb{R}_+}$ via the Stratonovich differential equation

$$\begin{cases} d\gamma(t) = \gamma(t) \odot dB_t \\ \gamma(0) = \mathrm{e}, \end{cases}$$

where e is the identity element in G. Let $\mathbf{P}(\mathsf{G}) = \mathscr{C}_0(\mathbb{R}_+,\mathscr{G})$ denote the space of continuous G-valued paths starting at e, with the image measure of the Wiener measure by $I : (B_t)_{t\in\mathbb{R}_+} \mapsto (\gamma(t))_{t\in\mathbb{R}_+}$. Here we take

$$\mathscr{S} = \{F = f(\gamma(t_1),\ldots,\gamma(t_n)) \quad : \quad f \in \mathscr{C}_b^\infty(\mathsf{G}^n)\},$$

and

$$\mathscr{U} = \left\{ \sum_{i=1}^n u_i F_i \quad : \quad F_i \in \mathscr{S},\ u_i \in L^2(\mathbb{R}_+;\mathscr{G}),\ i = 1,\ldots,n,\ n \geq 1 \right\}.$$

Next is the definition of the right derivative operator D.

Definition 6.1. For $F = f(\gamma(t_1),\ldots,\gamma(t_n)) \in \mathscr{S}$, $f \in \mathscr{C}_b^\infty(\mathsf{G}^n)$, we let $DF \in L^2(\Omega\times\mathbb{R}_+;\mathscr{G})$ be defined as

$$\langle DF, v\rangle = \frac{d}{d\varepsilon} f\left(\gamma(t_1)e^{\varepsilon\int_0^{t_1} v_s ds},\ldots,\gamma(t_n)e^{\varepsilon\int_0^{t_n} v_s ds}\right)_{|\varepsilon=0}, \qquad v \in L^2(\mathbb{R}_+,\mathscr{G}).$$

For $F \in \mathscr{S}$ of the form $F = f(\gamma(t_1),\ldots,\gamma(t_n))$, we also have

$$D_t F = \sum_{i=1}^n \partial_i f(\gamma(t_1),\ldots,\gamma(t_n))\mathbf{1}_{[0,t_i]}(t), \qquad t \geq 0.$$

The operator D is known to be closable and to admit an adjoint δ that satisfies

$$E[F\delta(v)] = E[\langle DF, v\rangle], \quad F \in \mathscr{S},\ v \in \mathscr{U}, \tag{6.3}$$

cf., e.g., [1]. Let $I\!D_{p,k}(X)$, $k \geq 1$, denote the completion of the space of smooth X-valued random variables under the norm

$$\|u\|_{I\!D_{p,k}(X)} = \sum_{l=0}^{k} \|D^l u\|_{L^p(W, X\otimes H^{\otimes l})}, \qquad p \in [1,\infty],$$

where $H = L^2(\mathbb{R}_+, \mathscr{G})$, and $X \otimes H$ denotes the completed symmetric tensor product of X and H. We also let $I\!D_{p,k} = I\!D_{p,k}(\mathbb{R})$, $p \in [1,\infty]$, $k \geq 1$.

Next we turn to the definition of the covariant derivative on the path space $\mathbf{P}(\mathsf{G})$, cf. [1].

Definition 6.2. Let the operator ∇ be defined on $u \in I\!D_{2,1}(H)$ as

$$\nabla_s u_t = D_s u_t + \mathbf{1}_{[0,t]}(s)\operatorname{ad} u_t \in \mathscr{G} \otimes \mathscr{G}, \qquad s,t \in \mathbb{R}_+. \tag{6.4}$$

When $h \in H$, we have

$$\nabla_s h_t = \mathbf{1}_{[0,t]}(s)\operatorname{ad} h_t, \qquad s,t \in \mathbb{R}_+,$$

and $\operatorname{ad} v \in \mathscr{G} \otimes \mathscr{G}$, $v \in \mathscr{G}$, is the matrix

$$(\langle e_j, \operatorname{ad}(e_i)v\rangle)_{1\leq i,j\leq d} = (\langle e_j, [e_i, v]\rangle)_{1\leq i,j\leq d}.$$

The operator $\operatorname{ad}(v)$ is antisymmetric on $\mathscr{G}$ because $(\cdot,\cdot)$ is Ad-invariant. In addition if $u = hF$, $h \in H$, $F \in I\!D_{2,1}$, we have

$$D_s u_t = D_s F \otimes h(t), \qquad \operatorname{ad} u_t = F \operatorname{ad} h(t), \qquad s,t \in \mathbb{R}_+,$$

and

$$\begin{aligned}
\langle e_i \otimes e_j, \nabla_s u_t\rangle &= \langle e_i \otimes e_j, \nabla_s (hF)(t)\rangle \\
&= \langle e_i \otimes e_j, D_s F \otimes h(t)\rangle + \mathbf{1}_{[0,t]}(s) F \langle e_i \otimes e_j, \operatorname{ad} h(t)\rangle \\
&= \langle h(t), e_j\rangle\langle e_i, D_s F\rangle + \mathbf{1}_{[0,t]}(s) F \langle e_j, \operatorname{ad}(e_i) h(t)\rangle \\
&= \langle h(t), e_j\rangle\langle e_i, D_s F\rangle + \mathbf{1}_{[0,t]}(s) F \langle e_j, [e_i, h(t)]\rangle,
\end{aligned}$$

$i,j = 1,\ldots,d$. In the commutative case, we have $\operatorname{ad}(v) = 0$, $v \in \mathscr{G}$, hence $\nabla = D$.

By (6.4), we have

$$(\nabla_v u)(t) := (\nabla u) v_t = \int_0^t (\nabla_s u_t) v_s ds, \qquad t \in \mathbb{R}_+,$$

is the covariant derivative of $u \in \mathscr{U}$ in the direction $v \in L^2(\mathbb{R}_+;\mathscr{G})$, with $\nabla_v u \in L^2(\mathbb{R}_+;\mathscr{G})$, cf. [1] and Lemma 3.4 in [4].

It is known that D and ∇ satisfy the commutation relation

$$D\delta(u) = u + \delta(\nabla^* u), \tag{6.5}$$

for $u \in I\!D_{2,1}(H)$ such that $\nabla^* u \in I\!D_{2,1}(H \otimes H)$, cf., e.g., [1]. On the other hand, the commutation relation (6.5) shows that the Skorohod isometry [9]

$$E[\delta(u)\delta(v)] = E[\langle u, v\rangle] + E\left[\text{trace}\,(\nabla u)(\nabla v)\right], \qquad u, v \in I\!D_{2,1}(H), \tag{6.6}$$

holds as a consequence of (6.5), cf. [1] and Theorem 3.3 in [4], where

$$\text{trace}\,(\nabla u)(\nabla v) = \langle \nabla u, \nabla^* v\rangle_{H\otimes H} = \int_0^\infty \int_0^\infty \langle \nabla_s u_t, \nabla_t^\dagger v_s\rangle_{\mathbb{R}^d\otimes\mathbb{R}^d} ds dt,$$

and $\nabla_t^\dagger v_s$ denotes the transpose of the matrix $\nabla_t v_s$, $s, t \in \mathbb{R}_+$. Note also that we have

$$\nabla_s u_t = D_s u_t, \qquad s > t, \tag{6.7}$$

Hence the Skorohod isometry (6.6) can be rewritten as

$$E[\delta(u)\delta(v)] = E[\langle u, v\rangle] + E\left[\text{trace}\,(\nabla u)(Dv)\right], \qquad u, v \in I\!D_{2,1}(H), \tag{6.8}$$

as a consequence of the following lemma. Note that for $u \in I\!D_{2,1}(H)$ and $v \in H$, we have

$$(\nabla u)^k v(t) = \int_0^\infty \cdots \int_0^\infty (\nabla_{t_k} u_t \nabla_{t_{k-1}} u_{t_k} \cdots \nabla_{t_1} u_{t_2}) v_{t_1} dt_1 \cdots dt_k, \qquad t \in \mathbb{R}_+,$$

and

$$\begin{aligned}\text{trace}\,(\nabla u)^k &= \langle \nabla^\dagger u, (\nabla u)^{k-1}\rangle \\ &= \int_0^\infty \cdots \int_0^\infty \langle \nabla_{t_k}^\dagger u_{t_1}, \nabla_{t_{k-1}} u_{t_k} \cdots \nabla_{t_1} u_{t_2}\rangle dt_1 \cdots dt_k,\end{aligned}$$

$k \geq 1$.

Lemma 6.1. *For all $u, v \in I\!D_{2,1}(H)$, we have*

$$\text{trace}\,(\nabla u)^k(\nabla v) = \text{trace}\,(\nabla u)^k(Dv), \qquad k \geq 1.$$

Proof. We have

$$\begin{aligned}\text{trace}\,(\nabla u)^k(\nabla v) &= \int_0^\infty \int_0^\infty \langle (\nabla_s u_t)^k, \nabla_t^\dagger v_s\rangle_{\mathbb{R}^d\otimes\mathbb{R}^d} ds dt \\ = \int_0^\infty \int_0^t \langle (\nabla_s u_t)^k, \nabla_t^\dagger v_s\rangle_{\mathbb{R}^d\otimes\mathbb{R}^d} ds dt &+ \int_0^\infty \int_t^\infty \langle (\nabla_s^\dagger u_t)^k, \nabla_t v_s\rangle_{\mathbb{R}^d\otimes\mathbb{R}^d} ds dt\end{aligned}$$

$$
\begin{aligned}
&= \int_0^\infty \int_0^t \langle (\nabla_s u_t)^k, \nabla_t^\dagger v_s \rangle_{\mathbb{R}^d \otimes \mathbb{R}^d} ds dt + \int_0^\infty \int_0^t \langle (\nabla_t^\dagger u_s)^k, \nabla_s v_t \rangle_{\mathbb{R}^d \otimes \mathbb{R}^d} ds dt \\
&= \int_0^\infty \int_0^t \langle (\nabla_s u_t)^k, D_t^\dagger v_s \rangle_{\mathbb{R}^d \otimes \mathbb{R}^d} ds dt + \int_0^\infty \int_0^t \langle D_s v_t, (\nabla_t^\dagger u_s)^k \rangle_{\mathbb{R}^d \otimes \mathbb{R}^d} ds dt \\
&= \int_0^\infty \int_0^t \langle (\nabla_s u_t)^k, D_t^\dagger v_s \rangle_{\mathbb{R}^d \otimes \mathbb{R}^d} ds dt + \int_0^\infty \int_t^\infty \langle D_t v_s, (\nabla_s^\dagger u_t)^k \rangle_{\mathbb{R}^d \otimes \mathbb{R}^d} ds dt \\
&= \int_0^\infty \int_0^t \langle (\nabla_s u_t)^k, D_t^\dagger v_s \rangle_{\mathbb{R}^d \otimes \mathbb{R}^d} ds dt + \int_0^\infty \int_t^\infty \langle D_t^\dagger v_s, (\nabla_s u_t)^k \rangle_{\mathbb{R}^d \otimes \mathbb{R}^d} ds dt \\
&= \operatorname{trace} (\nabla u)^k (Dv).
\end{aligned}
$$

□

In addition we have the following lemma, which will be used to apply our invariance results to adapted processes.

Lemma 6.2. *Assume that the process $u \in I\!D_{2,1}(H)$ is adapted with respect to the Brownian filtration $(\mathscr{F}_t)_{t\in\mathbb{R}_+}$. Then we have*

$$
\operatorname{trace} (\nabla u)^k = 0, \qquad k \geq 2..
$$

Proof. For almost all $t_1, \dots, t_{k+1} \in \mathbb{R}_+$ there exists $i \in \{1, \dots, k+1\}$ such that $t_i > t_{i+1 \bmod k+1}$, and (6.7) yields

$$
\begin{aligned}
\nabla_{t_i} u_{t_{i+1 \bmod k+1}} &= D_{t_i} u_{t_{i+1 \bmod k+1}} + \mathbf{1}_{[0, t_{i+1 \bmod k+1}]}(t_i) \\
&= D_{t_i} u_{t_{i+1 \bmod k+1}} \\
&= 0,
\end{aligned}
$$

since $(u_t)_{t\in\mathbb{R}_+}$ is adapted. □

We close this section with three lemmas that will be used in the sequel.

Lemma 6.3. *For any $u \in I\!D_{2,1}(H)$, we have*

$$
\langle (\nabla u) v, u \rangle = \frac{1}{2} \langle v, D \langle u, u \rangle \rangle, \qquad v \in H.
$$

Proof. We have

$$
\begin{aligned}
(\nabla^* u) u_t &= \int_0^\infty (\nabla_t u_s)^\dagger u_s ds \\
&= \int_0^\infty (D_t u_s)^\dagger u_s ds + \int_0^\infty \mathbf{1}_{[0,s]}(t) (\operatorname{ad} u_s)^\dagger u_s ds
\end{aligned}
$$

$$= \int_0^\infty (D_t u_s)^\dagger u_s ds - \int_0^\infty \mathbf{1}_{[0,s]}(t)\mathrm{ad}(u_s)u_s ds$$
$$= \int_0^\infty (D_t u_s)^\dagger u_s ds$$
$$= (D^* u)u_t..$$

Next, the relation $D\langle u, u\rangle = 2(D^*u)u$ shows that

$$\begin{aligned}\langle(\nabla u)v, u\rangle &= \langle(\nabla^* u)u, v\rangle \\ &= \langle(D^* u)u, v\rangle \\ &= \frac{1}{2}\langle v, D\langle u, u\rangle\rangle..\end{aligned}$$

□

Lemma 6.4. *For all $u \in \mathbb{D}_{2,2}(H)$ and $v \in \mathbb{D}_{2,1}(H)$, we have*

$$\langle\nabla^* u, D((\nabla u)^k v)\rangle = \mathrm{trace}\,((\nabla u)^{k+1}\nabla v) + \sum_{i=2}^{k+1} \frac{1}{i}\langle(\nabla u)^{k+1-i} v, D\,\mathrm{trace}\,(\nabla u)^i\rangle, \quad k \in \mathbf{N}.$$

Proof. Note that we have the commutation relation $\nabla D = D\nabla$, and as a consequence for all $1 \le k \le n$, we have

$$\langle\nabla^* u, D((\nabla u)^k v)\rangle = \int_0^\infty \cdots \int_0^\infty \langle\nabla_{t_k}^\dagger u_{t_{k+1}}, D_{t_{k+1}}(\nabla_{t_{k-1}} u_{t_k} \cdots \nabla_{t_0} u_{t_1} v_{t_0})\rangle dt_0 \cdots dt_{k+1}$$

$$= \int_0^\infty \cdots \int_0^\infty \langle\nabla_{t_k}^\dagger u_{t_{k+1}}, \nabla_{t_{k-1}} u_{t_k} \cdots \nabla_{t_0} u_{t_1} D_{t_{k+1}} v_{t_0}\rangle dt_0 \cdots dt_{k+1}$$

$$+ \int_0^\infty \cdots \int_0^\infty \langle\nabla_{t_k}^\dagger u_{t_{k+1}}, D_{t_{k+1}}(\nabla_{t_{k-1}} u_{t_k} \cdots \nabla_{t_0} u_{t_1}) v_{t_0}\rangle dt_0 \cdots dt_{k+1}$$

$$= \mathrm{trace}\,((\nabla u)^{k+1} Dv) + \sum_{i=0}^{k-1} \int_0^\infty \cdots \int_0^\infty$$

$$\langle\nabla_{t_k}^\dagger u_{t_{k+1}}, \nabla_{t_{k+1}} u_{t_{k+2}} \cdots \nabla_{t_i+1} u_{t_i+2}(\nabla_{t_i} D_{t_{k+1}} u_{t_i+1})\nabla_{t_{i-1}} u_{t_i} \cdots \nabla_{t_0} u_{t_1} v_{t_0}\rangle dt_0 \cdots dt_{k+1}$$

$$= \mathrm{trace}\,((\nabla u)^{k+1} Dv) + \sum_{i=0}^{k-1} \frac{1}{k+1-i} \int_0^\infty \cdots \int_0^\infty$$

$$\langle\nabla_{t_i}\langle\nabla_{t_k}^\dagger u_{t_{k+1}}, \nabla_{t_{k+1}} u_{t_{k+2}} \cdots \nabla_{t_i+1} u_{t_i+2} \nabla_{t_{k+1}} u_{t_i+1}\rangle, \nabla_{t_{i-1}} u_{t_i} \cdots \nabla_{t_0} u_{t_1} v_{t_0}\rangle dt_0 \cdots dt_{k+1}$$

$$= \mathrm{trace}\,((\nabla u)^{k+1} Dv) + \sum_{i=0}^{k-1} \frac{1}{k+1-i}\langle(\nabla u)^i v, D\,\mathrm{trace}\,(\nabla u)^{k+1-i}\rangle,$$

and we conclude by Lemma 6.1. □

Lemma 6.5. *For all* $u \in I\!D_{2,2}(H)$ *and* $v \in I\!D_{2,1}(H)$ *such that* $\|\nabla u\|_{L^\infty(\Omega;H\otimes H)} < 1$, *we have*

$$\langle \nabla^* u, D((I-\nabla u)^{-1}v)\rangle = \operatorname{trace}(\nabla u)(I-\nabla u)^{-1}(\nabla v) - \langle (I-\nabla u)^{-1}v, D\log\det{}_2(I-\nabla u)\rangle..$$

Proof. By Lemma 6.4, we have

$$\begin{aligned}
\langle \nabla^* u, D((I-\nabla u)^{-1}v)\rangle &= \sum_{n=0}^{\infty}\langle \nabla^* u, D((\nabla u)^n v))\rangle \\
&= \sum_{n=0}^{\infty}\operatorname{trace}((\nabla u)^{n+1}Dv) + \sum_{n=0}^{\infty}\sum_{i=2}^{n+1}\frac{1}{i}\langle (\nabla u)^{n+1-i}v, D\operatorname{trace}(\nabla u)^i\rangle \\
&= \operatorname{trace}(\nabla u)(I-\nabla u)^{-1}(Dv) + \sum_{i=2}^{\infty}\frac{1}{i}\sum_{n=0}^{\infty}\langle (\nabla u)^n v, D\operatorname{trace}(\nabla u)^i\rangle \\
&= \operatorname{trace}(\nabla u)(I-\nabla u)^{-1}(Dv) + \sum_{i=2}^{\infty}\frac{1}{i}\langle (I-\nabla u)^{-1}v, D\operatorname{trace}(\nabla u)^i\rangle \\
&= \operatorname{trace}(\nabla u)(I-\nabla u)^{-1}(\nabla v) - \langle (I-\nabla u)^{-1}u, D\log\det{}_2(I-\nabla u)\rangle,
\end{aligned}$$

by Lemma 6.1 and since $\det_2(I-\lambda\nabla u)$ satisfies

$$\det{}_2(I-\lambda\nabla u) = \exp\left(-\sum_{i=2}^{\infty}\frac{\lambda^i}{i}\operatorname{trace}(\nabla u)^i\right), \tag{6.9}$$

cf. [8] page 108, which shows that

$$D\log\det{}_2(I-\lambda\nabla u) = -\sum_{i=2}^{\infty}\frac{\lambda^i}{i}D\operatorname{trace}(\nabla u)^i..$$

□

6.3 Moment Identities on Path Space

The following moment identity extends Theorem 2.1 of [5] to the path space setting. The Wiener case is obtained by taking $\nabla = D$.

Proposition 6.1. *For any* $n \geq 1$ *and* $u \in I\!D_{n+1,2}(H)$, $v \in I\!D_{n+1,1}(H)$, *we have*

$$\begin{aligned}
E[\delta(u)^n\delta(v)] &= nE\left[\delta(u)^{n-1}\langle u, v\rangle\right] \qquad (6.10)\\
&+\frac{1}{2}\sum_{k=2}^{n}\frac{n!}{(n-k)!}E\left[\delta(u)^{n-k}\langle(\nabla u)^{k-2}v, D\langle u,u\rangle\rangle\right]\\
&+\sum_{k=1}^{n}\frac{n!}{(n-k)!}E\left[\delta(u)^{n-k}\left(\operatorname{trace}((\nabla u)^{k+1}\nabla v) + \sum_{i=2}^{k}\frac{1}{i}\langle(\nabla u)^{k-i}v, D\operatorname{trace}(\nabla u)^i\rangle\right)\right]..
\end{aligned}$$

For $n = 1$ the above identity (6.10) coincides with the Skorohod isometry (6.8).

When $\langle u, u\rangle$ is deterministic, $u \in I\!D_{2,1}(H)$, and $\operatorname{trace}(\nabla u)^k = 0$ a.s., $k \geq 2$, Proposition 6.1 yields

$$E[\delta(u)^{n+1}] = n\langle u, u\rangle E\left[\delta(u)^{n-1}\right], \qquad n \geq 1,$$

and by induction we have

$$E[\delta(u)^{2m}] = \frac{(2m)!}{2^m m!}\langle u, u\rangle^m, \qquad 0 \leq 2m \leq n+1,$$

and $E[\delta(u)^{2m+1}] = 0$, $0 \leq 2m \leq n$, while $E[\delta(u)] = 0$ for all $u \in I\!D_{2,1}(H)$, hence the following corollary of Proposition 6.1.

Corollary 6.1. *Let $u \in I\!D_{\infty,2}(H)$ such that $\langle u, u\rangle$ is deterministic and*

$$\operatorname{trace}(\nabla u)^k = 0, \quad a.s., \quad k \geq 2. \tag{6.11}$$

Then $\delta(u)$ has a centered Gaussian distribution with variance $\langle u, u\rangle$.

In particular, under the conditions of Corollary 6.1, $\delta(Rh)$ has a centered Gaussian distribution with variance $\langle h, h\rangle$ when $u = Rh$, $h \in H$, and R is a random mapping with values in the isometries of H, such that $Rh \in \bigcap_{p>1} I\!D_{p,2}(H)$ and $\operatorname{trace}(DRh)^k = 0$, $k \geq 2$. In the Wiener case this recovers Theorem 2.1-b) of [12], cf. also Corollary 2.2 of [5].

In addition, Lemma 6.2 shows that Condition (6.11) holds when the process u is adapted with respect to the Brownian filtration.

Next we prove Proposition 6.1 based on Lemmas 6.3, 6.4, and on Lemma 6.6 below.

Proof of Proposition 6.1. Let $n \geq 1$ and $u \in I\!D_{n+1,2}(H)$. We show that for any $n \geq 1$ and $u \in I\!D_{n+1,2}(H)$, $v \in I\!D_{n+1,1}(H)$, we have

$$E[\delta(u)^n\delta(v)] = \sum_{k=1}^{n}\frac{n!}{(n-k)!}E\left[\delta(u)^{n-k}\left(\langle(\nabla u)^{k-1}v, u\rangle + \langle\nabla^* u, D((\nabla u)^{k-1}v)\rangle\right)\right].. \tag{6.12}$$

We have $(\nabla u)^{k-1}v \in I\!D_{(n+1)/k,1}(H)$, $\delta(u) \in I\!D_{(n+1)/(n-k+1),1}$, and by Lemma 6.6 below applied to $F = 1$ we get

$$\begin{aligned}
&E\left[\delta(u)^l\langle(\nabla u)^i v, D\delta(u)\rangle\right] - lE\left[\delta(u)^{l-1}\langle(\nabla u)^{i+1}v, D\delta(u)\rangle\right] \\
&= E\left[\delta(u)^l\langle(\nabla u)^i v, u\rangle\right] + E\left[\delta(u)^l\langle(\nabla u)^i v, \delta(\nabla^* u)\rangle\right] \\
&\quad - lE\left[\delta(u)^{l-1}\langle(\nabla u)^{i+1}v, u\rangle\right] - lE\left[\delta(u)^{l-1}\langle(\nabla u)^{i+1}v, \delta(\nabla^* u)\rangle\right] \\
&= E\left[\delta(u)^l\langle(\nabla u)^i v, u\rangle\right] + E[\delta(u)^l\langle\nabla^* u, D((\nabla u)^i v)\rangle],
\end{aligned}$$

and applying this formula to $l = n - k$ and $i = k - 1$ yields

$$E[\delta(u)^n\delta(v)] = E[\langle v, D\delta(u)^n\rangle] = nE[\delta(u)^{n-1}\langle v, D\delta(u)\rangle]$$
$$= \sum_{k=1}^{n}\frac{n!}{(n-k)!}\left(E\left[\delta(u)^{n-k}\langle(\nabla u)^{k-1}v, D\delta(u)\rangle\right]-(n-k)E\left[\delta(u)^{n-k-1}\langle(\nabla u)^{k}v, D\delta(u)\rangle\right]\right)$$
$$= \sum_{k=1}^{n}\frac{n!}{(n-k)!}\left(E\left[\delta(u)^{n-k}\langle(\nabla u)^{k-1}v, u\rangle\right]+E\left[\delta(u)^{n-k}\langle\nabla^* u, D((\nabla u)^{k-1}v)\rangle\right]\right)..$$

We conclude by applying Lemmas 6.3 and 6.4. The next lemma extends the argument of Lemma 3.1 in [5] pages 120–121 to the path space case, including an additional random variable $F \in I\!D_{2,1}$.

Lemma 6.6. *Let $F \in I\!D_{2,1}$, $u \in I\!D_{n+1,2}(H)$, and $v \in I\!D_{n+1,1}(H)$. For all $k, i \geq 0$ we have*

$$E[F\delta(u)^k\langle(\nabla u)^i v, \delta(\nabla^* u)\rangle] - kE[F\delta(u)^{k-1}\langle(\nabla^* u)^{i+1}v, \delta(\nabla^* u)\rangle]$$
$$= kE[F\delta(u)^{k-1}\langle(\nabla u)^{i+1}v, u\rangle] + E[\delta(u)^k\langle(\nabla u)^{i+1}v, DF\rangle] + E[F\delta(u)^k\langle\nabla^* u, D((\nabla u)^i v)\rangle]..$$

Proof. We have

$$\begin{aligned}
&E[F\delta(u)^k\langle(\nabla u)^i v, \delta(\nabla^* u)\rangle] - iE[F\delta(u)^{k-1}\langle(\nabla^* u)^{i+1}v, \delta(\nabla^* u)\rangle]\\
&= E[\langle\nabla^* u, D(F\delta(u)^k(\nabla u)^i v)\rangle] - kE[F\delta(u)^{k-1}\langle(\nabla^* u)^{i+1}v, \delta(\nabla^* u)\rangle]\\
&= kE[F\delta(u)^{k-1}\langle\nabla^* u, (\nabla u)^i v \otimes D\delta(u)\rangle] - kE[F\delta(u)^{k-1}\langle(\nabla^* u)^{i+1}v, \delta(\nabla^* u)\rangle]\\
&\quad + E[\delta(u)^k\langle\nabla^* u, D(F(\nabla u)^i v)\rangle]\\
&= kE[F\delta(u)^{k-1}\langle\nabla^* u, (\nabla u)^i v \otimes u\rangle] + kE[F\delta(u)^{k-1}\langle\nabla^* u, (\nabla u)^i v \otimes \delta(\nabla^* u)\rangle]\\
&\quad - kE[F\delta(u)^{k-1}\langle(\nabla^* u)^{i+1}v, \delta(\nabla^* u)\rangle] + E[\delta(u)^k\langle\nabla^* u, D(F(\nabla u)^i v)\rangle]\\
&= kE[F\delta(u)^{k-1}\langle(\nabla u)^{i+1}v, u\rangle] + E[\delta(u)^k\langle(\nabla u)^{i+1}v, DF\rangle]\\
&\quad + E[F\delta(u)^k\langle\nabla^* u, D((\nabla u)^i v)\rangle],
\end{aligned}$$

where we used the commutation relation (6.5). □

The case of the left derivative D^L defined as

$$\langle D^L F, v\rangle = \frac{d}{d\varepsilon}f\left(e^{\varepsilon\int_0^{t_1}v_s ds}\gamma(t_1), \ldots, e^{\varepsilon\int_0^{t_n}v_s ds}\gamma(t_n)\right)_{|\varepsilon=0}, \quad v \in L^2(\mathbb{R}_+, \mathscr{G}),$$

for $F = f(\gamma(t_1), \ldots, \gamma(t_n)) \in \mathscr{S}$, $f \in \mathscr{C}_b^\infty(\mathrm{G}^n)$, can be dealt with by application of the existing results on the flat Wiener space, using the expression of its adjoint the left divergence δ^L which can be written as

$$\delta^L(u) = \hat{\delta}(\mathrm{Ad}\,\gamma.u.)$$

using the Skorohod integral operator $\hat{\delta}$ on the flat space $\mathbb{R}^d$, cf. [3,10], and Sect. 13.1 of [11].

6.4 Random Hermite Polynomials on Path Space

In this section we extend the results of [6] on the expectation of Hermite polynomials to the path space framework. This also allows us to recover the invariance results of Sect. 6.3 in Corollary 6.2 and to derive a Girsanov identity in Corollary 6.3 as a consequence of the derivation formula stated in Proposition 6.2.

It is well known that the Gaussianity of X is not required for $E[H_n(X,\sigma^2)]$ to vanish when σ^2 is allowed to be random. Indeed, such an identity also holds in the random adapted case under the form

$$E\left[H_n\left(\int_0^\infty u_t dB_t, \int_0^\infty |u_t|^2 dt\right)\right] = 0, \tag{6.13}$$

where $(u_t)_{t\in\mathbb{R}_+}$ is a square-integrable process adapted to the filtration generated by $(B_t)_{t\in\mathbb{R}_+}$, since $H_n\left(\int_0^\infty u_t dB_t, \int_0^\infty |u_t|^2 dt\right)$ is the n-th order iterated multiple stochastic integral of $u_{t_1}\cdots u_{t_n}$ with respect to $(B_t)_{t\in\mathbb{R}_+}$, cf. [7] and page 319 of [2].

In Theorem 6.1 below, we extend Relations (6.1) and (6.13) by computing the expectation of the random Hermite polynomial $H_n(\delta(u), \|u\|^2)$ in the Skorohod integral $\delta(u)$, $n \ge 1$. This also extends Theorem 3.1 of [6] to the setting of path spaces over Lie groups.

Theorem 6.1. *For any $n \ge 0$ and $u \in I\!D_{n+1,2}(H)$ we have*

$$E[H_{n+1}(\delta(u), \|u\|^2)]$$
$$= \sum_{l=0}^{n-1} \frac{n!}{l!} E\left[\delta(u)^l \sum_{0\le 2k\le n-1-l} \frac{(-1)^k}{k!} \frac{\|u\|^{2k}}{2^k} \langle \nabla^* u, D((\nabla u)^{n-2k-l-1} u)\rangle\right]..$$

Clearly, it follows from Theorem 6.1 that if $u \in I\!D_{n,2}(H)$ and

$$\langle \nabla^* u, D((\nabla u)^k u)\rangle = 0, \qquad 0 \le k \le n-2, \tag{6.14}$$

then we have

$$E[H_n(\delta(u), \|u\|^2)] = 0, \qquad n \ge 1, \tag{6.15}$$

which extends Relation (6.13) to the anticipating case. In addition, from Theorem 6.1 and Lemma 6.3 we have

$$E[H_{n+1}(\delta(u), \|u\|^2)]$$
$$= \sum_{l=0}^{n-1} \frac{n!}{l!} E\left[\delta(u)^l \sum_{0\le 2k\le n-1-l} \frac{(-1)^k}{k!} \frac{\|u\|^{2k}}{2^k} \operatorname{trace}((\nabla u)^{n+1-2k-l})\right] \tag{6.16}$$
$$+ \sum_{l=0}^{n-1} \frac{n!}{l!} E\left[\delta(u)^l \sum_{0\le 2k\le n-1-l} \frac{(-1)^k}{k!} \frac{\|u\|^{2k}}{2^k} \sum_{i=2}^{n-2k-l} \frac{1}{i} \langle (\nabla u)^{n-2k-l-i} u, D \operatorname{trace}(\nabla u)^i\rangle\right]..$$

As a consequence, Lemma 6.2 leads to the following corollary of Theorem 6.1, which extends Corollary 3.3 of [6] to the path space setting.

Corollary 6.2. *Let* $u \in I\!D_{n,2}(H)$ *such that* $\nabla u : H \to H$ *is a.s. quasi-nilpotent in the sense that*

$$\mathrm{trace}\,(\nabla u)^k = 0, \qquad k \geq 2, \tag{6.17}$$

or more generally that (6.14) *holds. Then for any* $n \geq 1$ *we have*

$$E[H_n(\delta(u), \|u\|^2)] = 0..$$

As above, Lemma 6.2 shows that Corollary 6.2 holds when the process $(u_t)_{t\in\mathbb{R}_+}$ is adapted with respect to the Brownian filtration, and this shows that (6.13) holds for the stochastic integral $\delta(u)$ on path space when the process $(u_t)_{t\in\mathbb{R}_+}$ is adapted.

We now turn to the proof of Theorem 6.1, which follows the same steps as the proof of Theorem 3.1 in [6], the main change being in the different roles played here by ∇ and D.

Proof of Theorem 6.1.

Step 1. We show that for any $n \geq 1$ and $u \in I\!D_{n+1,2}(H)$ we have

$$\begin{aligned} E[H_{n+1}(\delta(u), \|u\|^2)] = & \sum_{0\leq 2k\leq n-1} \frac{(-1)^k n!}{k! 2^k (n-2k-1)!} E[\delta(u)^{n-2k-1} \langle u, u\rangle^k \langle u, \delta(\nabla^* u)\rangle] \\ & + \sum_{1\leq 2k\leq n} \frac{(-1)^k n!}{k! 2^k (n-2k)!} E[\delta(u)^{n-2k} \langle u, D\langle u, u\rangle^k\rangle].. \end{aligned} \tag{6.18}$$

For $F \in I\!D_{2,1}$ and $k, l \geq 1$ we have

$$\begin{aligned} E[F\delta(u)^{l+1}] &= \frac{l+2k+1}{2k} E[F\delta(u)^{l+1}] - \frac{l+1}{2k} E[F\delta(u)^{l+1}] \\ &= \frac{l+2k+1}{2k} E[F\delta(u)^{l+1}] - \frac{l+1}{2k} E[\langle u, D(\delta(u)^l F)\rangle] \\ &= \frac{l+2k+1}{2k} E[F\delta(u)^{l+1}] - \frac{l(l+1)}{2k} E[F\delta(u)^{l-1}\langle u, D\delta(u)\rangle] - \frac{l+1}{2k} E[\delta(u)^l \langle u, DF\rangle] \\ &= \frac{l+2k+1}{2k} E[F\delta(u)^{l+1}] - \frac{l(l+1)}{2k} E[F\delta(u)^{l-1}\langle u, u\rangle] \\ &\quad - \frac{l(l+1)}{2k} E[F\delta(u)^{l-1}\langle u, \delta(\nabla^* u)\rangle] - \frac{l+1}{2k} E[\delta(u)^l \langle u, DF\rangle], \end{aligned}$$

i.e.

$$\begin{aligned} & E[F\delta(u)^{n-2k+1}] + \frac{(n-2k)(n-2k+1)}{2k} E[F\delta(u)^{n-2k-1}\langle u, u\rangle] \\ &= \frac{n+1}{2k} E[F\delta(u)^{n-2k+1}] - \frac{(n-2k)(n-2k+1)}{2k} E[F\delta(u)^{n-2k-1}\langle u, \delta(\nabla^* u)\rangle] \\ &\quad - \frac{n-2k+1}{2k} E[\delta(u)^{n-2k}\langle u, DF\rangle].. \end{aligned}$$

Hence, taking $F = \langle u, u\rangle^k$, we get

$$\begin{aligned}
E[\delta(u)^{n+1}] &= E[\langle u, D\delta(u)^n\rangle] \\
&= nE[\delta(u)^{n-1}\langle u, D\delta(u)\rangle] \\
&= nE[\delta(u)^{n-1}\langle u, u\rangle] + nE[\delta(u)^{n-1}\langle u, \delta(\nabla^* u)\rangle] \\
&= nE[\delta(u)^{n-1}\langle u, \delta(\nabla^* u)\rangle] \\
&\quad - \sum_{1\le 2k\le n+1} \frac{(-1)^k n!}{(k-1)!2^{k-1}(n+1-2k)!}\Bigg(E[\delta(u)^{n-2k+1}\langle u, u\rangle^k] \\
&\qquad + \frac{(n-2k+1)(n-2k)}{2k} E[\delta(u)^{n-2k-1}\langle u, u\rangle^{k+1}]\Bigg) \\
&= nE[\delta(u)^{n-1}\langle u, \delta(\nabla^* u)\rangle] \\
&\quad - \sum_{1\le 2k\le n+1} \frac{(-1)^k n!}{(k-1)!2^{k-1}(n+1-2k)!}\Bigg(\frac{n+1}{2k} E[\delta(u)^{n-2k+1}\langle u, u\rangle^k] \\
&\qquad - \frac{(n-2k)(n-2k+1)}{2k} E[\delta(u)^{n-2k-1}\langle u, u\rangle^k\langle u, \delta(\nabla^* u)\rangle] \\
&\qquad - \frac{n-2k+1}{2k} E[\delta(u)^{n-2k}\langle u, D\langle u, u\rangle^k\rangle]\Bigg) \\
&= - \sum_{1\le 2k\le n+1} \frac{(-1)^k (n+1)!}{k!2^k(n+1-2k)!} E[\delta(u)^{n-2k+1}\langle u, u\rangle^k] \\
&\quad + \sum_{0\le 2k\le n-1} \frac{(-1)^k n!}{k!2^k(n-2k-1)!} E[\delta(u)^{n-2k-1}\langle u, u\rangle^k\langle u, \delta(\nabla^* u)\rangle] \\
&\quad + \sum_{1\le 2k\le n} \frac{(-1)^k n!}{k!2^k(n-2k)!} E[\delta(u)^{n-2k}\langle u, D\langle u, u\rangle^k\rangle],
\end{aligned}$$

which yields (6.18) after using the identity (6.20).

Step 2. For $F \in I\!D_{2,1}$ and $k, i \ge 0$, by Lemma 6.6, we have

$$\begin{aligned}
&E[F\delta(u)^k\langle(\nabla u)^i u, \delta(\nabla^* u)\rangle] - kE[F\delta(u)^{k-1}\langle(\nabla^* u)^{i+1} u, \delta(\nabla^* u)\rangle] \\
&= kE[F\delta(u)^{k-1}\langle(\nabla u)^{i+1} u, u\rangle] + E[\delta(u)^k\langle(\nabla u)^{i+1} u, DF\rangle] \\
&\quad + E[F\delta(u)^k\langle\nabla^* u, D((\nabla u)^i u)\rangle]..
\end{aligned}$$

Hence, replacing k above with $l-i$, we get

$$E[F\delta(u)^l\langle u,\delta(\nabla^*u)\rangle] = l!E[F\langle(\nabla u)^l u,\delta(\nabla^*u)\rangle] + \sum_{i=0}^{l-1}\frac{l!}{(l-i)!}$$

$$\big(E[F\delta(u)^{l-i}\langle(\nabla u)^i u,\delta(\nabla^*u)\rangle] - (l-i)E[F\delta(u)^{l-i-1}\langle(\nabla^*u)^{i+1}u,\delta(\nabla^*u)\rangle]\big)$$

$$= l!E[F\langle(\nabla u)^l u,\delta(\nabla^*u)\rangle] + \sum_{i=0}^{l-1}\frac{l!}{(l-i-1)!}E[F\delta(u)^{l-i-1}\langle(\nabla u)^{i+1}u,u\rangle]$$

$$+\sum_{i=0}^{l-1}\frac{l!}{(l-i)!}E[\delta(u)^{l-i}\langle(\nabla u)^{i+1}u,DF\rangle]$$

$$+\sum_{i=0}^{l-1}\frac{l!}{(l-i)!}E[F\delta(u)^{l-i}\langle\nabla^*u,D((\nabla u)^i u)\rangle]$$

$$= l!E[\langle(\nabla u)^{l+1}u,DF\rangle] + \sum_{i=0}^{l-1}\frac{l!}{(l-i-1)!}E[F\delta(u)^{l-i-1}\langle(\nabla u)^{i+1}u,u\rangle]$$

$$+\sum_{i=0}^{l-1}\frac{l!}{(l-i)!}E[\delta(u)^{l-i}\langle(\nabla u)^{i+1}u,DF\rangle]$$

$$+\sum_{i=0}^{l}\frac{l!}{(l-i)!}E[F\delta(u)^{l-i}\langle\nabla^*u,D((\nabla u)^i u)\rangle]$$

$$= l!E[\langle(\nabla u)^{l+1}u,DF\rangle] + \sum_{i=0}^{l-1}\frac{l!}{(l-i-1)!}E[F\delta(u)^{l-i-1}\langle(\nabla u)^{i+1}u,u\rangle]$$

$$+\sum_{i=1}^{l}\frac{l!}{(l-i+1)!}E[\delta(u)^{l-i+1}\langle(\nabla u)^i u,DF\rangle]$$

$$+\sum_{i=0}^{l}\frac{l!}{(l-i)!}E[F\delta(u)^{l-i}\langle\nabla^*u,D((\nabla u)^i u)\rangle],$$

thus, letting $F=\langle u,u\rangle^k$ and $l=n-2k-1$ above, and using (6.18) in Step 1, we get

$$E[H_{n+1}(\delta(u),\|u\|^2)] = \sum_{0\le 2k\le n-1}\frac{(-1)^k n!}{k!2^k(n-2k-1)!}$$

$$E[\delta(u)^{n-2k-1}\langle u,u\rangle^k\langle u,\delta(\nabla^*u)\rangle]$$

$$+ \sum_{1\le 2k\le n} \frac{(-1)^k n!}{k!2^k(n-2k)!} E[\delta(u)^{n-2k}\langle u, D\langle u,u\rangle^k\rangle]$$

$$= \sum_{1\le 2k\le n} (-1)^k \frac{n!}{k!2^k} E[\langle(\nabla u)^{n-2k}u, D\langle u,u\rangle^k\rangle]$$

$$+ \sum_{0\le 2k\le n-2} \frac{(-1)^k}{k!2^k} \sum_{i=0}^{n-2k-2} \frac{n!}{(n-2(k+1)-i)!}$$

$$E[\langle u,u\rangle^k \delta(u)^{n-2(k+1)-i}\langle(\nabla u)^{i+1}u, u\rangle]$$

$$+ \sum_{1\le 2k\le n-1} \frac{(-1)^k}{k!2^k} \sum_{i=1}^{n-2k-1} \frac{n!}{(n-2k-i)!}$$

$$E[\delta(u)^{n-2k-i}\langle(\nabla u)^i u, D\langle u,u\rangle^k\rangle]$$

$$+ \sum_{0\le 2k\le n-1} \frac{(-1)^k}{k!2^k} \sum_{i=0}^{n-2k-1} \frac{n!}{(n-2k-1-i)!}$$

$$E[\langle u,u\rangle^k \delta(u)^{n-2k-i-1}\langle\nabla^* u, D((\nabla u)^i u)\rangle]$$

$$+ \sum_{1\le 2k\le n-1} \frac{(-1)^k n!}{k!2^k(n-2k)!} E[\delta(u)^{n-2k}\langle u, D\langle u,u\rangle^k\rangle]$$

$$= \sum_{1\le 2k\le n} (-1)^k \frac{n!}{k!2^k} E[\langle(\nabla u)^{n-2k}u, D\langle u,u\rangle^k\rangle]$$

$$- \sum_{0\le 2k\le n-2} \frac{(-1)^{k+1}}{(k+1)!2^{k+1}} \sum_{i=0}^{n-2k-2} \frac{n!}{(n-2(k+1)-i)!}$$

$$E[\delta(u)^{n-2(k+1)-i}\langle(\nabla u)^i u, D\langle u,u\rangle^{k+1}\rangle]$$

$$+ \sum_{1\le 2k\le n-1} \frac{(-1)^k}{k!2^k} \sum_{i=0}^{n-2k-1} \frac{n!}{(n-2k-i)!} E[\delta(u)^{n-2k-i}\langle(\nabla u)^i u, D\langle u,u\rangle^k\rangle]$$

$$+ \sum_{0\le 2k\le n-1} \frac{(-1)^k}{k!2^k} \sum_{i=0}^{n-2k-1} \frac{n!}{(n-2k-1-i)!}$$

$$E[\langle u,u\rangle^k \delta(u)^{n-2k-i-1}\langle\nabla^* u, D((\nabla u)^i u)\rangle]$$

$$= - \sum_{0\le 2k\le n-2} \frac{(-1)^{k+1}}{(k+1)!2^{k+1}} \sum_{i=0}^{n-2k-2} \frac{n!}{(n-2(k+1)-i)!}$$

$$E[\delta(u)^{n-2(k+1)-i}\langle(\nabla u)^i u, D\langle u,u\rangle^{k+1}\rangle]$$

$$+ \sum_{1\le 2k\le n} \frac{(-1)^k}{k!2^k} \sum_{i=0}^{n-2k} \frac{n!}{(n-2k-i)!} E[\delta(u)^{n-2k-i} \langle (\nabla u)^i u, D\langle u,u\rangle^k \rangle]$$

$$+ \sum_{0\le 2k\le n-1} \frac{(-1)^k}{k!2^k} \sum_{i=0}^{n-2k-1} \frac{n!}{(n-2k-1-i)!}$$

$$E[\langle u,u\rangle^k \delta(u)^{n-2k-i-1} \langle \nabla^* u, D((\nabla u)^i u)\rangle]$$

$$= \sum_{0\le 2k\le n-1} \frac{(-1)^k}{k!2^k} \sum_{i=0}^{n-2k-1} \frac{n!}{(n-2k-1-i)!}$$

$$E[\langle u,u\rangle^k \delta(u)^{n-2k-i-1} \langle \nabla^* u, D((\nabla u)^i u)\rangle],$$

where we applied Lemma 6.3 with $v = (\nabla u)^i u$, which shows that

$$\langle u,u\rangle^k \langle (\nabla u)^{i+1} u, u\rangle = \frac{1}{2} \langle u,u\rangle^k \langle (\nabla u)^i u, D\langle u,u\rangle\rangle = \frac{1}{2(k+1)} \langle (\nabla u)^i u, D\langle u,u\rangle^{k+1}\rangle..$$

6.5 Girsanov Identities on Path Space

In the sequel we let $ID_{\infty,2}(H) = \bigcap_{n\ge 1} ID_{n,2}(H)$. The next result follows from Theorem 6.1 and extends Corollary 4.1 of [6] with the same proof, which is omitted here.

Corollary 6.3. *Let $u \in ID_{\infty,2}(H)$ with $E[e^{|\delta(u)|+\|u\|^2/2}] < \infty$, and such that $\nabla u : H \to H$ is a.s. quasi-nilpotent in the sense of* (6.17) *or more generally that* (6.14) *holds. Then we have*

$$E\left[\exp\left(\delta(u) - \frac{1}{2}\|u\|^2\right)\right] = 1.. \tag{6.19}$$

Again, Relation (6.19) shows in particular that if $u \in ID_{\infty,2}(H)$ is such that $\|u\|$ is deterministic and (6.17) or more generally (6.14) holds, then we have

$$E\left[e^{\delta(u)}\right] = e^{-\frac{1}{2}\|u\|^2},$$

i.e., $\delta(u)$ has a centered Gaussian distribution with variance $\|u\|^2$.

As a consequence of Theorem 6.1, we also have the following derivation formula.

Proposition 6.2. *Let $u \in ID_{\infty,2}(H)$ such that $E[e^{a|\delta(u)|+a^2\|u\|^2}] < \infty$ for some $a > 0$. Then we have*

$$\frac{\partial}{\partial\lambda} E\left[e^{\lambda\delta(u) - \frac{\lambda^2}{2}\|u\|^2}\right] = \lambda E\left[e^{\lambda\delta(u) - \lambda^2\langle u,u\rangle/2} \langle \nabla^* u, D((I-\lambda\nabla u)^{-1} u)\rangle\right],$$

for all $\lambda \in (-a/2, a/2)$ such that $|\lambda| < \|\nabla u\|^{-1}_{L^\infty(\Omega;H\otimes H)}$.

Proof. From the identity

$$H_n(x,\mu) = \sum_{0\le 2k\le n} \frac{n!(-\mu/2)^k}{k!(n-2k)!} x^{n-2k}, \qquad x,\mu \in \mathbb{R}, \tag{6.20}$$

we get the bound

$$|H_n(x,\sigma^2)| \le \sum_{0\le 2k\le n} \frac{(-1)^k}{k!2^k} \frac{n!}{(n-2k)!} |x|^{n-2k}(-\sigma^2)^k = H_n(|x|,-\sigma^2),$$

hence

$$\begin{aligned} E\left[\sum_{n=0}^{\infty} \frac{|\lambda|^n}{n!} |H_{n+1}(\delta(u),\|u\|^2)|\right] &\le E\left[\sum_{n=0}^{\infty} \frac{|\lambda|^n}{n!} H_{n+1}(|\delta(u)|,-\|u\|^2)\right] \\ &= E\left[(|\delta(u)| + \lambda\|u\|^2) e^{|\lambda\delta(u)|+\lambda^2\|u\|^2/2}\right] \\ &= E\left[e^{2|\lambda\delta(u)|+4\lambda^2\|u\|^2}\right] \\ &< \infty, \end{aligned}$$

hence by the Fubini theorem we can exchange the infinite sum and the expectation to obtain

$$\frac{\partial}{\partial\lambda} E\left[e^{\lambda\delta(u)-\frac{\lambda^2}{2}\|u\|^2}\right] = \sum_{n=0}^{\infty} \frac{\lambda^n}{n!} E[H_{n+1}(\delta(u),\|u\|^2)]$$

$$= \sum_{n=0}^{\infty} \frac{\lambda^n}{n!} \sum_{l=0}^{n-1} \frac{n!}{l!} E\left[\delta(u)^l \sum_{0\le 2k\le n-1-l} \frac{(-1)^k}{k!} \frac{\|u\|^{2k}}{2^k} \langle \nabla^* u, D((\nabla u)^{n-2k-l-1}u)\rangle\right]$$

$$= \lambda E\left[e^{\lambda\delta(u)-\lambda^2\langle u,u\rangle/2} \langle \nabla^* u, D((I-\lambda\nabla u)^{-1}u)\rangle\right]..$$

□

In addition, Relation (6.16) yields the following result, in which $\det_2(I-\lambda\nabla u)$ denotes the Carleman–Fredholm determinant of $I-\lambda\nabla u$.

Corollary 6.4. *Let $u \in I\!D_{\infty,2}(H)$ such that $E[e^{a|\delta(u)|+a^2\|u\|^2}] < \infty$ for some $a > 0$. Then we have*

$$\begin{aligned} \frac{\partial}{\partial\lambda} E\left[e^{\lambda\delta(u)-\frac{\lambda^2}{2}\|u\|^2}\right] &= -E\left[e^{\lambda\delta(u)-\lambda^2\langle u,u\rangle/2} \frac{\partial}{\partial\lambda} \log\det_2(I-\lambda\nabla u)\right] \\ &\quad -\lambda E\left[e^{\lambda\delta(u)-\lambda^2\langle u,u\rangle/2} \langle (I-\lambda\nabla u)^{-1}u, D\log\det_2(I-\lambda\nabla u)\rangle\right], \end{aligned}$$

for all $\lambda \in (-a/2, a/2)$ such that $|\lambda| < \|\nabla u\|^{-1}_{L^\infty(\Omega;H\otimes H)}$.

Proof. From Lemma 6.5, we have

$$\begin{aligned}
&\lambda\langle\nabla^* u, D((I-\lambda\nabla u)^{-1}v)\rangle \\
&= \lambda\,\mathrm{trace}\,(\nabla u)(I-\lambda\nabla u)^{-1}\nabla u - \lambda\langle (I-\lambda\nabla u)^{-1}u, D\log\det_2(I-\lambda\nabla u)\rangle \\
&= -\frac{\partial}{\partial\lambda}\log\det_2(I-\lambda\nabla u) - \lambda\langle (I-\lambda\nabla u)^{-1}u, D\log\det_2(I-\lambda\nabla u)\rangle,
\end{aligned}$$

since (6.9) also shows that

$$\begin{aligned}
\frac{\partial}{\partial\lambda}\log\det_2(I-\lambda\nabla u) &= -\sum_{n=2}^{\infty}\lambda^{n-1}\,\mathrm{trace}\,(\nabla u)^n \\
&= -\lambda\sum_{n=0}^{\infty}\lambda^n\langle\nabla^* u, (\nabla u)^{n+1}\rangle \\
&= -\lambda\langle\nabla^* u, (I-\lambda\nabla u)^{-1}\nabla u\rangle \\
&= -\lambda\,\mathrm{trace}\,(\nabla u)(I-\lambda\nabla u)^{-1}(\nabla u).. \qquad \square
\end{aligned}$$

When (6.17) or more generally (6.14) holds, Proposition 6.2 and Corollary 6.4 show that

$$\frac{\partial}{\partial\lambda}E\left[e^{\lambda\delta(u)-\frac{\lambda^2}{2}\|u\|^2}\right] = 0,$$

for λ in a neighborhood of 0, which recovers the result of Corollary 6.3.

References

1. Fang, S., Franchi, J.: Platitude de la structure riemannienne sur le groupe des chemins et identité d'énergie pour les intégrales stochastiques. C. R. Acad. Sci. Paris Sér. I Math. **321**(10), 1371–1376 (1995)
2. Meyer, P.A.: Un cours sur les intégrales stochastiques. In: Séminaire de Probabilités, X (Seconde partie: Théorie des intégrales stochastiques, Univ. Strasbourg, Strasbourg, année universitaire 1974/1975), pp. 245–400. Lecture Notes in Math., vol. 511. Springer, Berlin (1976). Available at http://www.numdam.org/numdam-bin/item?id=SPS_1976__10__245_0
3. Pontier, M., Üstünel, A.S.: Analyse stochastique sur l'espace de Lie-Wiener. C. R. Acad. Sci. Paris Sér. I Math. **313**, 313–316 (1991)
4. Privault, N.: Quantum stochastic calculus applied to path spaces over Lie groups. In: Proceedings of the International Conference on Stochastic Analysis and Applications, pp. 85–94. Kluwer, Dordrecht (2004)
5. Privault, N.: Moment identities for Skorohod integrals on the Wiener space and applications. Electron. Comm. Probab. **14**, 116–121 (electronic) (2009)
6. Privault, N.: Random Hermite polynomials and Girsanov identities on the Wiener space. Infin. Dimens. Anal. Quant. Probab. Relat. Top. **13**(4), 663–675 (2010)
7. Segall, A., Kailath, T.: Orthogonal functionals of independent-increment processes. IEEE Trans. Inform. Theor. **IT-22**(3), 287–298 (1976)

8. Simon, B.: Trace ideals and their applications, vol. 35 of London Mathematical Society Lecture Note Series. Cambridge University Press, Cambridge (1979)
9. Skorokhod, A.V.: On a generalization of a stochastic integral. Theor. Probab. Appl. **XX**, 219–233 (1975)
10. Üstünel, A.S.: Stochastic analysis on Lie groups. In: Stochastic Analysis and Related Topics VI: The Geilo Workshop, Progress in Probability, pp. 129–158. Birkäuser, Basel (1996)
11. Üstünel, A.S.: Analysis on Wiener space and applications. Preprint arXiv:1003.1649v1 (2010)
12. Üstünel, A.S., Zakai, M.: Random rotations of the Wiener path. Probab. Theor. Relat. Fields **103**(3), 409–429 (1995)
13. Üstünel, A.S., Zakai, M.: Transformation of measure on Wiener space. Springer Monographs in Mathematics. Springer, Berlin (2000)
14. Zakai, M., Zeitouni, O.: When does the Ramer formula look like the Girsanov formula? Ann. Probab. **20**(3), 1436–1440 (1992)

Chapter 7
Derivatives of Solutions of Semilinear Parabolic PDEs and Variational Inequalities with Neumann Boundary Conditions

Denis Talay

Abstract This chapter is a survey of the results obtained by Bossy et al. [Ann. Inst. H. Poincaré Probab. Stat. **47**(2), 395–424 (2011)]. We explicit the derivative of the flows of one-dimensional reflected diffusion processes. This allows us to get stochastic representations for space derivatives of viscosity solutions of one-dimensional semilinear parabolic partial differential equations and parabolic variational inequalities with Neumann boundary conditions. These results are applied to estimate American options hedging errors resulting from artificial Neumann boundary conditions which are necessary to localize numerical resolutions in bounded domains.

Keywords Artificial boundary conditions • Backward stochastic differential equations • Reflected diffusions • Semilinear parabolic partial differential equations • Variational inequalities

AMS Classification. Primary: 60H10, 65U05.

Dedication

One of the first times I met Ali Süleyman was during the Filtering and Control of Random Processes Colloquium organized by H. Korezlioglu, G. Mazziotto and J. Szpirglas in Paris in 1983. I was quite impressed by his lecture on distributions-valued semimartingales [16]. Almost 30 years afterwards, a lot of speakers of this E.N.S.T.-C.N.E.T. Colloquium participated to the Conference in Ali Süleyman's honour at the same place. This cannot be by chance. In addition to be a great

---

D. Talay (✉)
INRIA, 2004 Route des Lucioles, B.P. 93, 06902 Sophia-Antipolis, France
e-mail: denis.talay@inria.fr

L. Decreusefond and J. Najim (eds.), *Stochastic Analysis and Related Topics*, Springer Proceedings in Mathematics & Statistics 22, DOI 10.1007/978-3-642-29982-7_7,

mathematician, Ali Süleyman is a wonderful colleague. I am happy to dedicate him this note which is related to some of his recent interests [17].

7.1 Motivation

Our main motivation comes from the hedging of American options and the hedging errors resulting from artificial Neumann boundary conditions which are necessary to localize the numerical resolution of variational inequalities in bounded domains. The price at time t of such an option is $V(t, S_t)$, and the hedging strategy (more or less formally) is $\partial_x V(t, S_t)$, where S_t is the spot price at time t of the stock, and the function $V(t,x)$ solves a variational inequality of the type

$$\begin{cases} \min\left\{V(t,x) - L(t,x); -\frac{\partial V}{\partial t}(t,x) - \mathscr{A}V(t,x) - r(t,x)V(t,x)\right\} = 0, \\ \qquad (t,x) \in [0,T) \times \mathbb{R}^d, \\ V(T,x) = g(x), \ x \in \mathbb{R}^d; \end{cases} \tag{7.1}$$

here g is the payoff function, $\mathscr{A}$ is the infinitesimal generator of the stock price, and r is the instantaneous interest rate. The numerical resolution of such a PDE in infinite domains is impossible. Therefore one introduces a boundary and artificial boundary conditions in order to reduce the computation to a bounded domain. We here consider the case of nonhomogeneous Neumann artificial boundary conditions.

From now on, we will consider a slightly more general variational inequality than (7.1):

$$\begin{cases} \min\left\{V(t,x) - L(t,x); -\frac{\partial V}{\partial t}(t,x) - \mathscr{A}V(t,x) - f(t,x,V(t,x),(\nabla V\sigma)(t,x))\right\} = 0, \\ \qquad (t,x) \in [0,T) \times \mathbb{R}^d, \\ V(T,x) = g(x), \ \ x \in \mathbb{R}^d. \end{cases} \tag{7.2}$$

Choosing a bounded domain $\mathscr{O}$ included in $\mathbb{R}^d$ and homogeneous Neumann artificial boundary conditions leads to the following equation:

$$\begin{cases} \min\left\{v(t,x) - L(t,x); -\frac{\partial v}{\partial t}(t,x) - \mathscr{A}v(t,x) - f(t,x,v(t,x),(\nabla v\sigma)(t,x))\right\} = 0, \\ \qquad (t,x) \in [0,T) \times \mathscr{O}, \\ v(T,x) = g(x), \ x \in \overline{\mathscr{O}}, \\ (\nabla v(t,x) \, ; \, n(x)) + h(t,x) = 0, \ (t,x) \in [0,T) \times \partial\mathscr{O}, \end{cases} \tag{7.3}$$

where, for all x in $\partial\mathscr{O}$, $n(x)$ denotes the inward unit normal vector at point x.

Of course, the problems (7.2) and (7.3) coincide in the domain $\mathcal{O}$ if the function $h(x)$ precisely satisfies

$$(\nabla V(t,x)\ ;\ n(x)) + h(t,x) = 0,\ (t,x) \in [0,T) \times \partial\mathcal{O}.$$

In practice, as the function $\nabla V(t,x)$ is unknown, one deduces hopefully good choices for $h(t,x)$ from the physical or economical properties of the solution to the Eq. (7.2), or from a priori estimates on this solution. A key issue then is to estimate the localization error $V(t,x) - v(t,x)$ in terms of the mis-specification of $(\nabla V(t,x)\ ;\ n(x))$ at the parabolic boundary $[0,T) \times \partial\mathcal{O}$. Backward stochastic differential equations (BSDE) appear to be a useful tool to get sharp localization error estimates.

Berthelot et al. [2] have estimated $|V(t,x) - v(t,x)|$ by using BSDE. By extending El Karoui et al. [6] and Ma and Cvitanić's [9] methods, they proved the existence and uniqueness of the continuous viscosity solution to (7.3), which allowed them to obtain the following estimate: under smoothness conditions on the coefficients and on $\partial\mathcal{O}$, there exists $C > 0$ such that, for all $0 \le t \le T$ and $x \in \mathcal{O}$,

$$|V(t,x) - v(t,x)| \le C \left\{ \mathbb{E} \sup_{t \le s \le T} |(\nabla V(s, X_s^{t,x}) + h(s, X_s^{t,x}); \eta(X_s^{t,x}))|^4 \mathbb{I}_{\{X_s^{t,x} \in \partial\mathcal{O}\}} \right\}^{1/4}, \tag{7.4}$$

where $(X_s^{t,x})$ solves the reflected forward stochastic differential equation (SDE)

$$\begin{cases} X_s^{t,x} = x + \int_t^s b(X_\theta^{t,x})d\theta + \int_t^s \sigma(X_\theta^{t,x})dW_\theta + K_s^{t,x},\ 0 \le t \le s \le T, \\ K_s^{t,x} = \int_t^s \eta(X_\theta^{t,x})d|K|_\theta^{t,x} \text{ with } |K|_s^{t,x} = \int_t^s \mathbb{I}_{\{X_\theta^{t,x} \in \partial\mathcal{O}\}} d|K|_\theta^{t,x}. \end{cases} \tag{7.5}$$

Notice that the estimate (7.4) takes into account the time spent by the process (X_t) at the boundary $\partial\mathcal{O}$.

In [4], limiting themselves to the one-dimensional case $d = 1$, Bossy et al. have obtained estimates for $|\partial_x V(t,x) - \partial_x v(t,x)|$, where the derivatives are understood in the sense of the distributions. We here survey the results in this latter paper, that is: the differentiability of the stochastic flow $X^{t,x}$; the probabilistic interpretation, in terms of BSDEs, of the derivative in the sense of the distributions $\partial_x v(t,x)$; stochastic representations and estimates for $|\partial_x V(t,x) - \partial_x v(t,x)|$.

We emphasize that, unfortunately, so far the results concern one-dimensional problems only because of the need to explicitly represent the stochastic flow derivative $\partial_x X^{t,x}$ (Malliavin derivatives were also explicited by Lépingle et al. [8] in the one-dimensional case only); in addition, in order to get stochastic representations for $\partial_x v(t,x)$ when $h \neq 0$, an integration by parts technique which is used seems limited to the one-dimensional case.

7.2 Our Key Tool: Derivatives of Flows of 1D Reflected Diffusions

7.2.1 Derivative of the Flow of the Reflected Diffusion X: Examples

Our first example is the Brownian motion reflected at 0. Let $x > 0$. The resolution of the Skorokhod problem shows that the adapted increasing process

$$k_s^x(\omega) := \sup\{0, -x + \sup_{0\le r\le s} W_r(\omega)\}$$

is such that the process $X_s^x := x - W_s + k_s^x$ is positive and satisfies

$$\int_0^T \mathbb{I}_{(0,\infty)}(X_s^x)dk_s^x = 0.$$

We obviously have

$$\partial_x X_s^x = 1 + \frac{\partial}{\partial x} k_s^x = \mathbb{I}_{\inf_{0\le r\le s} X_r^x > 0}.$$

Our second example is the Brownian motion doubly reflected at points d and d'. Let x be in (d, d'). Kruk et al. [7] have explicitly solved the Skorokhod problem corresponding to a two-sided reflection. To simplify the notation we suppose here that $d = 0$.

Consider $X_s^x := x - W_s + \tilde{k}_s^x$, where the increasing process $\tilde{k}_s^s$ is defined by

$$-\tilde{k}_s^x := \left[0 \wedge \inf_{0\le r\le s}(x - W_r)\right] \bigvee \sup_{0\le r\le s}\left[(x - W_r - d') \wedge \inf_{r\le\theta\le s}(x - W_\theta)\right].$$

On the event

$$\mathscr{E}_s^{0,x} := \left\{\omega \in \Omega : 0 < \inf_{r\in[0,s]} X_r^x(\omega) \le \sup_{r\in[0,s]} X_r^x(\omega) < d'\right\}, \tag{7.6}$$

the process $(\tilde{k}_r^x, r \le s)$ is null and thus $\frac{\partial}{\partial x} X_s^x = 1$.

On $\Omega - \mathscr{E}_s^{0,x}$, one has $-\tilde{k}_s^x = x + G_s$ for some r.v. G_s independent of x, and thus $\frac{\partial}{\partial x} X_s^x = 0$.

7.2.2 *Derivative of the Flow of the Reflected Diffusion X: The General 1D Case*

Consider the 1D doubly reflected SDE

$$\begin{cases} X_s^{t,x} = x + \int_t^s b(X_r^{t,x})dr + \int_t^s \sigma(X_r^{t,x})dW_r + K_s^{t,x}, \\ K_s^{t,x} = \int_t^s \eta(X_r^t)d|K|_r^{t,x} \text{ with } |K|_s^{t,x} = \int_t^s \mathbb{I}_{\{X_r^{t,x}\in\{d,d'\}\}} d|K|_r^{t,x}, \end{cases}$$

where $\eta(d) = 1$ and $\eta(d') = -1$. Menaldi [12] has proved the following theorem in its multidimensional version.

Theorem 7.1. *For $n \geq 1$ define the function β_n by*

$$\beta_n(y) := \begin{cases} -n(y-d') & \text{if } y \geq d', \\ 0 & \text{if } d \leq y \leq d', \\ n(d-y) & \text{if } y \leq d. \end{cases}$$

Then the penalized process $X^{t,x,n}$ solution to

$$X_s^{t,x,n} = x + \int_t^s b(X_r^{t,x,n})dr + \int_t^s \sigma(X_r^{t,x,n})dW_r + \int_t^s \beta_n(X_r^{t,x,n})dr \tag{7.7}$$

satisfies, for all $p \geq 1$ and all $t \leq T$,

$$\lim_{n\to\infty} \sup_{x\in(d,d')} \mathbb{E} \sup_{t\leq s\leq T} |X_s^{t,x} - X_s^{t,x,n}|^p = 0.$$

Let $D := H^1(d,d')$, $\widetilde{\Omega} := (d,d') \times \Omega$, and $d\widetilde{\mathbb{P}} := dx \otimes d\mathbb{P}$.

Let $\widetilde{D}_1$ be the space of functions $\gamma(x,\omega)$ such that there exists a Borel measurable function $\widetilde{\gamma}$ from $\widetilde{\Omega}$ to $\mathbb{R}$ satisfying $\gamma = \widetilde{\gamma}$ $\widetilde{\mathbb{P}}$ -a.e. and, for all (x,ω), the map $y \longrightarrow \widetilde{\gamma}(x+y,\omega)$ is locally absolutely continuous. For $\gamma \in \widetilde{D}_1$, set

$$\nabla\gamma(x,\omega) := \liminf_{y\to 0} \frac{\widetilde{\gamma}(x+y,\omega) - \widetilde{\gamma}(x,\omega)}{y}.$$

Bouleau and Hirsch [5] have shown that this definition is proper. Finally, set

$$\widetilde{D} := \left\{\gamma \in \mathbb{L}^2(\widetilde{\mathbb{P}}) \bigcap \widetilde{D}_1;\ \nabla\gamma \in \mathbb{L}^2(\widetilde{\mathbb{P}})\right\}.$$

Theorem 7.2. *Suppose that b and σ are bounded Lipschitz functions. Suppose the ellipticity condition $\inf_{x\in\mathbb{R}} \sigma(x) > 0$. Set*

$$\mathscr{E}_s^{t,x} := \left\{\omega \in \Omega : d < \inf_{r\in[t,s]} X_r^{t,x}(\omega) \leq \sup_{r\in[t,s]} X_r^{t,x}(\omega) < d'\right\}.$$

Then the flow $X^{t,x}$ belongs to $\widetilde{\mathbb{D}}$ and

$$\partial_x X_s^{t,x} = J_s^{t,x}\, \mathbb{I}_{\mathscr{E}_s^{t,x}},\ \widetilde{\mathbb{P}}-a.s., \tag{7.8}$$

where, denoting by b' and σ' arbitrary versions of the a.e. derivatives of b and σ,

$$J_s^{t,x} := \exp\left\{\int_t^s \sigma'(X_r^{t,x})dW_r + \int_s^t (b'(X_r^{t,x}) - \tfrac{1}{2}\sigma'^2(X_r^{t,x}))dr\right\}.$$

Remark. Lépingle et al. [8] have exhibited a similar formula for Malliavin derivatives of $X_s^{t,x}$.

Sketch of the proof.

The first step in the proof consists in reducing the study to the one-sided reflection case. The main arguments (a cut and paste of the paths procedure inspired by Lépingle et al. [8], and the comparison lemma for one-dimensional diffusion processes) make intensively use of the one-dimensional framework.

The second step consists in solving the case of the reflection at the sole point d. Consider the penalized processes

$$\widehat{X}_s^{t,x,n} = x + \int_t^s b(\widehat{X}_r^{t,x,n})dr + \int_t^s \sigma(\widehat{X}_r^{t,x,n})dW_r + \int_t^s n(d - \widehat{X}_r^{t,x,n})^+ dr,$$

and the one-sided reflected process

$$\widehat{X}_s^{t,x} = x + \int_t^s b(\widehat{X}_r^{t,x})dr + \int_t^s \sigma(\widehat{X}_r^{t,x})dW_r + \Lambda_s^d(\widehat{X}^{t,x}),$$

where $\Lambda^d(\widehat{X}^{t,x})$ is the local time at point d of the semi-martingale $\widehat{X}^{t,x}$. We have

$$\lim_{n\to\infty} \sup_{x\in(d,d')} \mathbb{E} \sup_{t\le s\le T} |\widehat{X}_s^{t,x} - \widehat{X}_s^{t,x,n}|^p = 0.$$

In view of Theorem 1 in [5], the stochastic flow $\widehat{X}^{t,x,n}$ is differentiable in the sense of the distributions, and its derivative, denoted by $\partial_x \widehat{X}_s^{t,x,n}$, satisfies, $\widetilde{\mathbb{P}}$-a.s.,

$$\partial_x \widehat{X}_s^{t,x,n} = \exp\left\{\int_t^s \sigma'(\widehat{X}_r^{t,x,n})dW_r + \int_t^s (b'(\widehat{X}_r^{t,x,n}) - \tfrac{1}{2}\sigma'^2(\widehat{X}_r^{t,x,n}))dr\right\}$$
$$\times \exp\left\{-n\int_t^s \mathbb{I}_{\widehat{X}_r^{t,x,n}<d}\,dr\right\}.$$

A straightforward adaptation of the inequality (3.23) in [12] readily leads to: for all $t \leq s \leq T$, $\widetilde{\mathbb{P}}$-a.s., the sequence $(\partial_x \widehat{X}^{t,x,n})$ converges weakly to some process denoted by $\partial_x \widehat{X}^{t,x}$. Here, the weak convergence means: for all stochastic flow U_s^x such that $\partial_x U_s^x$ is well defined $\widetilde{\mathbb{P}}$-a.s. and

$$\int_d^{d'} \mathbb{E}|U_s^x|^2 dx + \int_d^{d'} \mathbb{E}\left(\partial_x U_s^x\right)^2 dx < +\infty,$$

then

$$\int_d^{d'} \mathbb{E}\left[U_s^x\left(\partial_x \widehat{X}_s^{t,x,n} - \partial_x \widehat{X}_s^{t,x}\right) + \partial_x U_s^x (\partial_x \widehat{X}_s^{t,x,n} - \partial_x \widehat{X}_s^{t,x})\right] dx \xrightarrow[n\to+\infty]{} 0.$$

In addition, the flow $\widehat{X}^{t,x}$ is in $\widetilde{D}$.

Then one easily checks that, to prove that

$$\partial_x \widehat{X}_s^{t,x} = \exp\left\{\int_t^s \sigma'(\widehat{X}_r^{t,x}) dW_r + \int_t^s (b'(\widehat{X}_r^{t,x}(\omega)) - \tfrac{1}{2}\sigma'^2(\widehat{X}_r^{t,x}))dr\right\} \mathbb{I}_{\widehat{\mathscr{E}}_s^{t,x}}, \tag{7.9}$$

where

$$\widehat{\mathscr{E}}_s^{t,x} := \left\{\omega \in \Omega,\ \inf_{t\leq r\leq s} \widehat{X}_r^{t,x}(\omega) > d\right\},$$

it suffices to prove: for all x in (d, d'), $\mathbb{P}$-a.s.,

$$\exp\left\{-n\int_t^s \mathbb{I}_{\widehat{X}_r^{t,x,n}(\omega)<d}\right\} \mathbb{I}_{\widehat{\mathscr{E}}_s^{t,x}} \xrightarrow[n\to+\infty]{} \mathbb{I}_{\widehat{\mathscr{E}}_s^{t,x}},$$

and

$$\exp\left\{-n\int_t^s \mathbb{I}_{\widehat{X}_r^{t,x,n}(\omega)<d}\right\} \mathbb{I}_{\Omega-\widehat{\mathscr{E}}_s^{t,x}} \xrightarrow[n\to+\infty]{} 0.$$

To this end, a key calculation consists in checking that, for some x_0 depending on x and d,

$$\int_0^{s-t} \mathbb{E}_{x_0} \exp\left\{-n\int_0^{s-t-\theta} \mathbb{I}_{W_r\leq x_0} dr\right\} d\mathbb{P}^W_{\tau_{x_0}}(\theta) \xrightarrow[n\to+\infty]{} 0,$$

where $\tau_{x_0} := \inf\{r \geq 0, W_r = x_0\}$. This is done by using the formula (1.5.3) in [3, p. 160]: this integral is equal to

$$\int_0^{s-t} I_0\left(\frac{n(s-t-\theta)}{2}\right) \exp\left(-\frac{n}{2}(s-t-\theta)\right) \frac{|x_0|}{\sqrt{2\pi\theta^3}} \exp\left(-\frac{x_0^2}{2\theta}\right) d\theta,$$

where I_0 is a Bessel function which satisfies

$$I_0(y)e^{-y} \leq 1 \text{ for all } y \geq 0$$

and

$$I_0(y) \approx \frac{e^y}{\sqrt{2\pi y}} \text{ as } y \to +\infty$$

(see, e.g., [1, pp. 375 and 638]). ■

7.3 Stochastic Representations of Derivatives of Solutions of Semilinear Parabolic PDEs

In this section we consider a semilinear parabolic PDE in an interval with a Neumann boundary condition:

$$\begin{cases} \frac{\partial u}{\partial t}(t,x) + \mathscr{A}u(t,x) + f(t,x,u(t,x),\sigma(x)\frac{\partial}{\partial x}u(t,x)) = 0, \, (t,x) \in [0,T)\times(d'd'), \\ u(T,x) = g(x), \; x \in \{d,d'\}, \\ \frac{\partial u}{\partial x}(t,x) + h(t,x) = 0, \; (t,x) \in [0,T) \times \{d,d'\}, \end{cases} \tag{7.10}$$

where

$$\mathscr{A} = \frac{1}{2}\sigma^2(x)\frac{\partial^2}{\partial x^2} + b(x)\frac{\partial}{\partial x}.$$

Consider the BSDE

$$\underline{Y}_s^{t,x} = g(X_T^{t,x}) + \int_s^T h(r, X_r^{t,x}) dK_r^{t,x} + \int_s^T f(r, X_r^{t,x}, \underline{Y}_r^{t,x}, \underline{Z}_r^{t,x}) dr - \int_s^T \underline{Z}_r^{t,x} dW_r. \tag{7.11}$$

Pardoux and Zhang [15] have shown that the function $u(t,x) := \underline{Y}_t^{t,x}$ is the unique viscosity solution to the above semilinear parabolic PDE (7.10). We aim to get probabilistic representations for the derivative of $u(t,x)$.

Remark.

All the terms in (7.11) can be formally differentiated w.r.t. x except the integral $\int_s^T h(r, X_r^{t,x}) dK_r^{t,x}$ which leads to specific difficulties.

7.3.1 A Representation of $\partial_x u(t,x)$ Involving g' and ∇f

Theorem 7.3. *Let b and σ be as in theorem 7.2. Suppose that the function f is in $\mathscr{C}([0,T] \times [d,d'] \times \mathbb{R} \times \mathbb{R})$, bounded and uniformly Lipschitz w.r.t. the space variables. In addition, suppose that the function h is continuous on $[0,T] \times [d,d']$ and that the functions $h(t,d)$ and $h(t,d')$ are continuously differentiable on $[0,T]$. Suppose also that $g'(x) = -h(T,x)$ for $x = d$ or $x = d'$.*

Then the function $u(t,x) := \underline{Y}_t^{t,x}$ *is in the Sobolev space* $H^1(d,d')$ *for all* $0 \le t \le T$ *and, for almost all* x *in* (d,d'),

$$\partial_x u(t,x) = \mathbb{E}\left\{g'(X_T^{t,x})\partial_x X_T^{t,x}\mathbb{I}_{\{\tau^{t,x}>T\}} - h(\tau^{t,x}, X_{\tau^{t,x}}^{t,x})J_{\tau^{t,x}}^{t,x}\mathbb{I}_{\{\tau^{t,x}\le T\}}\right.$$
$$\left. + \int_t^{T\wedge\tau^{t,x}} \left[\partial_x f(r,\underline{\Theta}_r^{t,x})\partial_x X_r^{t,x} + \partial_y f(r,\underline{\Theta}_r^{t,x})\underline{\Psi}_r^{t,x} + \partial_z f(r,\underline{\Theta}_r^{t,x})\underline{\Gamma}_r^{t,x}\right]dr\right\}, \tag{7.12}$$

where $\underline{\Theta}_s^{t,x} := (X_s^{t,x}, \underline{Y}_s^{t,x}, \underline{Z}_s^{t,x})$ *solves (7.5) and (7.11),* $J_r^{t,x}$ *is as in (7.8), and* $(\underline{\Psi}_s^{t,x}, \underline{\Gamma}_s^{t,x})$ *is the unique adapted process satisfying*

$$\mathbb{E}\sup_{t\le s\le T}|\underline{\Psi}_s^{t,x}|^2 + \mathbb{E}\int_0^T |\underline{\Gamma}_r^{t,x}|^2 dr < \infty,$$

and, for all $0 \le s \le T$,

$$\underline{\Psi}_s^{t,x} = g'(X_T^{t,x})\partial_x X_T^{t,x}\mathbb{I}_{\{\tau^{t,x}>T\}} - h(\tau^{t,x}, X_{\tau^{t,x}}^{t,x})J_{\tau^{t,x}}^{t,x}\mathbb{I}_{\{\tau^{t,x}\le T\}}$$
$$+ \int_{s\wedge\tau^{t,x}}^{T\wedge\tau^{t,x}} \left[\partial_x f(r,\underline{\Theta}_r^{t,x})\partial_x X_r^{t,x} + \partial_y f(r,\underline{\Theta}_r^{t,x})\underline{\Psi}_r^{t,x} + \partial_z f(r,\underline{\Theta}_r^{t,x})\underline{\Gamma}_r^{t,x}\right]dr$$
$$- \int_{s\wedge\tau^{t,x}}^{T\wedge\tau^{t,x}} \underline{\Gamma}_r^{t,x}\, dW_r. \tag{7.13}$$

Sketch of the proof.

The first step in the proof consists in reducing the study to the homogeneous case $h \equiv 0$. This is done by interpolating the functions $h(t,d)$ and $h(t,d')$ by a smooth function h and by observing that if $(\widehat{Y}^{t,x}, \widehat{Z}_s^{t,x})$ solves (7.11) with the coefficients

$$\begin{cases} \widehat{h}(x) & := 0, \\ \widehat{g}(x) & := g(x) + H(T,x), \\ \widehat{f}(t,x,y,z) & := f(t,x,y-H(t,x), z-h(t,x)\sigma(x)) - \dfrac{\partial H}{\partial r}(r,x) \\ & \qquad +b(x)h(r,x) + \dfrac{1}{2}\sigma^2(x)\dfrac{\partial h}{\partial x}(r,x), \end{cases}$$

where $H(r,x) := \int_d^x h(r,\xi)d\xi$, then

$$\begin{cases} \underline{Y}_s^{t,x} & := H(s, X_s^{t,x}) - \widehat{Y}_s^{t,x}, \\ \underline{Z}_s^{t,x} & := h(s, X_s^{t,x})\sigma(X_s^{t,x}) - \widehat{Z}_s^{t,x} \end{cases}$$

solves (7.11).

The second step of the proof starts with considering the penalized process defined as in (7.7). From N'Zi et al. [14, Theorem 3.2], $\widetilde{\mathbb{P}}$-a.s.,

$$\partial_x X_s^{t,x,n} = \exp\left\{\int_t^s \sigma'(X_r^{t,x,n})dW_r + \int_t^s (b_n'(X_r^{t,x,n}) - \tfrac{1}{2}(\sigma')^2(X_r^{t,x,n}))dr\right\}; \tag{7.14}$$

the BSDE

$$\Psi_s^{t,x,n} = g'(X_T^{t,x,n})\partial_x X_T^{t,x,n} + \int_s^T \Big[\partial_x f(r,\Theta_r^{t,x,n})\Phi_r^{t,x,n} + \partial_y f(r,\Theta_r^{t,x,n})\Psi_r^{t,x,n} + \partial_z f(r,\Theta_r^{t,x,n})\Gamma_r^{t,x,n}\Big]dr - \int_s^T \Gamma_r^{t,x,n} dW_r \tag{7.15}$$

has a unique solution, and $\partial_x Y^{t,x,n} = \Psi^{t,x,n}$. Tedious calculations then allow one to let n tend to infinity and get (7.13). ∎

7.3.2 A Representation of $\partial_x u(t,x)$ Without g' and ∇f

Inspired by the results in [11] (see also [10]), we now aim to prove a formula of Elworthy's type for $\partial_x u(t,x)$ which does not suppose that the function f is everywhere differentiable.

Set

$$N_s^{t,x} := \frac{1}{s-t}\int_t^s \sigma^{-1}(X_r^{t,x})\partial_x X_r^{t,x} dW_r.$$

Theorem 7.4. *Under the hypotheses of Theorem 7.3 one has*

$$\partial_x u(t,x) = \mathbb{E}\Big[g(X_T^{t,x})N_T^{t,x} - h(\tau^{t,x}, X_{\tau^{t,x}}^{t,x})J_{\tau^{t,x}}^{t,x}\mathbb{I}_{\tau^{t,x}\le T} + \int_t^T f(r, X_r^{t,x}, \underline{Y}_r^{t,x}, \underline{Z}_r^{t,x})N_r^{t,x}dr\Big].$$

A key point in the proof is the following integration by parts formula:

Lemma 7.1. *For all differentiable function ϕ with bounded derivative and all $t \le r \le T$,*

$$\mathbb{E}\left[\phi'(X_r^{t,x})\partial_x X_r^{t,x}\mathbb{I}_{r<\tau^{t,x}}\right] = \mathbb{E}[\phi(X_r^{t,x})N_r^{t,x}].$$

Sketch of the proof.

Consider the event $\mathscr{E}_{\theta,r}^{t,x}$ defined as in (7.6). Lépingle et al. [8] have shown that the Malliavin derivative of $X_r^{t,x}$ satisfies:

$$\forall t \leq \theta < r, \; D_\theta X_r^{t,x} = \sigma(X_\theta^{t,x}) \frac{J_r^{t,x}}{J_\theta^{t,x}} \mathbb{I}_{\mathscr{E}_{\theta,r}^{t,x}}.$$

Therefore

$$\frac{1}{\sigma(X_\theta^{t,x})} D_\theta \phi(X_r^{t,x}) J_\theta^{t,x} = \phi'(X_r^{t,x}) J_r^{t,x} \mathbb{I}_{\mathscr{E}_{\theta,r}^{t,x}}.$$

On then slightly modifies the proof of Elworthy's formula (see, e.g., [13]) by integrating the previous equality w.r.t θ between times t and $r \wedge \tau^{t,x}$. ■

7.4 Stochastic Representations of Derivatives of Solutions of 1D Variational Inequalities

Consider the system

$$\begin{cases} \underline{\mathscr{Y}}_s^{t,x} = g(X_T^{t,x}) + \int_s^T f(r, X_r^{t,x}, \underline{\mathscr{Y}}_r^{t,x}, \underline{\mathscr{Z}}_r^{t,x}) dr + \int_s^T h(r, X_r^{t,x}) dK_r^{t,x} \\ \qquad + \underline{\mathscr{R}}_T^{t,x} - \underline{\mathscr{R}}_s^{t,x} - \int_s^T \underline{\mathscr{Z}}_r^{t,x} dW_r, \\ \underline{\mathscr{Y}}_s^{t,x} \geq L(s, X_s^{t,x}) \text{ for all } 0 \leq t \leq s \leq T, \\ (\underline{\mathscr{R}}_s^{t,x}, 0 \leq t \leq s \leq T) \text{ is a continuous increasing process such that} \\ \int_t^T (\underline{\mathscr{Y}}_r^{t,x} - L(r, X_r^{t,x})) d\underline{\mathscr{R}}_r^{t,x} = 0. \end{cases} \tag{7.16}$$

Using El Karoui et al. [6] and Ma and Cvitanić's [9] methods, Berthelot et al. [2] have shown that the function $v(t,x) := \underline{\mathscr{Y}}_t^{t,x}$ is the unique (in an appropriate space of functions) viscosity solution of the following parabolic system with a nonhomogeneous Neumann boundary condition:

$$\begin{cases} \min\left\{ v(t,x) - L(t,x); -\frac{\partial v}{\partial t}(t,x) - \mathscr{A} v(t,x) - f(t,x,v(t,x), \partial_x v(t,x)\sigma(x)) \right\} = 0, \\ \qquad (t,x) \in [0,T) \times (d,d'), \\ v(T,x) = g(x), \; x \in [d,d'], \\ \partial_x v(t,x) + h(t,x) = 0, \; (t,x) \in [0,T) \times \{d,d'\}. \end{cases} \tag{7.17}$$

The following stochastic representation of $\partial_x v(t,x)$ holds true:

Theorem 7.5. *Suppose that the function L is in $\mathscr{C}^{1,2}([0,T] \times \mathbb{R}; \mathbb{R})$, bounded with bounded derivatives. Under the assumptions of Theorem 7.3, it holds that, for all t in $[0,T]$ and almost all x in (d,d'),*

$$\partial_x v(t,x) = \mathbb{E}\left[N_T^{t,x} g(X_T^{t,x}) - h(\tau^{t,x}, X_{\tau^{t,x}}^{t,x}) J_{\tau^{t,x}}^{t,x} \mathbb{I}_{\{\tau^{t,x} \le T\}} + \int_t^T f(r, X_r^{t,x}, \mathscr{Y}_r^{t,x}, \mathscr{Z}_r^{t,x}) N_r^{t,x} dr + \int_t^T N_r^{t,x} d\mathscr{R}_r^{t,x}\right]. \tag{7.18}$$

Beginning of the proof.

As above, the first step consists in reducing the analysis to the homogeneous case $h(t,x) \equiv 0$.

The second step consists in introducing the BSDEs

$$\begin{cases} \mathscr{Y}_s^{t,x,n} = g(X_T^{t,x,n}) + \int_s^T f(r, X_r^{t,x,n}, \mathscr{Y}_r^{t,x,n}, \mathscr{Z}_r^{t,x,n}) dr - \int_s^T \mathscr{Z}_r^{t,x,n} dW_r \\ \qquad + \mathscr{R}_T^{t,x,n} - \mathscr{R}_s^{t,x,n}, \\ \forall t \le s \le T,\ \mathscr{Y}_s^{t,x,n} \ge L(s, X_s^{t,x,n}), \\ \{\mathscr{R}_s^{t,x,n}, t \le s \le T\} \text{ is an increasing continuous process such that} \\ \int_t^T (\mathscr{Y}_s^{t,x,n} - L(s, X_s^{t,x,n})) d\mathscr{R}_s^{t,x,n} = 0, \end{cases}$$

where $X^{t,x,n}$ is the solution to (7.7). Set $v^n(t,x) := \mathscr{Y}_t^{t,x,n}$. Ma and Zhang [11] have shown that, for almost all x in (d,d'),

$$\begin{aligned} \partial_x v^n(t,x) &= \mathscr{Z}_t^{t,x,n} \sigma^{-1}(x) \\ &= \mathbb{E}\Bigg(g(X_T^{t,x,n}) N_T^{t,x,n} + \int_t^T f(r, X_r^{t,x,n}, \mathscr{Y}_r^{t,x,n}, \mathscr{Z}_r^{t,x,n}) N_r^{t,x,n} dr \\ &\qquad + \int_t^T N_r^{t,x,n} d\mathscr{R}_r^{t,x,n} \Bigg). \end{aligned}$$

Technical and lengthy calculations allow one to prove that the right-hand side of the preceding equality tends to the right-hand side of (7.18) with $h(t,x) \equiv 0$. The following key estimates are necessary and interesting by themselves. First, there exist $0 < \beta < 1$ and $C > 0$ such that, for all x in (d,d'),

$$\mathbb{E}\left| \int_t^T N_r^{t,x} d|K|_r^{t,x} \right| \le \frac{C}{[(x-d) \wedge (d'-x)]^{\frac{14}{11}}} (T-t)^{\beta}. \tag{7.19}$$

Second, denoting by $(\mathscr{Y}^{t,x}, \mathscr{Z}^{t,x}, \mathscr{R}^{t,x})$ the solution to (7.16) for $h(t,x) \equiv 0$, one has: for all $t \le r \le T$,

$$
\begin{aligned}
d\mathscr{R}_r^{t,x} \leq \mathbb{I}_{\mathscr{Y}_r^{t,x}=L(r,X_r^{t,x})} \Big[& \frac{\partial L}{\partial r}(r, X_r^{t,x}) + \frac{\partial L}{\partial x}(r, X_r^{t,x}) b(X_r^{t,x}) \\
& + \frac{1}{2}\frac{\partial^2 L}{\partial x^2}(r, X_r^{t,x})\sigma^2(X_r^{t,x}) + f(r, X_r^{t,x}, \mathscr{Y}_r^{t,x}, \mathscr{Z}_r^{t,x}))\Big]^- dr \\
+ \mathbb{I}_{\mathscr{Y}_r^{t,x}=L(r,X_r^{t,x})} & \left[\left(\frac{\partial L}{\partial x}(r, X_r^{t,x})\right)\eta(X_r^{t,x})\right]^- d|K|_r^{t,x}.
\end{aligned}
\tag{7.20}
$$

■

7.5 Conclusion

Coming back to our original motivation described in the Introduction, we deduce from Theorem 7.5 a tractable estimate of the error induced by the artificial Neumann boundary condition $h(t,x)$. In this section, we suppose that $\partial_x V(t,d)$ and $\partial_x V(t,d')$ are well defined for all times $t \in [0,T]$. For example, if in addition of assumptions of Theorem 7.5, we suppose that b and σ are differentiable with bounded derivatives, Ma and Zhang [11, Theorem 5.1] have shown that $\partial_x V(\cdot,\cdot)$ is a bounded continuous function on $[0,T] \times \mathbb{R}$.

The following quantity represents the order of magnitude of the mis-specification at the boundary $\{d, d'\}$:

$$
\begin{aligned}
\epsilon(h) := & \sup_{t\leq r\leq T} \left(|V(r,d) - v(r,d)| + |V(r,d') - v(r,d')|\right) \\
& + \sup_{t\leq r\leq T} \left(|\partial_x V(r,d) + h(r,d)| + |\partial_x V(r,d') + h(r,d')|\right).
\end{aligned}
$$

We are in a position to prove the following estimate for the error induced by the artificial Neumann boundary condition $h(t,x)$:

Theorem 7.6. *Suppose that $\partial_x V(r,d)$ and $\partial_x V(r,d')$ are well defined for all times $r \in [t,T]$. Under the hypotheses of Theorem 7.5, there exists C independent of h such that, for all $\rho < \frac{1}{2}$,*

$$
\int_d^{d'} |\partial_x V(t,x) - \partial_x v(t,x)|^2 dx \leq C\epsilon(h)^\rho \wedge \epsilon(h).
$$

Remark.
A better estimate can be derived for semilinear PDEs, namely,

$$
\int_d^{d'} |\partial_x V(t,x) - \partial_x v(t,x)|^2 dx \leq C\epsilon(h).
$$

Sktech of the proof.

As shown in [2], the viscosity solution of (7.2) restricted to (d, d') satisfies $V(t,x) = \check{\mathscr{Y}}_t^{t,x}$, where $(\check{\mathscr{Y}}, \check{\mathscr{Z}}, \check{\mathscr{R}})$ is the unique solution to the reflected BSDE

$$\begin{cases} \check{\mathscr{Y}}_s^{t,x} = g(X_T^{t,x}) + \int_s^T f(r, X_r^{t,x}, \check{\mathscr{Y}}_r^{t,x}, \check{\mathscr{Z}}_r^{t,x})dr - \int_s^T \partial_x V(r, X_r^{t,x}) dK_r^{t,x} \\ \qquad + \check{\mathscr{R}}_T^{t,x} - \check{\mathscr{R}}_s^{t,x} - \int_s^T \check{\mathscr{Z}}_r^{t,x} dW_r, \\ \check{\mathscr{Y}}_s^{t,x} \geq L(s, X_s^{t,x}) \text{ for all } 0 \leq t \leq s \leq T, \\ (\check{\mathscr{R}}_s^{t,x}, 0 \leq t \leq s \leq T) \text{ is a continuous increasing process such that} \\ \int_t^T (\check{\mathscr{Y}}_r^{t,x} - L(r, X_r^{t,x})) d\check{\mathscr{R}}_r^{t,x} = 0. \end{cases} \tag{7.21}$$

In view of Theorem 7.5 we deduce that, for all x in (d, d'),

$$\begin{aligned} \partial_x V(t,x) - \partial_x v(t,x) = \mathbb{E}\Big[& \big(\partial_x V(\tau^{t,x}, X_{\tau^{t,x}}^{t,x}) + h(\tau^{t,x}, X_{\tau^{t,x}}^{t,x})\big) J_{\tau^{t,x}}^{t,x} \mathbb{I}_{\{\tau^{t,x} \leq T\}} \\ & + \int_t^T N_r^{t,x} \Big(f(r, X_r^{t,x}, \check{\mathscr{Y}}_r^{t,x}, \check{\mathscr{Z}}_r^{t,x}) \\ & - f(r, X_r^{t,x}, \underline{\mathscr{Y}}_r^{t,x}, \underline{\mathscr{Z}}_r^{t,x}) \Big) dr \\ & + \int_t^T N_r^{t,x} \Big(d\check{\mathscr{R}}_r^{t,x} - d\underline{\mathscr{R}}_r^{t,x} \Big) \Big]. \end{aligned}$$

The key difficulty is then to show that

$$\int_t^T \frac{1}{(r-t)^{\frac{3}{2}}} \left\{ \mathbb{E} |\check{\mathscr{R}}_r^{t,x} - \underline{\mathscr{R}}_r^{t,x}|^2 \right\}^{\frac{1}{2}} dr \leq \frac{C\epsilon(h)^{2\gamma}}{((x-d) \wedge (d'-x))^{1-2\gamma}} \int_t^T \frac{1}{(r-t)^{\frac{1}{2}+2\gamma}} dr,$$

and to deduce that

$$\begin{aligned} \left| \mathbb{E}\left[\int_t^T N_r^{t,x} (d\check{\mathscr{R}}_r^{t,x} - d\underline{\mathscr{R}}_r^{t,x}) \right] \right| & \leq \mathbb{E}\left[|N_T^{t,x}| |\check{\mathscr{R}}_T^{t,x} - \underline{\mathscr{R}}_T^{t,x}| \right] \\ & \quad + \mathbb{E}\left[\int_t^T \frac{|N_r^{t,x}|}{(r-t)} |\check{\mathscr{R}}_r^{t,x} - \underline{\mathscr{R}}_r^{t,x}| dr \right] \\ & \leq C\epsilon(h) + C\epsilon(h)^{2\gamma} \end{aligned}$$

for some real number C uniform w.r.t. the real number x in $[d, d']$ and the function $h(t,x)$. ■

Two challenging questions, which are important issues for applications, need now to be tackled for multidimensional PDEs or variational inequalities: first, the extension of our work to the multidimensional case and hypo-elliptic diffusions (see [17]); second, given a desired accuracy on the approximation of $\partial_x V(t,x)$ or the hedging strategy of an American option, the relevant choice of a function $h(t,x)$ and of an artificial boundary.

References

1. Abramowitz, M., Stegun, I.A.: Handbook of Mathematical Functions, with Formulas, Graphs, and Mathematical Tables. Dover Publications, New York (1964)
2. Berthelot, C., Bossy, M., Talay, D.: Numerical analysis and misspecifications in Finance: From model risk to localisation error estimates for nonlinear PDEs. In: Stochastic Processes and Applications to Mathematical Finance, pp. 1–25. World Scientific Publishing, River Edge (2004)
3. Borodin, A.N., Salminen, P.: Handbook of Brownian Motion- Facts and Formulae. Probability and Its applications. Birkhaüser, Basel (2002)
4. Bossy, M., Cissé, M., Talay, D.: Stochastic representations of derivatives of solutions of one dimensional parabolic variational inequalities with Neumann boundary conditions. Ann. Inst. H. Poincaré Probab. Stat. **47**(2), 395–424 (2011)
5. Bouleau, N., Hirsch, F.: On the derivability, with respect to the initial data, of solution of a stochastic differential equation with Lipschitz coefficients. In: Séminaire de Théorie du Potentiel, Paris, No. 9. Lecture Notes Math., vol. 1393, pp. 39–57 (1989)
6. EL Karoui, N., Kapoudjian, C., Pardoux, E., Peng, S., Quenez, M.-C.: Reflected solution of backward SDE's, and related obstacle problem for PDE's. Ann. Probab. **25**(2), 702–737 (1997)
7. Kruk, L., Lehoczky, J., Ramanan, K., Shreve, S.: An explicit formula for the Skorokhod map on $[0,a]$. Ann. Probab. **35**(5), 1740–1768 (2007)
8. Lépingle, D., Nualart, D., Sanz, M.: Dérivation stochastique de diffusions réfléchies. Ann. I.H.P., sec B, t.**25**(3), 283–305 (1989)
9. Ma, J., Cvitanić, J.: Reflected forward-backward SDE's and obstacle problems with boundary conditions. J. Appl. Stoch. Anal. **14**(2), 113–138 (2001)
10. Ma, J., Zhang, J.: Representation theorems for backward stochastic differential equations. Ann. Appl. Probab. **12**(4), 1390–1418 (2002)
11. Ma, J., Zhang, J.: Representations and regularities for solutions to BSDEs with reflections. Stoch. Proc. Appl. **115**, 539–569 (2005)
12. Menaldi, J.L.: Stochastic variational inequality for reflected diffusion. Indiana Univ. Math. J. **32**(5), 733–744 (1983)
13. Nualart, D.: The Malliavin Calculus and Related Topics. Springer, Berlin (2006)
14. N'Zi, M., Ouknine, Y., Sulem, A.: Regularity and representation of viscosity solutions of partial differential equations via backward stochastic differential equations. Stoch. Proc. Appl. **116**(9), 1319–1339 (2006)
15. Pardoux, E., Zhang, S.: Generalized BSDEs and nonlinear Neumann boundary value problems. Probab. Theor. Relat. Fields **110**, 535–558 (1998)
16. Üstünel, A.S.: Distribution valued semimartingales and applications to control and filtering. In: Korezlioglu, H., Mazziotto, G. Szpirglas, J. (eds.) Filtering and Control of Random Processes Colloquium. Lecture Notes Control Inf. Sciences, vol. 61, pp. 314–325. Springer, Berlin (1984)
17. Üstünel, A.S.: Probabilistic solution of the American options. J. Funct. Anal. **256**(3), 3091–3105 (2009)

Chapter 8
Some Differential Systems Driven by a fBm with Hurst Parameter Greater than 1/4

Samy Tindel and Iván Torrecilla

Abstract This note is devoted to show how to push forward the algebraic integration setting in order to treat differential systems driven by a noisy input with Hölder regularity greater than $1/4$. After recalling how to treat the case of ordinary stochastic differential equations, we mainly focus on the case of delay equations. A careful analysis is then performed in order to show that a fractional Brownian motion with Hurst parameter $H > 1/4$ fulfills the assumptions of our abstract theorems.

Keywords Fractional Brownian motion; Rough paths theory; Stochastic delay equations

AMS Classification. 60H05, 60H07, 60G15.

Dedicated to Ali Süleyman Üstünel on occasion of his 60th birthday

8.1 Introduction

A differential equation driven by a d-dimensional fractional Brownian motion $B = (B^1, \ldots, B^d)$ is generically written as:

S. Tindel (✉)
Institut Élie Cartan Nancy, B.P. 239, 54506 Vandœuvre-lès-Nancy Cedex, France
e-mail: tindel@iecn.u-nancy.fr

I. Torrecilla
Facultat de Matemàtiques, Universitat de Barcelona, Gran Via 585, 08007 Barcelona, Spain

Facultat de Ciències Econòmiques i Empresarials, Universitat Pompeu Fabra, C/Ramon Trias Fargas, 25–27, 08005 Barcelona, Spain
e-mail: itorrecillatarantino@gmail.com

L. Decreusefond and J. Najim (eds.), *Stochastic Analysis and Related Topics*, Springer Proceedings in Mathematics & Statistics 22, DOI 10.1007/978-3-642-29982-7_8,

$$y_t = a + \int_0^t \sigma(y_s)\, dB_s, \quad t \in [0, T], \tag{8.1}$$

where a is an initial condition in $\mathbb{R}^n$, $\sigma : \mathbb{R}^n \to \mathbb{R}^{n,d}$ is a smooth enough function, and T is an arbitrary positive constant. The recent developments in rough paths analysis [4, 12, 17] have allowed to solve this kind of differential equation when the Hurst parameter H of the fractional Brownian motion is greater than 1/4, by first giving a natural meaning to the integral $\int_0^t \sigma(y_s)\, dB_s$ above. It should also be stressed that a great amount of information has been obtained about these systems, ranging from support theorems [10] to the existence of a density for the law of y_t at a fixed instant t (see [2, 3]).

In a parallel but somewhat different direction, the algebraic integration theory (introduced in [13]), is meant as an alternative and complementary method of generalized integration with respect to a rough path. It relies on some more elementary and explicit formulae, and its main advantage is that it allows to develop rather easily an intuition about the way to handle differential systems beyond the diffusion case given by (8.1). This fact is illustrated by the study of delay [19] and Volterra [6] type equations, as well as an attempt to handle partial differential equations driven by a rough path [7, 15]. In each of those cases, the main underlying idea consists in changing slightly the basic structures allowing a generalized integration theory (discrete differential operator δ, sewing map Λ, controlled processes) in order to adapt them to the context under consideration. While the technical details might be long and tedious, let us insist on the fact that the changes in the structures we have alluded to are always natural and (almost) straightforward. Some twisted Lévy areas also enter into the game in a natural manner.

However, all the results contained in the references mentioned above concern a fractional Brownian motion B with Hurst parameter $H > 1/3$, while the usual rough path theory enables to handle any $H > 1/4$ (see [4, 12] for the explicit application to fBm). This chapter can then be seen as a step in order to fill this gap, and we shall deal mainly with two kind of systems: first of all, we will show how to solve Eq. (8.1) when $1/4 < H \le 1/3$, thanks to the algebraic integration theory. The results we will obtain are not new, and the algebraic integration formalism has been extended to a much broader context in [14] by means of a tree-based expansion (let us mention again that the case $H > 1/4$ is also covered by the usual rough path theory). This study is thus included here as a preliminary step, where the changes in the structures (new definition of a controlled path, introduction of a Lévy *volume*) can be exhibited in a simple enough manner.

Then, in the second part of the chapter, we show how to adapt our formalism in order to deal with delay equations of the form:

$$\begin{cases} dy_t = \sigma(y_t, y_{t-r_1}, \dots, y_{t-r_q})\, dB_t & t \in [0, T], \\ y_t = \xi_t, & t \in [-r_q, 0], \end{cases} \tag{8.2}$$

where y is a $\mathbb{R}^n$-valued continuous process, q is a positive integer, $\sigma : \mathbb{R}^{n(q+1)} \to \mathbb{R}^{n,d}$ is a smooth enough function, B is a d-dimensional fractional Brownian motion with Hurst parameter $H > 1/4$, and T is an arbitrary positive constant. The delay in our equation is represented by the family $0 < r_1 < \ldots < r_q < \infty$, and the initial condition ξ is taken as a regular enough deterministic function on $[-r_q, 0]$. Though this kind of system is implicitly considered in [16] in the usual Brownian case, and in [8] for a Hurst parameter $H > 1/2$, the rough paths techniques have only been used in this context (to the best of our knowledge) in [19], where a delay equation driven by a fractional Brownian motion with Hurst parameter $H > 1/3$ is considered. Our chapter is thus an extension of this last result, and we shall obtain an existence and uniqueness theorem for Eq. (8.2) in the case $H > 1/4$, under reasonable regularity conditions on σ and ξ.

From our point of view the example of delay equations, which is interesting in its own right because of its potential physical applications, is also worth studying in order to measure the flexibility of the rough paths formalism. In case of a delay equation driven by a rough path, we get the satisfaction to see that the reduction to ordinary differential systems can be done in a reasonably simple way, once we assume the existence of some delayed area and volume based on the driving process. Let us point out that the infinite dimensional setting of [18] is avoided here, and that all our considerations only involve paths taking values in a finite dimensional space.

Let us also mention that, as in other examples of fractional differential systems, an important part of our work consists in verifying that the fractional Brownian motion satisfies the assumptions of our abstract theorems. The main available tools we are aware of for this kind of task are based on Russo–Vallois approximations [24], regularization procedures of the fBm path [4, 12], or Malliavin calculus. We have chosen here to work under this latter framework, since it leads to reasonably short calculations, and also because it allows us to build on the previous results obtained in [19], where this formalism was also adopted. It is a pleasure for us to recall at this point that the stochastic analysis of fBm has been initiated in the pathbreaking paper [5] by Decreusefond and stnel, to which we are obviously indebted for the current chapter.

Here is how our chapter is structured: Sect. 8.2 is devoted to recall the basic ingredients of the algebraic integration setting. The abstract results concerning ordinary and delayed systems are given at Sect. 8.3, and the bulk of the computations concerning delay systems can be found at Sect. 8.3.5. Finally, the application to fractional Brownian motion is given at Sect. 8.4.

8.2 Increments

To begin with, let us present the very basic algebraic structures which will allow to define a pathwise integral with respect to irregular functions. These elements are mainly borrowed from [13, 15].

8.2.1 Basic Notions of Algebraic Integration

For an arbitrary real number $T > 0$, a vector space V, and an integer $k \geq 1$, we denote by $\mathscr{C}_k(V)$ the set of functions $g : [0,T]^k \to V$, $g(t_1,\ldots,t_k) = g_{t_1\ldots t_k}$ such that $g_{t_1\cdots t_k} = 0$ whenever $t_i = t_{i+1}$ for some $1 \leq i \leq k-1$. Such a function will be called a $(k-1)$-*increment*, and we will set $\mathscr{C}_*(V) = \cup_{k\geq 1}\mathscr{C}_k(V)$.

On $\mathscr{C}_k(V)$ we introduce the operator δ defined as follows:

$$\delta : \mathscr{C}_k(V) \to \mathscr{C}_{k+1}(V), \qquad (\delta g)_{t_1\cdots t_{k+1}} = \sum_{i=1}^{k+1}(-1)^{k-i} g_{t_1\cdots \hat{t}_i \cdots t_{k+1}}, \tag{8.3}$$

where $\hat{t}_i$ means that this particular argument is omitted. A fundamental property of δ, which is easily verified, is that $\delta \circ \delta = 0$. We will denote $\mathscr{Z}\mathscr{C}_k(V) = \mathscr{C}_k(V) \cap \mathrm{Ker}\delta$ and $\mathscr{B}\mathscr{C}_k(V) = \mathscr{C}_k(V) \cap \mathrm{Im}\delta$.

Throughout the chapter we will mainly deal with actions of δ on $\mathscr{C}_i$, $i = 1,2$. That is, consider $g \in \mathscr{C}_1$ and $h \in \mathscr{C}_2$. Then, for any $s,u,t \in [0,T]$, we have

$$(\delta g)_{st} = g_t - g_s, \quad \text{and} \quad (\delta h)_{sut} = h_{st} - h_{su} - h_{ut}. \tag{8.4}$$

Furthermore, it is easily checked that $\mathscr{Z}\mathscr{C}_{k+1}(V) = \mathscr{B}\mathscr{C}_k(V)$ for any $k \geq 1$. In particular, we have the following property:

Lemma 8.1. *Let $k \geq 1$ and $h \in \mathscr{Z}\mathscr{C}_{k+1}(V)$. There exists a (non unique) $f \in \mathscr{C}_k(V)$ such that $h = \delta f$.*

Lemma 8.1 implies that all the elements $h \in \mathscr{C}_2(V)$ such that $\delta h = 0$ can be written as $h = \delta f$ for some (nonunique) $f \in \mathscr{C}_1(V)$. Thus we have a heuristic interpretation of $\delta|_{\mathscr{C}_2(V)}$ as a measure of how much a given 1-increment is far from being an exact increment of a function, i.e., a finite difference.

Remark 8.1. Here is a first elementary but important link between these algebraic structures and integration theory. Let f and g be two smooth real valued functions on $[0,T]$. Define $I \in \mathscr{C}_2$ by

$$I_{st} = \int_s^t \left(\int_s^v dg_w\right) df_v, \quad \text{for} \quad s,t \in [0,T].$$

Then, $(\delta I)_{sut} = [g_u - g_s][f_t - f_u] = (\delta g)_{su}(\delta f)_{ut}$. Hence we see that the operator δ transforms iterated integrals into products of increments, and we will be able to take advantage of both regularities of f and g in these products of the form $\delta g\, \delta f$.

Let us concentrate now on the case $V = \mathbb{R}^d$, and notice that our future discussions will mainly rely on k-increments with $k \leq 2$, for which we will use some analytical assumptions. Namely, we measure the size of these increments by Hölder-type norms defined in the following way. For $f \in \mathscr{C}_2(V)$ and $\mu \in (0,\infty)$, let

$$\|f\|_\mu = \sup_{s,t\in[0,T]} \frac{|f_{st}|}{|t-s|^\mu}, \tag{8.5}$$

and set $\mathcal{C}_2^\mu(V) = \{f \in \mathcal{C}_2(V);\ \|f\|_\mu < \infty\}$.

The usual Hölder spaces $\mathcal{C}_1^\mu(V)$ will be determined in the following way. For a continuous function $g \in \mathcal{C}_1(V)$, we simply set

$$\|g\|_\mu = \|\delta g\|_\mu, \tag{8.6}$$

where the right-hand side of this equality is defined after (8.5); we will say that $g \in \mathcal{C}_1^\mu(V)$ iff $\|g\|_\mu$ is finite. Notice that $\|\cdot\|_\mu$ is only a semi-norm on $\mathcal{C}_1(V)$. However, we will generally work on spaces of the type

$$\mathcal{C}_{1,a}^\mu(V) = \{g : [0,T] \to V;\ g_0 = a,\ \|g\|_\mu < \infty\},$$

for a given $a \in V$, on which $\|g\|_\mu$ then becomes a norm. For $h \in \mathcal{C}_3(V)$, we set

$$\|h\|_{\gamma,\rho} = \sup_{s,u,t\in[0,T]} \frac{|h_{sut}|}{|u-s|^\gamma |t-u|^\rho}$$

$$\|h\|_\mu = \inf\left\{\sum_i \|h_i\|_{\rho_i,\mu-\rho_i};\ h = \sum_i h_i,\ 0 < \rho_i < \mu\right\},$$

where the last infimum is taken over all sequences $\{h_i \in \mathcal{C}_3(V)\}$ such that $h = \sum_i h_i$. Then $\|\cdot\|_\mu$ is easily seen to be a norm on $\mathcal{C}_3(V)$, and we set

$$\mathcal{C}_3^\mu(V) := \{h \in \mathcal{C}_3(V);\ \|h\|_\mu < \infty\}.$$

Eventually, let $\mathcal{C}_j^{1+}(V) = \cup_{\mu>1}\mathcal{C}_j^\mu(V)$, $j = 1,2,3$, and remark that the same kind of norms can be considered on the spaces $\mathcal{Z}\mathcal{C}_3(V)$, leading to the definition of some spaces $\mathcal{Z}\mathcal{C}_3^\mu(V)$ and $\mathcal{Z}\mathcal{C}_3^{1+}(V)$.

With these notations in mind, the crucial point in our approach to pathwise integration of irregular processes is that, under mild smoothness conditions, the operator δ can be inverted. This inverse is called Λ and is defined in the following proposition, whose proof can be found in [13, 15]:

Proposition 8.1. *There exists a unique linear map $\Lambda : \mathcal{Z}\mathcal{C}_3^{1+}(V) \to \mathcal{C}_2^{1+}(V)$ such that*

$$\delta\Lambda = Id_{\mathcal{Z}\mathcal{C}_3^{1+}(V)} \quad \text{and} \quad \Lambda\delta = Id_{\mathcal{C}_2^{1+}(V)}.$$

In other words, for any $h \in \mathcal{C}_3^{1+}(V)$ such that $\delta h = 0$ there exists a unique $g = \Lambda(h) \in \mathcal{C}_2^{1+}(V)$ such that $\delta g = h$. Furthermore, for any $\mu > 1$, the map Λ is continuous from $\mathcal{Z}\mathcal{C}_3^\mu(V)$ to $\mathcal{C}_2^\mu(V)$ and we have

$$\|\Lambda h\|_\mu \le \frac{1}{2^\mu - 2}\|h\|_\mu, \qquad h \in \mathscr{Z}\mathscr{C}_3^\mu(V).$$

It is worth mentioning at this point that Λ gives rise to a kind of generalized Young integral, which is a second link between the algebraic structures introduced so far and a theory of generalized integration:

Corollary 8.1. *For any 1-increment $g \in \mathscr{C}_2(V)$ such that $\delta g \in \mathscr{C}_3^{1+}$,*

$$(Id - \Lambda\delta)g = \lim_{|\Pi_{st}|\to 0} \sum_{i=0}^{n} g_{t_i\, t_{i+1}},$$

where the limit is over any partition $\Pi_{st} = \{t_0 = s, \dots, t_n = t\}$ of $[s,t]$, whose mesh tends to zero. Thus by setting $\delta f = (Id - \Lambda\delta)g$, the 1-increment δf is the indefinite integral of the 1-increment g.

We can now explain heuristically how our generalized integral will be defined.

Remark 8.2. Let f and g be two real valued smooth functions, and define $I \in \mathscr{C}_2$ like in Remark 8.1. Thanks to this remark and Proposition 8.1, the following decomposition–recomposition for $I = \int df \int dg$ holds true:

$$\int dg \int df \;\xrightarrow{\delta}\; (\delta g)\,(\delta f) \;\xrightarrow{\Lambda}\; \int dg \int df \;,$$

where for the second step of this construction, we have only used the fact that the product of increments $(\delta g)\,(\delta f)$, considered as an element of $\mathscr{Z}\mathscr{C}_3$, is smooth enough. This simple procedure allows then to extend the notion of iterated integral to a non-smooth situation, by just applying the operator Λ to $(\delta g)\,(\delta f)$ whenever we are allowed to do it.

8.2.2 Some Further Notations

We summarize in this section some of the notation which will be used throughout the chapter.

A multilinear operator A of order l, from $\mathbb{R}^{d_1} \times \ldots \times \mathbb{R}^{d_l}$ to $\mathbb{R}^n$, is denoted as an element $A \in \mathbb{R}^{n,d_1,\dots,d_l}$. In order to avoid tricky matrix notations, we have decided to expand all our computations in coordinates, and use Einstein's convention on summations over repeated indices. Notice that we will also use the notation $A \in \mathbb{R}^{d_1,d_2,d_3,d_4}$ for a linear operator from $\mathbb{R}^{d_3,d_4}$ to $\mathbb{R}^{d_1,d_2}$. We hope that this convention won't lead to any ambiguity. The transposed of a matrix $M \in \mathbb{R}^{d_1,d_2}$ is written as M^*.

We shall meet two kind of products of increments: first, for $g \in \mathscr{C}_n(\mathbb{R}^{l,d})$ and $h \in \mathscr{C}_m(\mathbb{R}^d)$ we set gh for the element of $\mathscr{C}_{n+m-1}(\mathbb{R}^l)$ defined by

$$(gh)_{t_1,\dots,t_{m+n-1}} = g_{t_1,\dots,t_n} h_{t_n,\dots,t_{m+n-1}}, \quad t_1,\dots,t_{m+n-1} \in [0,T]. \tag{8.7}$$

If now $g \in \mathcal{C}_n(\mathbb{R}^{l,d})$ and $h \in \mathcal{C}_n(\mathbb{R}^d)$, we set $g \cdot h$ for the element of $\mathcal{C}_n(\mathbb{R}^l)$ defined by

$$(g \cdot h)_{t_1,\dots,t_n} = g_{t_1,\dots,t_n} h_{t_1,\dots,t_n}, \quad t_1,\dots,t_n \in [0,T]. \tag{8.8}$$

In order to avoid ambiguities, we shall denote by $\mathcal{N}[f;\mathcal{C}_j^\kappa]$ the κ-Hölder norm on the space $\mathcal{C}_j$, for $j = 1,2,3$. For $\zeta \in \mathcal{C}_1(V)$, we also set $\mathcal{N}[\zeta;\mathcal{C}_1^\infty(V)] = \sup_{0\le s\le T} |\zeta^i|_V$.

The integral of a real valued function f with respect to another real valued function g, when properly defined, is written indistinctly as $\int f dg$ or $\mathcal{J}(f dg)$.

8.3 Abstract Results

In this section, we will first recall the basic steps which allow to define rigorously and solve an equation of the form:

$$y_t = a + \int_0^t \sigma(y_s)\, dx_s, \quad t \in [0,T], \tag{8.9}$$

where a is an initial condition in $\mathbb{R}^n$, $\sigma : \mathbb{R}^n \to \mathbb{R}^{n,d}$ is a smooth enough function, T is an arbitrary positive constant, and x is a generic d-dimensional noisy input with Hölder regularity $\gamma > 1/4$. In the algebraic integration setting [13, 14], this task amounts to perform the following steps:

1. Definition of an incremental operator δ and its inverse Λ.
2. Definition of a suitable notion of controlled processes, and integration of those processes with respect to x.
3. Resolution of the equation, thanks to a fixed point procedure in the space of controlled processes.

Having dealt with the first of those points at Sect. 8.2.1, we turn now to the second part of this strategy. Then we shall show how to reduce differential delay equations to ordinary ones by increasing the dimension of the system.

8.3.1 Weakly Controlled Processes

Before giving the formal definition of a weakly controlled process in the context of Eq. (8.9), let us recall that when the regularity of the noise is $\gamma > 1/4$, the rough path setting relies on the a priori existence of an area (resp. volume) element $\mathbf{x}^2$ (resp. $\mathbf{x}^3$) satisfying the so-called Chen's relations:

Hypothesis 8.1. *The path $\mathbb{R}^d$-valued x is γ-Hölder continuous with $\gamma > 1/4$, and admits a Lévy area and a* volume element, *that is two increments $\mathbf{x^2} \in \mathscr{C}_2^{2\gamma}(\mathbb{R}^{d,d})$ and $\mathbf{x^3} \in \mathscr{C}_2^{3\gamma}(\mathbb{R}^{d,d,d})$ (which represent respectively $\mathscr{J}(dxdx)$ and $\mathscr{J}(dxdxdx)$, with the conventions of Sect. 8.2.2) satisfying:*

$$\delta\mathbf{x^2} = \delta x \otimes \delta x, \quad i.e.\ (\delta\mathbf{x}^{\mathbf{2},ij})_{sut} = (\delta x^i)_{su}(\delta x^j)_{ut}$$

$$\delta\mathbf{x^3} = \mathbf{x^2} \otimes \delta x + \delta x \otimes \mathbf{x^2}, \quad i.e.\ (\delta\mathbf{x}^{\mathbf{3},ijk})_{sut} = \mathbf{x}^{\mathbf{2},ij}_{su}(\delta x^k)_{ut} + (\delta x^i)_{su}\mathbf{x}^{\mathbf{2},jk}_{ut},$$

for any $s,u,t \in [0,T]$, and any $i,j,k \in \{1,\dots,d\}$.

The geometrical assumption for rough paths (which is satisfied by the fractional Brownian motion in the Stratonovich setting) also states that products of increments should be expressed in terms of iterated integrals:

Hypothesis 8.2. *Let $\mathbf{x^2}$ be the area process defined at Hypothesis 8.1, and denote by $\mathbf{x^{2,s}}$ the symmetric part of $\mathbf{x^2}$, i.e., $\mathbf{x^{2,s}} = \frac{1}{2}(\mathbf{x^2} + (\mathbf{x^2})^*)$. Then we suppose that for $0 \le s < t \le T$, we have:*

$$\mathbf{x}^{\mathbf{2,s}}_{st} = \frac{1}{2}(\delta x)_{st} \otimes (\delta x)_{st}.$$

With these hypotheses in mind, the natural class of processes which will be integrated against x are processes whose increments can be expressed simply enough in terms of the increments of x:

Definition 8.1. Let z be a process in $\mathscr{C}_1^\kappa(\mathbb{R}^l)$ with $\kappa \le \gamma$ and $3\kappa + \gamma > 1$, such that $z_0 = a \in \mathbb{R}^l$. We say that z is a weakly controlled path based on x if $\delta z \in \mathscr{C}_2^\kappa(\mathbb{R}^l)$ can be decomposed into

$$\delta z^i = \zeta^{1,ij}\delta x^j + \zeta^{2,ijk}\mathbf{x}^{\mathbf{2},kj} + r^i, \text{ i.e. } (\delta z^i)_{st} = \zeta^{1,ij}_s(\delta x^j)_{st} + \zeta^{2,ijk}_s\mathbf{x}^{\mathbf{2},kj}_{st} + r^i_{st}, \tag{8.10}$$

for any $1 \le i \le l$, $1 \le j,k \le d$. In the previous decomposition, we further assume that $\zeta^1 \in \mathscr{C}_1^\kappa(\mathbb{R}^{l,d})$ is a path with a given initial condition $\zeta^1_0 = b \in \mathbb{R}^{l,d}$, such that $\delta\zeta^1$ can be decomposed itself into:

$$\delta\zeta^{1,ij} = \zeta^{2,ijk}\delta x^k + \rho^{ij}, \text{ i.e. } (\delta\zeta^{1,ij})_{st} = \zeta^{2,ijk}_s(\delta x^k)_{st} + \rho^{ij}_{st},$$

for all $s,t \in [0,T]$, where ζ^2 is the increment which already appears in (8.10). As far as regularities of the increments at stake are concerned, we suppose that $\zeta^1 \in \mathscr{C}_1^\kappa(\mathbb{R}^{l,d})$, ζ^2 is an element of $\mathscr{C}_1^\kappa(\mathbb{R}^{l,d,d})$, and r and ρ are understood as regular remainders, such that $r \in \mathscr{C}_2^{3\kappa}(\mathbb{R}^l)$ and $\rho \in \mathscr{C}_2^{2\kappa}(\mathbb{R}^{l,d})$.

The space of weakly controlled paths will be denoted by $\mathscr{Q}_{\kappa,a,b}(\mathbb{R}^l)$, and a process $z \in \mathscr{Q}_{\kappa,a,b}(\mathbb{R}^l)$ can be considered in fact as a triple (z,ζ^1,ζ^2). The natural semi-norm on $\mathscr{Q}_{\kappa,a,b}(\mathbb{R}^l)$ is given by

$$\begin{aligned}\mathcal{N}[z;\mathcal{Q}_{\kappa,a,b}(\mathbb{R}^l)] &= \mathcal{N}[z;\mathcal{C}_1^{\kappa}(\mathbb{R}^l)] + \mathcal{N}[\zeta^1;\mathcal{C}_1^{\infty}(\mathbb{R}^{l,d})] + \mathcal{N}[\zeta^1;\mathcal{C}_1^{\kappa}(\mathbb{R}^{l,d})] \\ &\quad + \mathcal{N}[\zeta^2;\mathcal{C}_1^{\infty}(\mathbb{R}^{l,d,d})] + \mathcal{N}[\zeta^2;\mathcal{C}_1^{\kappa}(\mathbb{R}^{l,d,d})] \\ &\quad + \mathcal{N}[\rho;\mathcal{C}_2^{2\kappa}(\mathbb{R}^{l,d})] + \mathcal{N}[r;\mathcal{C}_2^{3\kappa}(\mathbb{R}^l)],\end{aligned}$$

where the notations $\mathcal{N}[g;\mathcal{C}_1^{\kappa}(V)]$ and $\mathcal{N}[\zeta;\mathcal{C}_1^{\infty}(V)]$ have been introduced at Sect. 8.2.2.

Remark 8.3. With respect to the case $\gamma > 1/3$, the link between ζ^1 and ζ^2 in the definition of controlled processes is new. This *cascade* relation between z, ζ^1, and ζ^2 is reminiscent of the Heinsenberg group structure of Lyons' theory, and is really natural for computational purposes.

We can now study the stability of controlled processes by composition with a regular function.

8.3.2 Composition of Controlled Processes

The results of this section can be summarized into the following:

Proposition 8.2. *Assume Hypothesis 8.2 holds true. Let* $z \in \mathcal{Q}_{\kappa,a,b}(\mathbb{R}^l)$ *with decomposition (8.10), consider a regular function* $\varphi \in \mathcal{C}_b^3(\mathbb{R}^l;\mathbb{R})$ *and set* $\hat{z} = \varphi(z)$, $\hat{a} = \varphi(a)$, $\hat{b} = \partial_i\varphi(a)b^i$. *Then* $\hat{z} \in \mathcal{Q}_{\kappa,\hat{a},\hat{b}}(\mathbb{R})$, *and the latter path admits the decomposition*

$$\delta\hat{z} = \hat{\zeta}^{1,j}\delta x^j + \hat{\zeta}^{2,jk}\mathbf{x}^{\mathbf{2},kj} + \hat{r}, \tag{8.11}$$

with

$$\hat{\zeta}^{1,j} = [\partial_i\varphi(z)\cdot\zeta^{1,ij}], \quad \hat{\zeta}^{2,jk} = [\partial_i\varphi(z)\cdot\zeta^{2,ijk}] + [\partial_{i_1i_2}\varphi(z)\cdot\zeta^{1,i_1j}\cdot\zeta^{1,i_2k}],$$

and where $\hat{r}$ *can be further decomposed into* $\hat{r} = \hat{r}^1 + \hat{r}^2 + \hat{r}^3$, *with:*

$$\begin{aligned}\hat{r}^1 &= \partial_i\varphi(z)r^i, \\ \hat{r}^2 &= \frac{1}{2}[\partial_{i_1i_2}\varphi(z)\cdot\zeta^{2,i_1j_1k_1}\cdot\zeta^{2,i_2j_2k_2}][\mathbf{x}^{\mathbf{2},k_1j_1}\cdot\mathbf{x}^{\mathbf{2},k_2j_2}] + \frac{1}{2}\partial_{i_1i_2}\varphi(z)[r^{i_1}\cdot r^{i_2}] \\ &\quad + [\partial_{i_1i_2}\varphi(z)\cdot\zeta^{1,i_1j_1}\cdot\zeta^{2,i_2jk}][\delta x^{j_1}\cdot(\mathbf{x}^{\mathbf{2}})^{kj}] \\ &\quad + [\partial_{i_1i_2}\varphi(z)\cdot\zeta^{1,i_1j}][\delta x^j\cdot r^{i_2}] + [\partial_{i_1i_2}\varphi(z)\cdot\zeta^{2,i_1jk}][\mathbf{x}^{\mathbf{2},kj}\cdot r^{i_2}], \\ \hat{r}^3 &= \delta\varphi(z) - \partial_i\varphi(z)\delta z^i - \frac{1}{2}\partial_{ij}\varphi(z)[\delta z^i\cdot\delta z^j].\end{aligned}$$

As far as $\hat{\zeta}^{1,j}$ *is concerned, for* $1 \le j \le d$, *it can be decomposed into*

$$\delta\hat{\zeta}^{1,j} = \hat{\zeta}^{2,jk}\delta x^k + \hat{\rho}^j \tag{8.12}$$

where the remainder $\hat{\rho}^j$ *can be expressed as* $\hat{\rho}^j = (\hat{\rho}^1)^j + (\hat{\rho}^2)^j$, *with:*

$$\begin{aligned}
\hat{\rho}^{1,j} &= \partial_i\varphi(z)\rho^{ij} + [\delta[\partial_i\varphi(z)] \cdot \delta\zeta^{1,ij}] \\
&\quad + [\partial_{i_1 i_2}\varphi(z) \cdot \zeta^{1,i_1 j} \cdot \zeta^{2,i_2 j_2 k_2}](\mathbf{x}^{\mathbf{2}})^{k_2 j_2} + [\partial_{i_1 i_2}\varphi(z) \cdot \zeta^{1,i_1 j}]r^{i_2}, \\
\hat{\rho}^{2,j} &= \zeta^{1,i_1 j}\delta[\partial_{i_1}\varphi(z)] - [\zeta^{1,i_1 j} \cdot \partial_{i_1 i_2}\varphi(z)]\delta z^{i_2}.
\end{aligned}$$

Finally, the following cubical bound holds true for the norm of $\hat{z}$*:*

$$\mathcal{N}[\hat{z}; \mathcal{Q}_{\kappa,\hat{a},\hat{b}}(\mathbb{R})] \le c_{\varphi,x,T}(1 + \mathcal{N}^3[z; \mathcal{Q}_{\kappa,a,b}(\mathbb{R}^l)]). \tag{8.13}$$

Proof. This proof is a matter of long and tedious Taylor expansions, and we shall omit most of the details. Let us just mention that we start from the relation:

$$\begin{aligned}
(\delta\hat{z})_{st} &= \varphi(z_t) - \varphi(z_s) = \partial_i\varphi(z_s)(\delta z^i)_{st} + \frac{1}{2}\partial_{i_1 i_2}\varphi(z_s)(\delta z^{i_1})_{st}(\delta z^{i_2})_{st} \\
&\quad + \varphi(z_t) - \varphi(z_s) - \partial_i\varphi(z_s)(\delta z^i)_{st} - \frac{1}{2}\partial_{i_1 i_2}\varphi(z_s)(\delta z^{i_1})_{st}(\delta z^{i_2})_{st}.
\end{aligned}$$

The desired decomposition (8.11) is then obtained by plugging relation (8.10) into the last identity and expanding further. It should also be noticed that some cancelations occur due to Hypothesis 8.2. Relation (8.12) is obtained in the same manner, and our bound (8.13) is a matter of standard computations once the expressions (8.11) and (8.12) are known. □

8.3.3 Integration of Controlled Paths

It is of course of fundamental importance for our purposes to be able to integrate a controlled process with respect to the driving signal x. This is achieved in the following proposition:

Proposition 8.3. *For fixed* $1/4 < \kappa \le \gamma$, *let* x *be a process satisfying Hypothesis 8.1. Let also* $m \in \mathcal{Q}_{\kappa,b,c}(\mathbb{R}^{1,d})$ *with decomposition* $m_0 = b \in \mathbb{R}^{1,d}$ *and*

$$(\delta m^i)_{st} = \mu_s^{1,ij}(\delta x^j)_{st} + \mu_s^{2,ijk}\mathbf{x}_{st}^{\mathbf{2},kj} + r_{st}^i, \quad 1 \le i \le d, \tag{8.14}$$

where $\mu^1 \in \mathcal{C}_1^\kappa(\mathbb{R}^{d,d})$, $\mu_0^1 = c \in \mathbb{R}^{d,d}$, *and where* $\delta\mu^1 \in \mathcal{C}_2^\kappa(\mathbb{R}^{d,d})$ *can be decomposed into*

$$(\delta\mu^{1,ij})_{st} = \mu_s^{2,ijk}(\delta x^k)_{st} + \rho_{st}^{ij}, \tag{8.15}$$

with $\mu^2 \in \mathscr{C}_1^\kappa(\mathbb{R}^{d,d,d})$, $\rho \in \mathscr{C}_2^{2\kappa}(\mathbb{R}^{d,d})$, $r \in \mathscr{C}_2^{3\kappa}(\mathbb{R}^{1,d})$. *Define then* z *by* $z_0 = a \in \mathbb{R}$ *and*

$$\delta z = m^i \delta x^i + \mu^{1,ij}\mathbf{x}^{2,ji} + \mu^{2,ijk}\mathbf{x}^{3,kji} + \Lambda\big(r^i\delta x^i + \rho^{ij}\mathbf{x}^{2,ji} + \delta\mu^{2,ijk}\mathbf{x}^{3,kji}\big). \tag{8.16}$$

Finally, set

$$\mathscr{J}(m^i dx^i) = \delta z. \tag{8.17}$$

Then,

(i) z *is well defined as an element of* $\mathscr{Q}_{\kappa,a,b}(\mathbb{R})$, *and* $\mathscr{J}(m^i dx^i)$ *coincides with a Riemann integral in case of some smooth processes* m *and* x.

(ii) The semi-norm of z *in* $\mathscr{Q}_{\kappa,a,b}(\mathbb{R})$ *can be estimated as*

$$\mathscr{N}[z; \mathscr{Q}_{\kappa,a,b}(\mathbb{R})] \le c_{x,T}\{1 + |b|_{\mathbb{R}^{1,d}} + T^{\gamma-\kappa}(|b|_{\mathbb{R}^{1,d}} + \mathscr{N}[m; \mathscr{Q}_{\kappa,b,c}(\mathbb{R}^{1,d})])\}. \tag{8.18}$$

Furthermore, we obtain

$$\|\delta z\|_\kappa \le c_{T,x} T^{\gamma-\kappa}(|b|_{\mathbb{R}^{1,d}} + \mathscr{N}[m; \mathscr{Q}_{\kappa,b,c}(\mathbb{R}^{1,d})]). \tag{8.19}$$

(iii) It holds

$$\mathscr{J}_{st}(m^i dx^i) = \lim_{|\Pi_{st}\to 0|} \sum_{q=0}^{n} [m_{t_q}^i(\delta x^i)_{t_q,t_{q+1}} + \mu_{t_q}^{1,ij}\mathbf{x}_{t_q,t_{q+1}}^{2,ji} + \mu_{t_q}^{2,ijk}\mathbf{x}_{t_q,t_{q+1}}^{3,kji}] \tag{8.20}$$

for any $0 \le s < t \le T$, *where the limit is taken over all partitions* $\Pi_{st} = \{t_0 = s, \ldots, t_n = t\}$ *of* $[s,t]$, *as the mesh of the partition goes to zero.*

Proof. Here again, the proof is long and cumbersome, and we prefer to avoid most of the technical details for sake of conciseness. Let us just try to justify the second part of the first assertion (about Riemann integrals).

Let us suppose then that x is a smooth function and that $m \in \mathscr{C}_1^\infty(\mathbb{R}^{1,d})$ admits the decomposition (8.14) with $\mu^1 \in \mathscr{C}_1^\infty(\mathbb{R}^{d,d})$, $\mu^2 \in \mathscr{C}_1^\infty(\mathbb{R}^{d,d,d})$, $\rho \in \mathscr{C}_2^\infty(\mathbb{R}^{d,d})$ and $r \in \mathscr{C}_2^\infty(\mathbb{R}^{1,d})$. Then $\mathscr{J}(m^i dx^i)$ is well defined, and we have

$$\int_s^t m_u^i dx_u^i = m_s^i[x_t^i - x_s^i] + \int_s^t [m_u^i - m_s^i]dx_u^i$$

for $s < t$, which can also be read as:

$$\mathscr{J}(m^i dx^i) = m^i\delta x^i + \mathscr{J}(\delta m^i dx^i). \tag{8.21}$$

Let us now plug the decomposition (8.14) into the expression (8.21). This yields

$$\mathcal{J}(m^i dx^i) = m^i \delta x^i + \mathcal{J}([\mu^{1,ij}\delta x^j]dx^i) + \mathcal{J}([\mu^{2,ijk}\mathbf{x}^{\mathbf{2},kj}]dx^i) + \mathcal{J}(r^i dx^i)$$
$$= m^i \delta x^i + \mu^{1,ij}\mathbf{x}^{\mathbf{2},ji} + \mu^{2,ijk}\mathbf{x}^{\mathbf{3},kji} + \mathcal{J}(r^i dx^i), \tag{8.22}$$

and observe that the terms $m^i\delta x^i$, $(\mu^1)^{ij}\mathbf{x}^{\mathbf{2},ji}$ and $\mu^{2,ijk}\mathbf{x}^{\mathbf{3},kji}$ in (8.22) are well defined provided that x, $\mathbf{x^2}$ and $\mathbf{x^3}$ are defined themselves. To push forward our analysis to the rough case, we still need to handle the term $\mathcal{J}(r^i dx^i)$. Owing to (8.22) we can write

$$\mathcal{J}(r^i dx^i) = \mathcal{J}(m^i dx^i) - m^i\delta x^i - \mu^{1,ij}\mathbf{x}^{\mathbf{2},ji} - \mu^{2,ijk}\mathbf{x}^{\mathbf{3},kji}, \tag{8.23}$$

and let us analyze this relation by applying δ to both sides of the last identity. Invoking standard rules on the operator δ, and the fact that x satisfies Hypothesis 8.1, we end up with:

$$\delta[\mathcal{J}(r^i dx^i)] = \delta\mu^{1,ij}\mathbf{x}^{\mathbf{2},ji} + \delta\mu^{2,ijk}\mathbf{x}^{\mathbf{3},kji} - \mu^{2,ijk}\delta x^k\mathbf{x}^{\mathbf{2},ji} + r^i\delta x^i,$$

and thanks to the fact that $\delta\mu^{1,ij} = \mu^{2,ijk}\delta x^k + \rho^{ij}$, we obtain:

$$\delta[\mathcal{J}(r^i dx^i)] = \rho^{ij}\mathbf{x}^{\mathbf{2},ji} + \delta\mu^{2,ijk}\mathbf{x}^{\mathbf{3},kji} + r^i\delta x^i. \tag{8.24}$$

Assuming now that $\rho^{ij}\mathbf{x}^{\mathbf{2},ji}$, $\delta\mu^{2,ijk}\mathbf{x}^{\mathbf{3},kji}$, $r^i\delta x^i \in \mathcal{C}_3^{\nu}$ with $\nu > 1$, then $\rho^{ij}\mathbf{x}^{\mathbf{2},ji} + \delta\mu^{2,ijk}\mathbf{x}^{\mathbf{3},kji} + r^i\delta x^i$ becomes an element of $\mathrm{Dom}(\Lambda)$. Thus, applying Λ to both sides of (8.24) and inserting the result into (8.21) we get the expression (8.16) of Proposition 8.3. This justifies the fact that (8.16) is a natural expression for $\mathcal{J}(m^i dx^i)$. □

As in [20], the previous proposition has a straightforward multidimensional extension, which we state in the following corollary:

Corollary 8.2. *Let x be a process satisfying Hypothesis 8.1 and let $m \in \mathcal{Q}_{\kappa,b,c}(\mathbb{R}^{l,d})$ with decomposition $m_0 = b \in \mathbb{R}^{l,d}$ and*

$$(\delta m^{ij})_{st} = \mu_s^{1,ijk}(\delta x^k)_{st} + \mu_s^{2,ijk_1k_2}\mathbf{x}_{st}^{\mathbf{2},k_2k_1} + r_{st}^{ij}\,; \quad \delta\mu_s^{1,ijk_1} = \mu_s^{2,ijk_1k_2}\delta x^{k_2} + \rho^{ijk_1}, \tag{8.25}$$

where $\mu^{1,ijk_1} \in \mathcal{C}_1^{\kappa}(\mathbb{R})$, $\mu^{2,ijk_1k_2} \in \mathcal{C}_1^{\kappa}(\mathbb{R})$, $\rho^{ijk_1} \in \mathcal{C}_1^{2\kappa}(\mathbb{R})$ and $r^{ij} \in \mathcal{C}_2^{3\kappa}(\mathbb{R})$, for $i = 1,\ldots,l$ and $j, k_1, k_2 = 1,\ldots,d$. Define z by $z_0 = a \in \mathbb{R}^l$ and

$$\delta z^i = \mathcal{J}(m^{ij}dx^j) \equiv m^{ij}\delta x^j + \mu^{1,ijk}\mathbf{x}^{\mathbf{2},kj}$$
$$+ \mu^{2,ijk_1k_2}\mathbf{x}^{\mathbf{3},k_2k_1j} + \Lambda\big(r^{ij}\delta x^j + \rho^{ijk}\mathbf{x}^{\mathbf{2},kj} + \delta\mu^{2,ijk_1k_2}\mathbf{x}^{\mathbf{3},k_2k_1j}\big). \tag{8.26}$$

Then the conclusions of Proposition 8.3 still hold in this context.

We also observe that our extended pathwise integral has a nice continuity property with respect to the driving path x, whose proof is also skipped here for sake of conciseness (see also [13, Proposition 4], and [20, Proposition 3.12]).

Proposition 8.4. *Let x be a function satisfying Hypotheses 8.1 and 8.2. Suppose that there exists a sequence $\{x^n;\ n \geq 1\}$ of piecewise C^1-functions from [0,T] to $\mathbb{R}^d$ such that*

$$\lim_{n\to\infty} \mathcal{N}[x^n - x; \mathcal{C}_1^{\gamma}(\mathbb{R}^d)] = 0, \quad \lim_{n\to\infty} \mathcal{N}[\mathbf{x}^{\mathbf{2},n} - \mathbf{x}^{\mathbf{2}}; \mathcal{C}_2^{2\gamma}(\mathbb{R}^{d,d})] = 0,$$

and $\lim_{n\to\infty} \mathcal{N}[\mathbf{x}^{\mathbf{3},n} - \mathbf{x}^{\mathbf{3}}; \mathcal{C}_2^{3\gamma}(\mathbb{R}^{d,d,d})] = 0$. *For $n \geq 1$, define $z^n \in \mathcal{C}_1^{\kappa}(\mathbb{R}^l)$ in the following way: set $z_0^n = b \in \mathbb{R}^l$ and assume that δz^n can be decomposed into:*

$$\delta z^{n,i} = \zeta^{1;n,ij}\delta x^j + \zeta^{2;n,ijk}\mathbf{x}^{\mathbf{2},kj} + r^{n,i}, \quad \delta\zeta^{1;n,ij} = \zeta^{2;n,ijk}\delta x^k + \rho^{n,ij},$$

for $1 \leq i \leq l$ and $1 \leq j,k \leq d$, where $\zeta^{1;n} \in \mathcal{C}_1^{\kappa}(\mathbb{R}^{l,d})$ satisfies $\zeta_0^{1;n} = c \in \mathbb{R}^{l,d}$, and $\zeta^{2;n} \in \mathcal{C}_1^{\kappa}(\mathbb{R}^{l,d,d})$, $\rho^n \in \mathcal{C}_2^{2\kappa}(\mathbb{R}^{l,d})$, and $r^n \in \mathcal{C}_2^{3\kappa}(\mathbb{R}^l)$. Let also z be a weakly controlled process with decomposition (8.10), such that $z_0 = b$, $\zeta_0^1 = c$, and suppose that

$$\begin{aligned}\lim_{n\to\infty} \{&\mathcal{N}[z^n - z; \mathcal{C}_1^{\kappa}(\mathbb{R}^l)] + \mathcal{N}[\zeta^{1;n} - \zeta^1; \mathcal{C}_1^{\infty}(\mathbb{R}^{l,d})] + \mathcal{N}[\zeta^{1;n} - \zeta^1; \mathcal{C}_1^{\kappa}(\mathbb{R}^{l,d})]\\ &+ \mathcal{N}[\zeta^{2;n} - \zeta^2; \mathcal{C}_1^{\infty}(\mathbb{R}^{l,d,d})] + \mathcal{N}[\zeta^{2;n} - \zeta^2; \mathcal{C}_1^{\kappa}(\mathbb{R}^{l,d,d})]\\ &+ \mathcal{N}[\rho^n - \rho; \mathcal{C}_2^{2\kappa}(\mathbb{R}^{l,d})] + \mathcal{N}[r^n - r; \mathcal{C}_2^{3\kappa}(\mathbb{R}^l)]\} = 0.\end{aligned}$$

Finally, let $\varphi : \mathbb{R}^l \to \mathbb{R}^{l',d}$ be a C_b^4-function. Then

$$\lim_{n\to\infty} \mathcal{N}[\mathcal{J}(\varphi(z^n)dx^n) - \mathcal{J}(\varphi(z)dx); \mathcal{C}_2^{\kappa}(\mathbb{R}^{l'})] = 0.$$

8.3.4 Rough Diffusion Equations

In this section, we shall apply the previous considerations to study differential equations driven by a rough signal, and recall that we first wish to solve simple equations of the form

$$dy_t = \sigma(y_t)dx_t, \quad y_0 = a, \tag{8.27}$$

where $t \in [0, T]$, y is a $\mathbb{R}^l$-valued continuous process, $\sigma : \mathbb{R}^l \to \mathbb{R}^{l,d}$ is a smooth enough function, x is a $\mathbb{R}^d$-valued path, and $a \in \mathbb{R}^l$ is a fixed initial condition.

In our algebraic setting, we rephrase Eq. (8.27) as follows: we shall say that y is a solution to (8.27), if $y_0 = a$, $y \in \mathcal{Q}_{\kappa,a,\sigma(a)}(\mathbb{R}^l)$ and for any $0 \leq s \leq t \leq T$ we have

$$(\delta y)_{st} = \mathcal{J}_{st}(\sigma(y)dx), \tag{8.28}$$

where the integral $\mathcal{J}(\sigma(y)dx)$ has to be understood in the sense of Corollary 8.2.

With these notations in mind, our existence and uniqueness result is the following:

Theorem 8.1. *Let x be a process satisfying Hypotheses 8.1 and 8.2, and $\sigma : \mathbb{R}^l \to \mathbb{R}^{l,d}$ be a C_b^4-function. Then*

(i) Equation (8.28) admits a unique solution y in $\mathcal{Q}_{\kappa,a,\sigma(a)}(\mathbb{R}^l)$ for any $\kappa < \gamma$ such that $3\kappa + \gamma > 1$.

(ii) The mapping $(a, x, \mathbf{x^2}, \mathbf{x^3}) \mapsto y$ is continuous from

$$\mathbb{R}^l \times \mathcal{C}_1^{\gamma}(\mathbb{R}^d) \times \mathcal{C}_2^{2\gamma}(\mathbb{R}^{d,d}) \times \mathcal{C}_2^{3\gamma}(\mathbb{R}^{d,d,d}) \text{ to } \mathcal{C}_1^{\kappa}(\mathbb{R}^l),$$

in the following sense: let z be the unique solution of (8.28) in $\mathcal{Q}_{\kappa,a,\sigma(a)}(\mathbb{R}^l)$ and $\tilde{z}$ the unique solution of (8.28) in $\mathcal{Q}_{\kappa,\tilde{a},\sigma(\tilde{a})}(\mathbb{R}^l)$, based on $x, \tilde{x}$, respectively. Then, there exists a positive constant $\hat{c}_{\sigma,x,\tilde{x}}$ depending only on $\sigma, x, \tilde{x}$ such that

$$\mathcal{N}[z - \tilde{z}; \mathcal{C}_1^{\kappa}(\mathbb{R}^l)] \le \hat{c}_{x,\tilde{x}}\{|a - \tilde{a}| + \mathcal{N}[x - \tilde{x}; \mathcal{C}_1^{\gamma}(\mathbb{R}^l)]$$
$$+ \mathcal{N}[\mathbf{x^2} - \tilde{\mathbf{x}}^2; \mathcal{C}_2^{2\gamma}(\mathbb{R}^{d,d})] + \mathcal{N}[\mathbf{x^3} - \tilde{\mathbf{x}}^3; \mathcal{C}_2^{3\gamma}(\mathbb{R}^{d,d,d})]\}.$$

Proof. As in [13, 15], we first identify the solution on a small interval $[0, \tau]$ as the fixed point of the map $\Gamma : \mathcal{Q}_{\kappa,a,\sigma(a)}(\mathbb{R}^l) \to \mathcal{Q}_{\kappa,a,\sigma(a)}(\mathbb{R}^l)$ defined by $\Gamma(z) = \hat{z}$ with $\hat{z}_0 = a$ and $\delta\hat{z} = \mathcal{J}(\sigma(z)dx)$. The first step in this direction is to show that the ball

$$B_M = \{z;\ z_0 = a,\ \mathcal{N}[z; \mathcal{Q}_{\kappa,a,\sigma(a)}([0,\tau];\ \mathbb{R}^l)] \le M\} \tag{8.29}$$

is invariant under Γ if τ is small enough and M is large enough. However, due to Corollary 8.2 and Proposition 8.2, invoking the fact that σ is bounded together with its derivatives and assuming $\tau \le 1$, we obtain

$$\begin{aligned}
&\mathcal{N}[\Gamma(z); \mathcal{Q}_{\kappa,a,\sigma(a)}(\mathbb{R}^l)] \\
&\le c_x\{1 + |\sigma(a)|_{\mathbb{R}^{l,d}} + \tau^{\gamma-\kappa}(|\sigma(a)|_{\mathbb{R}^{l,d}} + \mathcal{N}[\sigma(z); \mathcal{Q}_{\kappa,\hat{a},\hat{b}}(\mathbb{R}^{l,d})])\} \\
&\le c_{x,\sigma}\{1 + \tau^{\gamma-\kappa}\mathcal{N}[\sigma(z); \mathcal{Q}_{\kappa,\hat{a},\hat{b}}(\mathbb{R}^{l,d})]\} \\
&\le c_{x,\sigma}\{1 + \tau^{\gamma-\kappa}(1 + \mathcal{N}^3[z; \mathcal{Q}_{\kappa,a,\sigma(a)}(\mathbb{R}^l)])\} \\
&\le \tilde{c}_{x,\sigma}\{1 + \tau^{\gamma-\kappa}\mathcal{N}^3[z; \mathcal{Q}_{\kappa,a,\sigma(a)}(\mathbb{R}^l)]\},
\end{aligned} \tag{8.30}$$

where $\hat{a} = \sigma(a)$ and $\hat{b} = \partial_i\sigma(a)\sigma^{i\cdot}(a)$. Taking $M > \tilde{c}_{x,\sigma}$ and $\tau \le \tau_0 = \left(\frac{1}{M^2\tilde{c}_{x,\sigma}} - \frac{1}{M^3}\right)^{\frac{1}{\gamma-\kappa}} \wedge 1$, we obtain that $\tilde{c}_{x,\sigma}(1 + \tau^{\gamma-\kappa}M^3) \le M$. Therefore, the ball B_M defined at (8.29) is left invariant by Γ.

It is now a matter of standard considerations to settle a fixed point argument for Γ on $[0,\tau]$, and also to patch solutions on any interval of the form $[k\tau, (k+1)\tau]$ for $k \ge 1$. The details of this procedure are left to the reader. □

8.3.5 The Delay Equation Case

This section is devoted to show how to change the diffusion setting in order to cover the case of delayed systems, having in mind to solve an equation of the form:

$$\begin{cases} dy_t = \sigma(y_t, y_{t-r_1}, \ldots, y_{t-r_q})\, dx_t & t \in [0,T], \\ y_t = \xi_t, & t \in [-r_q, 0], \end{cases} \tag{8.31}$$

where x is $\mathbb{R}^d$-valued γ-Hölder continuous function with $\gamma > 1/4$, the function σ is smooth enough, ξ is a $\mathbb{R}^l$-valued 3γ-Hölder continuous function, and $0 < r_1 < \ldots < r_q < \infty$.

One possibility in order to solve Eq. (8.31) is to elaborate on the strategy followed in [19], which relied on the notion of *delayed controlled paths*. The reader is referred to [27] for a detailed account on this first strategy. However, we have chosen here to show that delay systems of the form (8.31), with a discrete family of delays $(r_1, \ldots, r_q)$, can be reduced to ordinary differential systems. This is obtained by a huge (though finite) dimension augmentation of the system we are considering and leads to shorter proofs, at the price of slightly nonoptimal hypothesis. It should also be noticed at this point that general delay systems are implicitly solved in [16] by considering an infinite dimensional rough path, namely the so-called frame process.

In order to specify our statements, let us introduce some additional notation. We assume, without loss of generality, that $T = m\, r_q$ for a certain $m \in \mathbb{N}$. Assume also for the moment that x is a smooth function, in which case Eq. (8.31) can be solved in the usual sense, and denote by y its solution. Notice that for notational convenience, we set $r_0 = 0$. For $(k_1, \ldots, k_q) \in \{0, \ldots, m-1\}^q$, with $k_1 \le k_2 \le \cdots \le k_q$, and $t \in [0, r_q]$, set

$$\hat{y}_t(k_1, \ldots, k_q) = y_{t+\sum_{j=1}^q k_j (r_j - r_{j-1})}, \quad \text{and} \quad \hat{x}_t(k_1, \ldots, k_q) = x_{t+\sum_{j=1}^q k_j (r_j - r_{j-1})}. \tag{8.32}$$

We shall consider $\hat{y}_t$ (resp. $\hat{x}_t$) as an element of $\mathbb{R}^{l\times\Delta}$ (resp. of $\mathbb{R}^{d\times\Delta}$), where

$$\Delta = \mathrm{Card}\{(k_1, \ldots, k_q) \in \{0, \ldots, m-1\}^q;\ k_1 \le k_2 \le \cdots \le k_q\} = \binom{m+q-1}{q}.$$

These notations being set, the representation of our delay equation as an ordinary differential system can be read as follows:

Proposition 8.5. *Let x be a regular $\mathbb{R}^d$-valued function, and σ a C_b^1-function. Then the delay equation (8.31) can be represented as an ordinary differential system of the form*

$$d\hat{y}_t = \hat{\sigma}(\hat{y}_t)\, d\hat{x}_t, \quad t \in [0, r_q], \tag{8.33}$$

where $\hat{x}, \hat{y}$ have been defined at (8.32) and for a certain matrix-valued function $\hat{\sigma}$ which inherits the regularity properties of σ. The initial condition of (8.33) is given by solving the equation in a sequential way.

Proof. Our claims can be deduced by elementary considerations. Indeed, Eq. (8.31) can be written as

$$\begin{aligned} d\,\hat{y}_t(k_1,\ldots,k_q) \\ &= \sigma\Big(\hat{y}_t(k_1,\ldots,k_q), \hat{y}_t((k_1-1),k_2,\ldots,k_q), \hat{y}_t((k_1-1),(k_2-1),\ldots,k_q), \\ &\qquad \ldots, \hat{y}_t((k_1-1),\ldots,(k_q-1))\Big)\, d\hat{x}_t(k_1,\ldots,k_q), \end{aligned} \tag{8.34}$$

from which Eq. (8.33) is easily deduced, with a block matrix $\hat{\sigma} \in \mathbb{R}^{l\times\Delta, d\times\Delta}$ whose blocks consist in Δ copies of σ.

One word should be said however about the order in which Eq. (8.34) has to be solved. Indeed, one has to be careful about the initial conditions in the differential system, since it involves the solution $\hat{y}$ itself if some $k_j \neq 0$. This forces us to solve Eq. (8.33) in a sequential way, according to a somewhat intricate algorithm. Our algorithm consists then in moving along tuples of the set $\{(k_1,\ldots,k_q) \in \{0,\ldots,m-1\}^q;\, k_1 \leq k_2 \leq \cdots \leq k_q\}$ following a path $(\mathbf{k}^N)_{N\leq\Delta}$, where $\mathbf{k}^N = (\mathbf{k}_1^N,\ldots,\mathbf{k}_q^N)$. The sequence $(\mathbf{k}^N)_{N\leq\Delta}$ is defined as follows:

(i) Start from $\mathbf{k}^0 = (0,\ldots,0)$.
(ii) If $\mathbf{k}_1^N = \cdots = \mathbf{k}_i^N$ and $\mathbf{k}_{i+1}^N > \mathbf{k}_i^N$ for $1 \leq i \leq q$, then set

$$\mathbf{k}^{N+1} = (\overbrace{0,\ldots,0}^{i-1}, \mathbf{k}_i^N + 1, \mathbf{k}_{i+1}^N,\ldots,\mathbf{k}_q^N).$$

It is easily seen that this algorithm allows to go from $(0,\ldots,0)$ (obtained for $N=0$) to $(m-1,\ldots,m-1)$ (obtained for $N=\Delta$). Furthermore, the following facts are also readily verified:

(1) For the index $(m-1,\ldots,m-1)$, we have been able to define

$$\hat{y}_0(m-1,\ldots,m-1) = y_{(m-1)r_q} = y_{T-r_q}, \quad \text{and} \quad \hat{y}_{r_q}(m-1,\ldots,m-1) = y_T.$$

This means that, once the index $(m-1,\ldots,m-1)$ is attained, the original delay equation is solved.

(2) For a given $N \leq \Delta$, any tuple of the form

$$(\mathbf{k}_1^N-1,\mathbf{k}_2^N,\ldots,\mathbf{k}_q^N), (\mathbf{k}_1^N-1,\mathbf{k}_2^N-1,\ldots,\mathbf{k}_q^N),\ldots,(\mathbf{k}_1^N-1,\mathbf{k}_2^N-1,\ldots,\mathbf{k}_q^N-1)$$

can be expressed as $\mathbf{k}^{N'}$ for another $N' < N$.

(3) By construction of our algorithm, the initial condition for each equation of our system is well defined in terms of previous elements of the system.

These three simple facts allow to solve the system (8.33) in an iterative way, which ends the proof. □

Now that we have been able to write Eq. (8.31) as the ordinary differential system (8.33), the strategy in order to solve it in the rough case is straightforward: just apply Theorem 8.1 to the process $\hat{x}$ defined by (8.32). This boils down to say that $\hat{x}$ must satisfy Hypotheses 8.1 and 8.2. Moreover, in case of a process $\hat{x}$ defined by the shift of one single process x, one can particularize those two assumptions into the following one:

Hypothesis 8.3. *The path $\mathbb{R}^d$-valued x is γ-Hölder continuous with $\gamma > 1/4$, and admits a delayed Lévy area and a delayed volume element, that is a collection of increments*

$$\mathbf{x}^2 = \left\{\mathbf{x}^{2,ij}(v_1, v_2);\ i, j \in \{1, \ldots, d\}\right\}$$
$$\mathbf{x}^3 = \left\{\mathbf{x}^{3,ijk}(v_1, v_2, v_3);\ i, j, k \in \{1, \ldots, d\}\right\},$$

where $v_1, v_2, v_3 \in \mathscr{S} := \left\{\sum_{j=1}^{q} k_j(r_j - r_{j-1});\ 0 \le k_1 \le k_2 \le \cdots \le k_q \le m-1\right\}$, such that $\mathbf{x}^2 \in \mathscr{C}_2^{2\gamma}(\mathbb{R}^{d^2 \times \Delta^2})$ and $\mathbf{x}^3 \in \mathscr{C}_2^{3\gamma}(\mathbb{R}^{d^3 \times \Delta^3})$. The families of increments $\mathbf{x}^2$ and $\mathbf{x}^3$ are also assumed to satisfy the following algebraic relations:

$$\delta \mathbf{x}^{2,ij}(v_1, v_2) = \delta x^i(v_1)\, \delta x^j(v_2)$$
$$\delta \mathbf{x}^{3,ijk}(v_1, v_2, v_3) = \mathbf{x}^{2,ij}(v_1, v_2)\, \delta x^k(v_3) + \delta x^i(v_1)\, \mathbf{x}^{2,jk}(v_2, v_3)$$

for any $i, j, k \in \{1, \ldots, d\}$ and $v_1, v_2, v_3 \in \mathscr{S}$, where we have set $\delta x_{st}(v) = \delta x_{s+v,t+v}$.

As far as the geometric property (analogous to Hypothesis 8.2) is concerned, it can be stated as follows:

Hypothesis 8.4. *Let $\mathbf{x}^2$ be the area process defined at Hypothesis 8.3. Then we suppose that for $0 \le s < t \le T$, $i, j \in \{1, \ldots, d\}$, and $v_1, v_2 \in \mathscr{S}$, we have:*

$$\mathbf{x}^{2,ij}_{st}(v_1, v_2) + \mathbf{x}^{2,ji}_{st}(v_2, v_1) = \delta x^i_{st}(v_1)\, \delta x^j_{st}(v_2).$$

Remark 8.4. Like in the non-delay case, if x is a smooth function, the increments $\mathbf{x}^2$ and $\mathbf{x}^3$ are simply given by

$$\mathbf{x}^{2,i_1 i_2}_{st}(v_1, v_2) = \int_{s<u_1<u_2<t} dx^{i_1}_{u_1+v_1}\, dx^{i_2}_{u_2+v_2}$$
$$\mathbf{x}^{3,i_1 i_2 i_3}_{st}(v_1, v_2, v_3) = \int_{s<u_1<u_2<u_3<t} dx^{i_1}_{u_1+v_1}\, dx^{i_2}_{u_2+v_2}\, dx^{i_3}_{u_3+v_3}.$$

We are now in a position to state the main result of this section. It can be stated and proved along the same lines as Theorem 8.1, so that the proof is skipped here for sake of conciseness.

Theorem 8.2. *Let x be a process satisfying Hypotheses 8.3 and 8.4, and $\sigma : \mathbb{R}^{l(q+1)} \to \mathbb{R}^{l,d}$ be a C_b^4-function. Then*

(i) Equation (8.33) admits a unique solution $\hat{y}$ in a space of controlled processes with respect to $\hat{x}$, of order κ for any $\kappa < \gamma$ such that $3\kappa + \gamma > 1$.
(ii) The mapping $(a, x, \mathbf{x}^2, \mathbf{x}^3) \mapsto y$ is continuous from

$$\mathbb{R}^l \times \mathscr{C}_1^\gamma(\mathbb{R}^d) \times \mathscr{C}_2^{2\gamma}(\mathbb{R}^{d^2 \times \Delta^2}) \times \mathscr{C}_2^{3\gamma}(\mathbb{R}^{d^3 \times \Delta^3}) \text{ to } \mathscr{C}_1^\kappa(\mathbb{R}^l),$$

in the following sense: let z be the unique solution of (8.33) and $\tilde{z}$ the unique solution of (8.33) in their respective controlled processes spaces, based on $x, \tilde{x}$, respectively. Then, there exists a positive constant $\hat{c}_{\sigma,x,\tilde{x}}$ depending only on $\sigma, x, \tilde{x}$ such that

$$\mathscr{N}[z - \tilde{z}; \mathscr{C}_1^\kappa(\mathbb{R}^l)] \le \hat{c}_{x,\tilde{x}}\{|a - \tilde{a}| + \mathscr{N}[x - \tilde{x}; \mathscr{C}_1^\gamma(\mathbb{R}^d)]$$
$$+ \mathscr{N}[\mathbf{x}^2 - \tilde{\mathbf{x}}^2; \mathscr{C}_2^{2\gamma}(\mathbb{R}^{d^2 \times \Delta^2})] + \mathscr{N}[\mathbf{x}^3 - \tilde{\mathbf{x}}^3; \mathscr{C}_2^{3\gamma}(\mathbb{R}^{d^3 \times \Delta^3})]\}.$$

8.4 Application to Fractional Brownian Motion

All the previous constructions rely on the specific assumptions that we have made on the process x. In this section, we prove how our results can be applied to fractional Brownian motion. More specifically, we first recall some basic definitions about fBm, and then define the delayed Lvy area $\mathbf{B}^2$. We shall then turn to the definition of the volume $\mathbf{B}^3$, which is the main difficulty in order to go from the case $H > 1/3$ treated in [19] to our rougher situation.

8.4.1 Basic Facts on Fractional Brownian Motion

Recall that a d-dimensional fBm with Hurst parameter $H \in (0, 1)$ defined on the real line is a centered Gaussian process

$$B = \{B_t = (B_t^1, \ldots, B_t^d);\ t \in \mathbb{R}\},$$

where $B^1, \ldots, B^d$ are d independent one-dimensional fBm, that is, each B^i is a centered Gaussian process with continuous sample paths and covariance function

$$R_H(t, s) = \mathbb{E}(B_t^i B_s^i) = \frac{1}{2}(|t|^{2H} + |s|^{2H} - |t - s|^{2H}), \tag{8.35}$$

for all $i \in \{1, \dots, d\}$. In the sequel, all the random variables we deal with are defined on a complete probability space $(\Omega, \mathscr{F}, \mathbb{P})$, and we assume that $\mathscr{F}$ is generated by the random variables $(B_t;\ t \in \mathbb{R})$. The fBm verifies the following two important properties:

- Scaling property: For any $c > 0$, $B^{(c)} = c^H B_{\cdot/c}$ is a fBm,
- Stationarity property: For any $h \in \mathbb{R}$, $B_{\cdot+h} - B_h$ is a fBm.

Notice that we work with a fBm indexed by $\mathbb{R}$ for sake of simplicity as in [19], since this allows some more elegant calculations for the definitions of the delayed Lévy area and volume, respectively. Furthermore, since the case $H > 1/2$ or the Brownian case $H = 1/2$ are less demanding than the rougher case, we shall mainly focus in this section on the range of parameter $H < 1/2$.

8.4.1.1 Gaussian Structure of *B*

Let us give a few facts about the Gaussian structure of fractional Brownian motion, following the pioneering article [5] and the review given in Chap. 5 of [22]. All the considerations in this direction will concern a one-dimensional fBm B, which will be enough for our applications.

Let $\mathscr{E}$ be the set of step-functions on $\mathbb{R}$ with values in $\mathbb{R}$. Consider the Hilbert space $\mathscr{H}$ defined as the closure of $\mathscr{E}$ with respect to the scalar product induced by

$$\left\langle \mathbb{1}_{[t,t']}, \mathbb{1}_{[s,s']} \right\rangle_{\mathscr{H}} = R_H(t', s') - R_H(t', s) - R_H(t, s') + R_H(t, s),$$

for any $-\infty < s < s' < +\infty$ and $-\infty < t < t' < +\infty$, and where $R_H(t, s)$ is given by (8.35). The mapping

$$\mathbb{1}_{[t,t']} \mapsto B_{t'} - B_t$$

can be extended to an isometry between $\mathscr{H}$ and the Gaussian space $H_1(B)$ associated with B. We denote this isometry by $\varphi \mapsto B(\varphi)$.

The spaces $\mathscr{H}$ and $H_1(B)$ can be characterized more precisely in the following way: first, we notice that a one-dimensional fBm defined on the real line, with $H \neq 1/2$, has the following integral representation in terms of a Wiener process W defined on $\mathbb{R}$ (see [26, Proposition 7.2.6] for details):

$$B_t = \frac{1}{C_1(H)} \int_{\mathbb{R}} \left[(t-s)_+^{H-1/2} - (-s)_+^{H-1/2} \right] dW_s, \quad t \in \mathbb{R}, \tag{8.36}$$

where

$$C_1(H) = \left(\int_0^{\infty} \left[(1+s)^{H-1/2} - s^{H-1/2} \right]^2 ds + \frac{1}{2H} \right)^{1/2}, \tag{8.37}$$

and where a_+ stands for the positive part of a real number a, namely $a_+ = \mathbb{1}_{\mathbb{R}_+}(a)\,a$. Using the representation (8.36), the authors in [23] define the following stochastic integral of a deterministic function with respect to a one-dimensional fBm B:

$$\int_{\mathbb{R}} f(u)dB_u = \frac{\Gamma(H+1/2)}{C_1(H)} \begin{cases} \int_{\mathbb{R}} \left(\mathscr{D}_-^{1/2-H} f\right)(u)dW_u, & H < 1/2, \\ \int_{\mathbb{R}} \left(\mathscr{I}_-^{H-1/2} f\right)(u)dW_u, & H > 1/2, \end{cases}$$

provided that the stochastic integral with respect to the Wiener process W makes sense, and where

$$\left(\mathscr{D}_-^{\alpha} f\right)(u) = \frac{\alpha}{\Gamma(1-\alpha)} \int_0^{\infty} \frac{f(r) - f(u+r)}{r^{1+\alpha}} dr, \tag{8.38}$$

$$\left(\mathscr{I}_-^{\alpha} f\right)(u) = \frac{1}{\Gamma(\alpha)} \int_u^{\infty} \frac{f(r)}{(r-u)^{1-\alpha}} dr, \tag{8.39}$$

for $0 < \alpha < 1$. The expressions (8.38) and (8.39) are respectively called right-sided fractional derivative and right-sided fractional integral on the whole real line. We remark that, in general,

$$\left(\mathscr{D}_-^{\alpha} f\right)(u) \equiv \lim_{\varepsilon \to 0} \frac{\alpha}{\Gamma(1-\alpha)} \int_{\varepsilon}^{\infty} \frac{f(r) - f(u+r)}{r^{1+\alpha}} dr.$$

We also notice that

$$\left(\mathscr{I}_-^{\alpha}\left(\mathscr{D}_-^{\alpha} f\right)\right)(u) = \left(\mathscr{D}_-^{\alpha}\left(\mathscr{I}_-^{\alpha} f\right)\right)(u) = f(u). \tag{8.40}$$

When f is a function defined on an interval $[a,b]$ with $-\infty < a < b < \infty$, extend f by setting $f^\star = f\,\mathbb{1}_{[a,b]}$. Define then

$$\left(\mathscr{D}_-^{\alpha} f^\star\right)(u) = \left(\mathscr{D}_{b-}^{\alpha} f\right)(u) = \frac{f(u)}{\Gamma(1-\alpha)(b-u)^{\alpha}} + \frac{\alpha}{\Gamma(1-\alpha)} \int_u^b \frac{f(u) - f(r)}{(r-u)^{1+\alpha}} dr, \tag{8.41}$$

$$\left(\mathscr{I}_-^{\alpha} f^\star\right)(u) = \left(\mathscr{I}_{b-}^{\alpha} f\right)(u) = \frac{1}{\Gamma(\alpha)} \int_u^b \frac{f(r)}{(r-u)^{1-\alpha}} dr, \tag{8.42}$$

for $0 < \alpha < 1$, $a < u < b$. The expressions (8.41) and (8.42) are respectively called right-sided fractional derivative and right-sided fractional integral on the interval $[a,b]$. In this context, as in the case of the whole line (see [25] for details and also [28]), the following relation holds true:

$$\left(\mathscr{D}_{b-}^{\alpha} f\right)(u) \equiv \frac{f(u)}{\Gamma(1-\alpha)(b-u)^{\alpha}} + \lim_{\varepsilon \to 0} \frac{\alpha}{\Gamma(1-\alpha)} \int_{u+\varepsilon}^b \frac{f(u) - f(r)}{(r-u)^{1+\alpha}} dr.$$

With these notations in hand, it is proved in [23] that the operator

$$(\mathcal{K} f)(u) \equiv \frac{\Gamma(H+1/2)}{C_1(H)} \begin{cases} \left(\mathcal{D}_-^{1/2-H} f\right)(u), & H < 1/2, \\ \left(\mathcal{I}_-^{H-1/2} f\right)(u), & H > 1/2, \end{cases}$$

is an isometry between $\mathcal{H}$ and a closed subspace of $L^2(\mathbb{R})$. In fact,

$$\langle \phi, \psi \rangle_{\mathcal{H}} = \langle \mathcal{K}\phi, \mathcal{K}\psi \rangle_{L^2(\mathbb{R})},$$

for all $\phi, \psi \in \mathcal{H}$. This also allows to write $B(\varphi)$ as $W(\mathcal{K}\varphi)$ for any $\varphi \in \mathcal{H}$, where $W(\mathcal{K}\varphi)$ has to be interpreted as a Wiener integral with respect to the Gaussian measure W. In particular, we have:

$$\mathbb{E}[|B(\varphi)|^2] = \|\varphi\|_{\mathcal{H}} = \|\mathcal{K}\varphi\|_{L^2(\mathbb{R})}. \tag{8.43}$$

8.4.1.2 Malliavin Calculus with Respect to the fBm B

Let $\mathcal{S}$ be the set of smooth cylindrical random variables of the form

$$F = f(B(\varphi_1), \ldots, B(\varphi_k)), \qquad \varphi_i \in \mathcal{H}, \quad i \in \{1, \ldots, k\},$$

where $f \in C^\infty(\mathbb{R}^{d,k}, \mathbb{R})$ is bounded with bounded derivatives. The derivative operator D of a smooth cylindrical random variable of the above form is defined as the $\mathcal{H}$-valued random variable

$$DF = \sum_{i=1}^{k} \frac{\partial f}{\partial x_i}(B(\varphi_1), \ldots, B(\varphi_k))\varphi_i.$$

This operator is closable from $L^p(\Omega)$ into $L^p(\Omega; \mathcal{H})$. As usual, $\mathbb{D}^{1,2}$ denotes the closure of the set of smooth random variables with respect to the norm

$$\|F\|_{1,2}^2 = \mathbb{E}|F|^2 + \mathbb{E}\|DF\|_{\mathcal{H}}^2.$$

In particular, considering a d-dimensional fBm $(B^1, \ldots, B^d)$, if $D^{B^i} F$ denotes the Malliavin derivative of $F \in \mathbb{D}_{B^i}^{1,2}$ with respect to B^i, where $\mathbb{D}_{B^i}^{1,2}$ denotes the corresponding Sobolev space, we have $D^{B^i} B_t^j = \delta_{i,j} \mathbb{1}_{(-\infty,t]}$ for $i, j = 1, \ldots, d$, where $\delta_{i,j}$ denotes the Kronecker symbol.

The divergence operator I is the adjoint of the derivative operator. If a random variable $\phi \in L^2(\Omega; \mathcal{H})$ belongs to $\mathrm{dom}(I)$, the domain of the divergence operator, then $I(\phi)$ is defined by the duality relationship

$$\mathbb{E}(F I(\phi)) = \mathbb{E}\langle DF, \phi \rangle_{\mathcal{H}}, \tag{8.44}$$

for every $F \in \mathbb{D}^{1,2}$. In addition, let us recall two useful properties verified by D and I:

- If $\phi \in \mathrm{dom}(I)$ and $F \in \mathbb{D}^{1,2}$ such that $F\phi \in L^2(\Omega;\mathscr{H})$, then we have the following integration by parts formula:

$$I(F\phi) = FI(\phi) - \langle DF, \phi\rangle_{\mathscr{H}}. \tag{8.45}$$

- If $\phi \in \mathbb{D}^{1,2}(\mathscr{H})$, $D_r\phi \in \mathrm{dom}(I)$ for all $r \in \mathbb{R}$ and $\{I(D_r\phi)\}_{r\in\mathbb{R}}$ is an element of $L^2(\Omega;\mathscr{H})$, then

$$D_r I(\phi) = \phi_r + I(D_r\phi). \tag{8.46}$$

One can relate the Malliavin derivatives with respect to B and W through the operator $\mathscr{K}$ defined above. Indeed, relation (8.40) shows that $\mathscr{K}$ is invertible. This allows to state, as in the case of a one-dimensional fBm B in an interval (see for example [22, Sect. 5.2] and also [1]), the following relations for the Malliavin derivative and divergence operators with respect to the processes B and W:

(i) For any $F \in \mathbb{D}^{1,2}_W = \mathbb{D}^{1,2}$, we have:

$$\mathscr{K} DF = D^W F,$$

where D^W denotes the derivative operator with respect to the process W, and $\mathbb{D}^{1,2}_W$ the corresponding Sobolev space.

(ii) $\mathrm{Dom}(I) = \mathscr{K}^{-1}(\mathrm{Dom}(I^W))$, and for any $\mathscr{H}$-valued random variable u in $\mathrm{Dom}(I)$ we have $I(u) = I^W(\mathscr{K} u)$, where I^W denotes the divergence operator with respect to the process W.

In addition, we have $\mathbb{D}^{1,2}(\mathscr{H}) = (\mathscr{K}^{-1})(\mathbb{L}^{1,2})$, where $\mathbb{L}^{1,2} = \mathbb{D}^{1,2}(L^2(\mathbb{R}))$, and this space is included in $\mathrm{dom}(I^W)$. Making use of the notations $I^W(\phi) = \int_{\mathbb{R}} \phi_u \, dW_u$ for any $\phi \in \mathrm{dom}(I^W)$, and $I(\phi) = \int_{\mathbb{R}} \phi_u \, dB_u$ for any $\phi \in \mathrm{dom}(I)$, we can write:

$$\int_{\mathbb{R}} \phi_u \, dB_u = \int_{\mathbb{R}} (\mathscr{K}\phi)(u) \, dW_u.$$

This kind of relation also holds when one considers functions defined on an interval. Indeed, for some fixed $-\infty < a < b < \infty$, and $H < 1/2$, relation (8.41) yields

$$\int_a^b \phi_u \, dB_u = \int_{\mathbb{R}} \phi_u 1\!\!1_{[a,b]}(u) \, dB_u = \int_{\mathbb{R}} (\mathscr{K}[\phi 1\!\!1_{[a,b]}])(u) \, dW_u = \int_{\mathbb{R}} (\mathscr{K}^{[a,b]}\phi)(u) \, dW_u,$$

where the operator $\mathscr{K}^{[a,b]}$ is defined by:

$$(\mathscr{K}^{[a,b]} f)(u) \equiv \frac{\Gamma(H+1/2)}{C_1(H)} \left(\mathscr{D}_{b-}^{1/2-H} f\right)(u), \quad \text{for} \quad a < u < b,$$

with $C_1(H)$ defined by (8.37). In case of an interval $[a,b]$, it should also be mentioned that an important subspace of integrable processes is the following: let $\mathscr{E}^{[a,b]}$ be the set of step-functions on $[a,b]$ with values in $\mathbb{R}$. As in [22, Sect. 5.2.3], we consider on this space the semi-norm

$$\|\varphi\|^2_{\mathscr{H}_K([a,b])} = \int_a^b \frac{\varphi_u^2}{(b-u)^{1-2H}} du + \int_a^b \left(\int_u^b \frac{|\varphi_r - \varphi_u|}{(r-u)^{3/2-H}} dr \right)^2 du.$$

Let $\mathscr{H}_K([a,b])$ be the Hilbert space defined as the closure of $\mathscr{E}^{[a,b]}$ with respect to the previous semi-norm. Then the space $\mathscr{H}_K([a,b])$ is continuously included in $\mathscr{H}$, and if $\phi \in \mathbb{D}^{1,2}(\mathscr{H}_K([a,b]))$, then $\phi \in \mathrm{Dom}(I)$.

8.4.1.3 Generalized Stochastic Integrals

The stochastic integrals we shall use in order to construct our delayed Lévy area and volume are defined, in a natural way, by Russo–Vallois' symmetric approximations, that is, for a given process ϕ:

$$\int_a^b \phi_w \, d^\circ B_w^i = L^2 - \lim_{\varepsilon \to 0} \frac{1}{2\varepsilon} \int_a^b \phi_w \left(B_{w+\varepsilon}^i - B_{w-\varepsilon}^i \right) dw,$$

provided the limit exists. It is well known that the Russo–Vallois symmetric integral coincides with Young's integral for $H > 1/2$, and with the classical Stratonovich integral in the Brownian case $H = 1/2$. Since these two cases are not very demanding from a technical point of view, we will focus our efforts on the case $1/4 < H < 1/2$. This being said, for $v_1, v_2, v_3 \in \mathscr{S}$, we will try to define the increments $\mathbf{B}^2$ and $\mathbf{B}^3$ as

$$\mathbf{B}^2_{st}(v_1,v_2) = \int_s^t d^\circ B_{u+v_2} \otimes \int_{s+v_1}^{u+v_1} d^\circ B_\tau, \text{ i.e. } (\mathbf{B}^2_{st}(v_1,v_2))^{ij} = \int_s^t d^\circ B^j_{u+v_2} \int_{s+v_1}^{u+v_1} d^\circ B^i_\tau$$

$$\mathbf{B}^3_{st}(v_1,v_2,v_3) = \int_s^t d^\circ B_{w+v_3} \otimes \int_s^w d^\circ B_{u+v_2} \otimes \int_s^u d^\circ B_{\tau+v_1},$$

$$\text{i.e. } (\mathbf{B}^3_{st}(v_1,v_2,v_3))^{ijk} = \int_s^t d^\circ B^k_{w+v_3} \int_s^w d^\circ B^j_{u+v_2} \int_s^u d^\circ B^i_{\tau+v_1}, \tag{8.47}$$

for all $i,j,k \in \{1,\ldots,d\}$, $0 \le s < t \le T < \infty$.

Interestingly enough, one can establish the existence of symmetric integrals, thanks to some Malliavin calculus criterions:

Proposition 8.6. *Let ϕ be a stochastic process such that $\phi \, \mathbb{1}_{[a,b]} \in \mathbb{D}^{1,2}(\mathscr{H}_K([a,b]))$, for all $-\infty < a < b < \infty$. Suppose also that*

$$\mathrm{Tr}_{[a,b]} D\phi := L^2 - \lim_{\varepsilon \to 0} \frac{1}{2\varepsilon} \int_a^b \langle D\phi_u, \mathbb{1}_{[u-\varepsilon, u+\varepsilon]} \rangle_{\mathscr{H}} du$$

is an almost surely finite random variable. Then $\int_a^b \phi_u d^\circ B^i_u$ *exists and verifies*

$$\int_a^b \phi_u d^\circ B^i_u = I(\phi \mathbb{1}_{[a,b]}) + \mathrm{Tr}_{[a,b]} D\phi.$$

Furthermore, the following algebraic relation is trivially satisfied for this kind of integrals:

Lemma 8.2. *Let* $\alpha = \{\alpha_w,\ w \in [a,b]\}$ *be a stochastic process such that its symmetric Russo–Vallois integral with respect to a one-dimensional fractional Brownian motion* B *exists, and let* F *be a random variable. Then* $F\alpha$ *is integrable with respect to* B *in the Russo–Vallois symmetric integral sense and* $\int_a^b F\alpha_w\, d^\circ B_w = F \int_a^b \alpha_w\, d^\circ B_w$.

We are now ready to show the existence of delayed areas and volumes with respect to fBm.

8.4.2 Delayed Lévy Areas

Before we turn to statements involving increments as functions of two parameters, let us deal first with fixed times s, t:

Proposition 8.7. *Let* B *be a* d*-dimensional fractional Brownian motion, with Hurst parameter* $H > 1/4$*. Then, for* $s, t \in [0, T]$*,* $v_1, v_2 \in \mathscr{S}$*, the delayed Lévy area, denoted by* $\mathbf{B}^2_{st}(v_1, v_2)$ *and defined by (8.47), is well defined. In addition, we have* $\mathbb{E}[|\mathbf{B}^2_{st}(v_1, v_2)|^2] \leq c|t-s|^{4H}$ *for a strictly positive constant* $c = c_{H,v_1,v_2,T} = c_{H,v_2-v_1,T}$*, exhibiting the following discontinuity phenomenon: we have* $\lim_{|v|\to 0} c_{H,v,T} = \infty$*, but* $c_{H,0,T}$ *is finite.*

Before we go to the core of the proof of this proposition, let us add a few comments to the result:

Remark 8.5. Instead of the Malliavin calculus tools we are invoking, one could have constructed the delayed Lvy area, thanks to the linearization methods contained in [4], plus some direct computations on the covariance of our underlying fBm. It is then rather easily checked that both methods would lead to the same area by means of convergence of Riemann sums (this fact has already been observed in [21] for a comparison between areas defined by linearization and analytic approximation respectively). Let us also notice that the methods of [4] have been greatly generalized in [11,12], ending up with a general criterion for the existence of a rough path above a multidimensional Gaussian process in terms of its covariance function. The reader might then wonder why we did not try to apply those criterions directly to the $2d$-Gaussian process $(B^1_{s+v}, \dots, B^d_{s+v}, B^1_{s+u}, \dots, B^d_{s+u})$. It turns out however that all components of the Gaussian process considered in [11, 12] are independent, while the dependence between B^j_{s+v} and B^j_{s+u} is an important feature of the computations below. We thus believe that the delayed area deserves a separate treatment.

Remark 8.6. The discontinuity result for $c_{H,v,T}$ alluded to in Proposition 8.7 is not a surprise, and had already been observed in [19]. A possible interesting development in this direction is the following: take up the program initiated by Ferrante and Rovira in [9], where (denoting by y^h the solution to our delayed system (8.31)) the dependence $h \mapsto y^h$ is investigated. For $H > 1/2$, Ferrante and Rovira establish the continuity of this map. However, a kind of surprising phase transition phenomenon might be observed at $H = 1/2$, insofar as we expect that for $H < 1/2$ (and under some non-degeneracy assumptions on the coefficent σ in (8.31)) the solution y^h will explode when $h \to 0$. We leave this problem for a subsequent communication.

Proof (Proof of Proposition 8.7). As mentioned before, the case $H \geq 1/2$ is rather easy to handle, and we thus focus on $1/4 < H < 1/2$. It should also be mentioned that Lvy areas can be constructed in a similar way to [19], though an extra attention has to be paid in order to treat irregular cases, when H approaches $1/4$. Considering the change of variable $u' = u + v_2$, we can write

$$\mathbf{B}^2_{st}(v_1, v_2) = \int_{s+v_2}^{t+v_2} (B_{u'-v_2+v_1} - B_{s+v_1}) \otimes d^\circ B_{u'}.$$

As a last preliminary remark, observe that due to the stationarity property of the fBm we shall work without loss of generality on the interval $[0, t-s]$ instead of $[s+v_2, t+v_2]$ in the sequel, that is, we can write

$$\mathbf{B}^2_{st}(v_1, v_2) = \int_0^{t-s} (B_{u-v} - B_{-v}) \otimes d^\circ B_u,$$

where $v = v_2 - v_1$. We denote this last integral by $\mathbf{B}^2_{0,t-s}(v)$, for notational convenience.

1. *Case* $i = j$ *and* $v \geq 0$. Consider the process $\phi = (B^i_{\cdot - v} - B^i_{-v})\, \mathbb{1}_{[0,t-s]}(\cdot)$. When $v \geq 0$, the arguments in [19, Proposition 5.2] for $1/3 < H < 1/2$ also hold for $1/4 < H \leq 1/3$. Thus

$$\mathbf{B}^{2,ii}_{0,t-s}(v) = I^{B^i}(\phi) + \mathrm{Tr}_{[0,t-s]} D^{B^i} \phi,$$

where $I^{B^i}(\phi)$ denotes the divergence integral of ϕ with respect to B^i and

$$\mathrm{Tr}_{[0,t-s]} D^{B^i} \phi = \begin{cases} \frac{1}{2}(t-s)^{2H}, & \text{if } v = 0, \\ -H v^{2H-1}(t-s) + \frac{1}{2}\left((t-s+v)^{2H} - v^{2H}\right), & \text{if } v > 0. \end{cases}$$

In addition, one can also prove, as in [19], that

$$\mathbb{E}\left|\mathbf{B}^{2,ii}_{0,t-s}(v)\right|^2 \le c_{H,v}|t-s|^{4H},$$

for any $v = v_2 - v_1$, with $v_1, v_2 \in \mathscr{S}$, where $\lim_{v\to 0} c_{H,v} = \infty$.

2. *Case* $i = j$ *and* $v < 0$. When $v < 0$, we will show that

$$\mathbf{B}^{2,ii}_{0,t-s}(v) = I^{B^i}(\phi) + \mathrm{Tr}_{[0,t-s]} D^{B^i}\phi, \tag{8.48}$$

where now

$$\mathrm{Tr}_{[0,t-s]} D^{B^i}\phi = H(-v)^{2H-1}(t-s) + \frac{1}{2}\left(|t-s+v|^{2H} - (-v)^{2H}\right). \tag{8.49}$$

Indeed, notice that $D^{B^i}_r \phi_u = 1\!\!1_{[-v,u-v]}(r)\, 1\!\!1_{[0,t-s]}(u)$ and furthermore, for $u \in [0, t-s]$ and $\varepsilon \in [0, -v]$, one can write

$$\begin{aligned}
&\langle 1\!\!1_{[-v,u-v]}, 1\!\!1_{[u-\varepsilon,u+\varepsilon]}\rangle_{\mathscr{H}} \\
&\quad = \frac{1}{2}\left(|-v+\varepsilon|^{2H} - |-v-\varepsilon|^{2H} + |-v-u-\varepsilon|^{2H} - |-v-u+\varepsilon|^{2H}\right) \\
&\quad = \frac{1}{2}\left((-v+\varepsilon)^{2H} - (-v-\varepsilon)^{2H} + |-v-u-\varepsilon|^{2H} - |-v-u+\varepsilon|^{2H}\right).
\end{aligned}$$

Performing now a Taylor expansion in a neighborhood of $\varepsilon = 0$, we get

$$(-v+\varepsilon)^{2H} - (-v-\varepsilon)^{2H} = 4H(-v)^{2H-1}\varepsilon + o\left(\varepsilon^2\right).$$

Thus, applying the dominated convergence theorem (details are left to the reader) we obtain

$$\lim_{\varepsilon\to 0}\int_0^{t-s} \frac{1}{4\varepsilon}\left((-v+\varepsilon)^{2H} - (-v-\varepsilon)^{2H}\right)du = H(-v)^{2H-1}(t-s). \tag{8.50}$$

Along the same lines, by separating the cases $-v \ge t-s$, $0 < u < -v < t-s$ and $-v \le u < t-s$, it can also be proved that

$$\lim_{\varepsilon\to 0}\int_0^{t-s} \frac{1}{4\varepsilon}\left(|-v-u-\varepsilon|^{2H} - |-v-u+\varepsilon|^{2H}\right)du = \frac{1}{2}\left(|t-s+v|^{2H} - (-v)^{2H}\right). \tag{8.51}$$

We now obtain (8.49) by putting together (8.50) and (8.51).

Let us bound now $\mathrm{Tr}_{[0,t-s]} D^{B^i}\phi$ from expression (8.49): in the case $-v \ge t-s$, invoking the fact that, for $0 < p < 1$ and $a \ge b > 0$, the inequality $a^p - b^p \le (a-b)^p$ holds true, we obtain

$$\left|\mathrm{Tr}_{[0,t-s]} D^{B^i}\phi\right| = H(-v)^{2H-1}(t-s) + \frac{1}{2}\left((-v)^{2H} - (-v-(t-s))^{2H}\right)$$
$$\leq H(t-s)^{2H} + \frac{1}{2}\left((-v)^{2H} - ((-v)^{2H} - (t-s)^{2H})\right) \leq (t-s)^{2H},$$

and in the case $-v < t-s$, we also have

$$\left|\mathrm{Tr}_{[0,t-s]} D^{B^i}\phi\right| = H(-v)^{2H-1}(t-s) + \frac{1}{2}\left|(t-s+v)^{2H} - (-v)^{2H}\right|$$
$$\leq H(-v)^{2H-1}T^{1-2H}(t-s)^{2H} + \frac{1}{2}\left((t-s)^{2H} + (-v)^{2H} + (-v)^{2H}\right)$$
$$\leq \left(H(-v)^{2H-1}T^{1-2H} + \frac{3}{2}\right)(t-s)^{2H}.$$

Thus, we have found

$$\left|\mathrm{Tr}_{[0,t-s]} D^{B^i}\phi\right| \leq \left(H(-v)^{2H-1}T^{1-2H} + \frac{3}{2}\right)(t-s)^{2H}, \tag{8.52}$$

for all $v = v_2 - v_1$, with $v_1, v_2 \in \mathcal{S}$.
We proceed now to bound the term $I^{B^i}(\phi)$ in (8.48): owing to (8.46), we have

$$D_r^{B^i} I^{B^i}(\phi) = (B^i_{r-v} - B^i_{-v})\mathbb{1}_{[0,t-s]}(r) + I^{B^i}\left(\mathbb{1}_{[-v,\cdot-v]}(r)\mathbb{1}_{[0,t-s]}(\cdot)\right)$$
$$= (B^i_{r-v} - B^i_{-v})\mathbb{1}_{[0,t-s]}(r) + I^{B^i}\left(\mathbb{1}_{[v+r,t-s]}(\cdot)\right)\mathbb{1}_{[-v,t-s-v]}(r)$$
$$= (B^i_{r-v} - B^i_{-v})\mathbb{1}_{[0,t-s]}(r) + (B^i_{t-s} - B^i_{v+r})\mathbb{1}_{[-v,t-s-v]}(r). \tag{8.53}$$

With this identity in hand and using the same arguments as in the proof of [19, Proposition 5.2], we obtain

$$\mathbb{E}|I^{B^i}(\phi)|^2 \leq c_H|t-s|^{4H}, \tag{8.54}$$

with a constant $c_H > 0$ independent of v.
Finally, (8.52) and (8.54) imply $\mathbb{E}|(\mathbf{B}^2_{0,t-s}(v))^{ii}|^2 \leq c_{H,v}|t-s|^{4H}$ for any $v = v_2 - v_1$, with $v_1, v_2 \in \mathcal{S}$, and thus, according to our stationarity argument:

$$\mathbb{E}\left|(\mathbf{B}^2_{st}(v_1, v_2))^{ii}\right|^2 \leq c_{H,v_1,v_2}|t-s|^{4H}, \tag{8.55}$$

for any $v_1, v_2 \in \mathcal{S}$.

3. *Case* $i \neq j$. This case can be treated similarly to [19, Proposition 5.2] and yields the same kind of inequality as in equation (8.55).

Our claim $\mathbb{E}[|\mathbf{B}^2_{st}(v_1, v_2)|^2] \leq c|t-s|^{4H}$ now stems easily from the inequalities we have obtained for the three cases $i = j$ and $v \geq 0$, $i = j$ and $v < 0$, and $i \neq j$. □

We can go one step further, and state a result concerning $\mathbf{B}^2$ as an increment.

Proposition 8.8. *Let $\mathbf{B}^2$ be the increment defined at Proposition 8.7. Then $\mathbf{B}^2$ satisfies Hypothesis 8.3 and 8.4.*

Proof. First, we have to ensure the almost sure existence of $\mathbf{B}^2_{st}(v_1, v_2)$ for all $s, t \in [0, T]$. This can be done by noticing that $\mathbf{B}^2_{st}(v_1, v_2)$ is a random variable in the second chaos of the fractional Brownian motion B, on which all L^p-norms are equivalent for $p > 1$. Hence we can write:

$$\mathbb{E}\left|\mathbf{B}^{2,ij}_{st}(v_1, v_2)\right|^p \le c_{H,v_1,v_2,p}|t-s|^{2pH}, \tag{8.56}$$

for any $i, j \in \{1, \dots, d\}$ and $p \ge 2$. With the same kind of calculations, one can also obtain the inequality

$$\mathbb{E}\left|\mathbf{B}^{2,ij}_{s_2t_2}(v_1, v_2) - \mathbf{B}^{2,ij}_{s_1t_1}(v_1, v_2)\right|^p \le c_{H,v_1,v_2,p}\left(|t_2-t_1|^{pH} + |s_2-s_1|^{pH}\right).$$

Then, a standard application of Kolmogorov's criterion yields the almost sure definition of the whole family $\{\mathbf{B}^2_{st}(v_1, v_2);\ s, t \in [0, T]\}$, and its continuity as a function of s and t.

Moreover, a direct application of Lemma 8.2 gives

$$\delta\mathbf{B}^2(v_1, v_2) = \delta B(v_1) \otimes \delta B(v_2), \tag{8.57}$$

and Fubini's theorem for Stratonovich integrals with respect to B also yield easily Hypothesis 8.4. Finally, it is readily checked that $\mathbf{B}^2(v_1, v_2) \in \mathscr{C}_2^{2\gamma}(\mathbb{R}^{d,d})$ for any $1/4 < \gamma < H$, $v_1, v_2 \in \mathscr{S}$ (separating the case $v_1 = v_2$). Indeed, it is sufficient to apply Corollary 4 in [13] (see also inequality (90) in [19]), having in mind the bound (8.56) and expression (8.57). □

8.4.3 Delayed Volumes

We study now the term $\mathbf{B}^3(v_1, v_2, v_3)$, starting from a similar statement as in Proposition 8.7:

Proposition 8.9. *Let B be a d-dimensional fractional Brownian motion, with Hurst parameter $H > 1/4$. Then, for $s, t \in [0, T]$, $v_1, v_2, v_3 \in \mathscr{S}$, the delayed volume, denoted by $\mathbf{B}^3(v_1, v_2, v_3)$ and defined by (8.47), is well defined. In addition, we have $\mathbb{E}[|\mathbf{B}^3(v_1, v_2, v_3)|^2] \le c|t-s|^{6H}$ for a strictly positive constant $c = c_{H,v_1,v_2,v_3,T}$ which tends to ∞ if $|v_2 - v_1| \to 0$ or $|v_3 - v_2| \to 0$, but which is also well defined if $v_1 = v_2 = v_3$.*

Proof. Here again, we focus on the case $1/4 < H < 1/2$. First, using the changes of variable $u' = u + v_2$ and $w' = w + v_3$, we can write

$$\mathbf{B}^3_{st}(v_1, v_2, v_3) = \int_{s+v_3}^{t+v_3}\left(\int_{s+v_2}^{w'-v_3+v_2}(B_{u'-v_2+v_1} - B_{s+v_1}) \otimes d^\circ B_{u'}\right) \otimes d^\circ B_{w'}.$$

Due to the stationarity property of the fBm, we shall work without loss of generality on the interval $[0, t-s]$ instead of $[s+v_3, t+v_3]$ in the sequel. For notational sake, we will also set $\tau = t - s$ in the remainder of the proof. We shall then evaluate

$$\mathbf{B}^3_\tau(v_1, v_2, v_3) = \int_0^\tau \left(\int_{-\eta_2}^{w-\eta_2} \left(B_{u-\eta_1} - B_{-\eta_2-\eta_1}\right) \otimes d^\circ B_u \right) \otimes d^\circ B_w,$$

where $\eta_1 = v_2 - v_1$ and $\eta_2 = v_3 - v_2$. Notice that $\int_{-\eta_2}^{w-\eta_2} \left(B_{u-\eta_1} - B_{-\eta_2-\eta_1}\right) \otimes d^\circ B_u$ behaves as $\mathbf{B}^2_{0w}(\eta_1)$.

1. Case $i = j = k$. Consider the process

$$\psi = \left(\int_{-\eta_2}^{\cdot-\eta_2} \left(B^i_{u-\eta_1} - B^i_{-\eta_2-\eta_1}\right) d^\circ B^i_u \right) 1\!\!1_{[0,\tau]}(\cdot).$$

We will define $\mathbf{B}^{3,iii}(\eta_1, \eta_2)$ as $\int_0^\tau \psi_u d^\circ B^i_u$, which amounts to show that $\psi \in \mathbb{D}^{1,2}(\mathcal{H}_K([0,T]))$ and to compute the trace of the process ψ.

With this aim in mind, let us first compute the Malliavin derivative of ψ: it is easily seen that

$$\begin{aligned} &D^{B^i}_r \psi_u \\ &= (B^i_{r-\eta_1} - B^i_{-\eta_2-\eta_1}) 1\!\!1_{[-\eta_2, u-\eta_2]}(r) 1\!\!1_{[0,\tau]}(u) + I^{B^i}\left(1\!\!1_{[-\eta_2-\eta_1, \cdot-\eta_1]}(r) 1\!\!1_{[-\eta_2, u-\eta_2]}(\cdot)\right) 1\!\!1_{[0,\tau]}(u) \\ &= (B^i_{r-\eta_1} - B^i_{-\eta_2-\eta_1}) 1\!\!1_{[-\eta_2, u-\eta_2]}(r) 1\!\!1_{[0,\tau]}(u) + (B^i_{u-\eta_2} - B^i_{r+\eta_1}) 1\!\!1_{[-\eta_2-\eta_1, u-\eta_2-\eta_1]}(r) 1\!\!1_{[0,\tau]}(u). \end{aligned} \tag{8.58}$$

From this identity, one can check that $\psi \in \mathbb{D}^{1,2}(\mathcal{H}_K([0,T]))$. We will now evaluate $\int_0^\tau \psi_u d^\circ B^i_u$ by separating the Skorokhod and the trace term in the symmetric integral.

(i) *Evaluation of the trace term.* We start by observing that $D^{B^i}\psi_u$ can also be written as:

$$\begin{aligned} D^{B^i}_r \psi_u = {} & I^{B^i}\left(1\!\!1_{[\cdot+\eta_1, u-\eta_2]}(r) 1\!\!1_{[-\eta_2-\eta_1, u-\eta_2-\eta_1]}(\cdot)\right) 1\!\!1_{[0,\tau]}(u) \\ & + I^{B^i}\left(1\!\!1_{[-\eta_2-\eta_1, \cdot-\eta_1]}(r) 1\!\!1_{[-\eta_2, u-\eta_2]}(\cdot)\right) 1\!\!1_{[0,\tau]}(u). \end{aligned} \tag{8.59}$$

Apply then Fubini's Theorem in order to get

$$\begin{aligned} &\int_0^\tau \langle D^{B^i}\psi_u, 1\!\!1_{[u-\varepsilon, u+\varepsilon]}\rangle_{\mathcal{H}} du \\ &= \int_{-\eta_2-\eta_1}^{\tau-\eta_2-\eta_1} \left(\int_{w+\eta_2+\eta_1}^\tau \langle 1\!\!1_{[w+\eta_1, u-\eta_2]}, 1\!\!1_{[u-\varepsilon, u+\varepsilon]}\rangle_{\mathcal{H}} du \right) dB^i_w \\ &\quad + \int_{-\eta_2}^{\tau-\eta_2} \left(\int_{w+\eta_2}^\tau \langle 1\!\!1_{[-\eta_2-\eta_1, w-\eta_1]}, 1\!\!1_{[u-\varepsilon, u+\varepsilon]}\rangle_{\mathcal{H}} du \right) dB^i_w, \end{aligned} \tag{8.60}$$

where the last two integrals have to be interpreted in the Wiener sense, and are well defined according to the criterions contained in [23].

We shall consider the case that $\eta_2 \geq 0$ and $\eta_1 + \eta_2 \geq 0$. The other cases can be obtained analogously. Let us evaluate the scalar product in (8.60): for a fixed $\eta_2 > 0$, $u \in [w + \eta_2 + \eta_1, \tau]$, $w \in [-\eta_2 - \eta_1, \tau - \eta_2 - \eta_1]$, and $\varepsilon \in [0, \eta_2]$, we can write

$$\begin{aligned}
&\left\langle \mathbb{1}_{[w+\eta_1, u-\eta_2]}, \mathbb{1}_{[u-\varepsilon, u+\varepsilon]} \right\rangle_{\mathcal{H}} \\
&= \frac{1}{2}\left(|-\eta_2+\varepsilon|^{2H} - |-\eta_2-\varepsilon|^{2H} + |w+\eta_1-u-\varepsilon|^{2H} - |w+\eta_1-u+\varepsilon|^{2H}\right) \\
&= \frac{1}{2}\left((\eta_2-\varepsilon)^{2H} - (\eta_2+\varepsilon)^{2H} + (u-w-\eta_1+\varepsilon)^{2H} - (u-w-\eta_1-\varepsilon)^{2H}\right) \\
&= 2H\left(-\eta_2^{2H-1} + (u-w-\eta_1)^{2H-1}\right)\varepsilon + o(\varepsilon^2).
\end{aligned}$$

If $\eta_2 = 0$, one can prove similarly that for ε small enough,

$$\left\langle \mathbb{1}_{[w+\eta_1, u-\eta_2]}, \mathbb{1}_{[u-\varepsilon, u+\varepsilon]} \right\rangle_{\mathcal{H}} = 2H(u-w-\eta_1)^{2H-1}\varepsilon + o(\varepsilon^2).$$

This yields easily the relation

$$\lim_{\varepsilon\to 0} \frac{1}{2\varepsilon} \langle \mathbb{1}_{[w+\eta_1, u-\eta_2]}, \mathbb{1}_{[u-\varepsilon, u+\varepsilon]} \rangle_{\mathcal{H}} = \begin{cases} H\left(-\eta_2^{2H-1} + (u-w-\eta_1)^{2H-1}\right) & \text{if } \eta_2 > 0, \\ H(u-w-\eta_1)^{2H-1} & \text{if } \eta_2 = 0. \end{cases}$$

The same kind of elementary arguments work for the scalar product in expression (8.60), and one obtains:

$$\lim_{\varepsilon\to 0} \frac{1}{2\varepsilon} \langle \mathbb{1}_{[-\eta_2-\eta_1, w-\eta_1]}, \mathbb{1}_{[u-\varepsilon, u+\varepsilon]} \rangle_{\mathcal{H}} = H\left(-(u-w+\eta_1)^{2H-1} + (\eta_2+\eta_1+u)^{2H-1}\right).$$

Thus, by an application of the dominated convergence theorem (whose details are left to the reader) we get, for a fixed $\eta_2 > 0$,

$$\begin{aligned}
&\mathrm{Tr}_{[0,\tau]} D^{B^i} \psi \\
&= \int_{-\eta_2-\eta_1}^{\tau-\eta_2-\eta_1} \left(-H\eta_2^{2H-1}(\tau-w-\eta_2-\eta_1) + \frac{1}{2}[(\tau-w-\eta_1)^{2H} - \eta_2^{2H}]\right) dB_w^i \\
&+ \frac{1}{2}\int_{-\eta_2}^{\tau-\eta_2} \left((\eta_2+\eta_1)^{2H} - (\tau-w+\eta_1)^{2H} + (\tau+\eta_2+\eta_1)^{2H} - (2\eta_2+\eta_1+w)^{2H}\right) dB_w^i,
\end{aligned} \tag{8.61}$$

and for $\eta_2 = 0$, we end up with:

$$\begin{aligned}
\mathrm{Tr}_{[0,\tau]} D^{B^i} \psi &= \frac{1}{2}\int_{-\eta_1}^{\tau-\eta_1} (\tau-w-\eta_1)^{2H} dB_w^i \\
&+ \frac{1}{2}\int_0^{\tau} \left(\eta_1^{2H} - (\tau-w+\eta_1)^{2H} + (\tau+\eta_1)^{2H} - (\eta_1+w)^{2H}\right) dB_w^i.
\end{aligned}$$

Some similar expressions, whose exact forms are skipped here for sake of conciseness, can be obtained for the remaining cases *(a)* $\eta_2 > 0$ and $\eta_1+\eta_2 < 0$, *(b)* $\eta_2 = 0$ and $\eta_1 < 0$, *(c)* $\eta_2 < 0$ and $\eta_1 + \eta_2 \geq 0$.

For the remainder of the chapter, the relation $a \lesssim b$ stands for $a \leq Cb$ with a universal constant C. Starting from Eq. (8.61), let us evaluate $\mathrm{Tr}_{[0,\tau]} D^{B^i} \psi$ for $\eta_2 > 0$ and $\eta_1 + \eta_2 \geq 0$. Observe first that one can write $\mathbb{E}[|\mathrm{Tr}_{[0,\tau]} D^{B^i} \psi|^2] \lesssim \sum_{l=1}^4 J_l$, where J_l can be decomposed itself as $J_l = \mathbb{E}[| \int_0^\tau F_l(w) dB^i_w|^2]$, with

$$F_1(w) = (\tau - w), \quad F_2(w) = (\tau + \eta_2 + \eta_1)^{2H} - (\eta_2 + \eta_1 + w)^{2H}$$

$$F_3(w) = (\eta_2 + \eta_1)^{2H} - (\tau - w + \eta_2 + \eta_1)^{2H}, \quad F_4(w) = (\tau - w + \eta_2)^{2H} - \eta_2^{2H}.$$

Thus, thanks to relation (8.43), we obtain:

$$J_l = \|F_l\|^2_{\mathscr{H}([0,\tau])} = c_H \left\| \mathscr{D}^{1/2-H}_{\tau-} F_l \right\|^2_{L^2([0,\tau])}.$$

Furthermore, each F_l is a power function, whose fractional derivative $\mathscr{D}^{1/2-H}_{\tau-} F_l$ can be computed explicitly. It is then easily shown that $\mathbb{E}|\mathrm{Tr}_{[0,\tau]} D^{B^i} \psi|^2 \leq c_{H,\eta_2,T}\,\tau^{6H}$, where $c_{H,\eta_2,T} = c_{H,T}\,\eta_2^{2(2H-1)} + c_H$. Analogously, for the other cases, we get $\mathbb{E}|\mathrm{Tr}_{[0,\tau]} D^{B^i} \psi|^2 \leq c_H \tau^{6H}$.

(ii) *Evaluation of the Skorokhod term.* We shall prove that $\mathbb{E}|I^{B^i}(\psi)|^2 \leq c_{H,\eta_1,T}\,\tau^{6H}$ and to this aim, let us decompose ψ into its Skorokhod and trace part. This gives $\mathbb{E}|I^{B^i}(\psi)|^2 \leq 2\mathbb{E}|I^{B^i}(\psi_1)|^2 + 2\mathbb{E}|I^{B^i}(\psi_2)|^2$, where

$$\psi_1(w) = \int_{-\eta_2}^{w-\eta_2} [B^i_{u-\eta_1} - B^i_{-\eta_2-\eta_1}] dB^i_u$$

$$\psi_2(w) = \mathrm{Tr}_{[0,w]} D^{B^i} \phi, \quad \text{with} \quad \phi = (B^i_{\cdot-\eta_1} - B^i_{-\eta_2-\eta_1}) \mathbb{1}_{[-\eta_2, w-\eta_2]}(\cdot).$$

The proof that

$$\mathbb{E}|I^{B^i}(\psi_2)|^2 \leq c_{H,\eta_1,T}\,\tau^{6H},$$

where $c_{H,\eta_1,T} \to \infty$ if $\eta_1 \to 0$ but is also well defined if $\eta_1 = 0$, can be obtained using the same arguments as for Step (i), and we then concentrate on the Skorokhod term $I^{B^i}(\psi_1)$.

To estimate $\mathbb{E}|I^{B^i}(\psi_1)|^2$, we use first identity (8.44), which can be read here as $\mathbb{E}|I^{B^i}(\psi_1)|^2 = \mathbb{E}[\langle \psi_1, D^{B^i} I^{B^i}(\psi_1) \rangle_{\mathscr{H}}]$. Taking into account relation (8.46), the expression (8.58) we have obtained for $D^{B^i}\psi_1$, and the isomorphism (8.43), we end up with

$$\mathbb{E}\left[|I^{B^i}(\psi_1)|^2\right] \lesssim Q_1 + Q_2 + Q_3, \tag{8.62}$$

where Q_1, Q_2, Q_3 are respectively defined by:

$$Q_1 = \mathbb{E}\left\| \mathscr{D}_{\tau-}^{1/2-H} \psi_1 \right\|_{L^2([0,\tau])}^2$$

$$Q_2 = \mathbb{E}\left\| \mathscr{D}_{(\tau-\eta_2)-}^{1/2-H} \left(\int_{\cdot+\eta_2}^{\tau} [B^i_{\cdot-\eta_1} - B^i_{-\eta_2-\eta_1}] dB^i_w \right) \right\|_{L^2([-\eta_2,\tau-\eta_2])}^2$$

$$Q_3 = \mathbb{E}\left\| \mathscr{D}_{(\tau-\eta_2-\eta_1)-}^{1/2-H} \left(\int_{\cdot+\eta_2+\eta_1}^{\tau} [B^i_{w-\eta_2} - B^i_{r+\eta_1}] dB^i_w \right) \right\|_{L^2([-\eta_2-\eta_1,\tau-\eta_2-\eta_1])}^2 .$$

We now estimate those three terms separately, starting with Q_1: invoking the very definition (8.41) of the fractional derivative $\mathscr{D}_{\tau-}^{1/2-H}$, it is easily seen that $Q_1 \lesssim A_1 + A_2$, where

$$A_1 = \mathbb{E}\int_0^{\tau} \left(\int_{-\eta_2}^{r-\eta_2} [B^i_{u-\eta_1} - B^i_{-\eta_2-\eta_1}] dB^i_u \right)^2 \frac{1}{(\tau-r)^{1-2H}} dr$$

$$A_2 = \mathbb{E}\int_0^{\tau} \left(\int_r^{\tau} \frac{\int_{-\eta_2}^{w-\eta_2} [B^i_{u-\eta_1} - B^i_{-\eta_2-\eta_1}] dB^i_u - \int_{-\eta_2}^{r-\eta_2} [B^i_{u-\eta_1} - B^i_{-\eta_2-\eta_1}] dB^i_u}{(w-r)^{3/2-H}} dw \right)^2 dr$$

The term A_1 is easily bounded: according to Fubini's theorem and to our previous bounds on $\mathbf{B^2}$, we have

$$\begin{aligned} A_1 &= \int_0^{\tau} \mathbb{E}\left(\int_{-\eta_2}^{r-\eta_2} [B^i_{u-\eta_1} - B^i_{-\eta_2-\eta_1}] dB^i_u \right)^2 \frac{1}{(\tau-r)^{1-2H}} dr \\ &\le c_H \int_0^{\tau} r^{4H} \frac{1}{(\tau-r)^{1-2H}} dr \le c_H \tau^{4H} \int_0^{\tau} \frac{1}{(\tau-r)^{1-2H}} dr = \frac{c_H}{2H} \tau^{6H}. \end{aligned}$$

The term A_2 is a little longer to treat. However, by resorting to the same kind of tools, one is able to prove that $A_2 \le c_H \tau^{6H}$, and gathering the estimates on A_1 and A_2, we obtain $Q_1 \le c_H \tau^{6H}$ as well. Finally, after some tedious computations which will be spared to the reader for sake of conciseness, we obtain the same kind of bound for Q_2 and Q_3.

Now one has to reverse our decomposition process: putting together our estimates on Q_1, Q_2, Q_3 and plugging them into (8.62), we get $\mathbb{E}[|I^{B^i}(\psi_1)|^2] \le c_H \tau^{6H}$, with a constant $c_H > 0$ independent of η_1, η_2. Finally, gathering the bounds on the Skorokhod and the trace term, one obtains $\mathbb{E}[|\mathbf{B}^{3,iii}(\eta_1,\eta_2)|^2] \le c|t-s|^{6H}$.

2. *Other cases.* The previous arguments and computations can be simplified to obtain the desired result for the case $i = k \neq j$ and $j = k \neq i$. The cases $i = j \neq k$ and $i \neq j \neq k$ can be treated by means of Wiener integrals estimations. This finishes the proof of our claim $\mathbb{E}[|\mathbf{B}^3(v_1,v_2,v_3)|^2] \le c|t-s|^{6H}$. □

As in the case of delayed Lévy areas, and with exactly the same kind of arguments, one can push forward the analysis in order to deal with $\mathbf{B}^3$ as an increment:

Theorem 8.3. *Let* $\mathbf{B}^3$ *be the increment defined at Proposition 8.9. Then* $\mathbf{B}^3$ *satisfies Hypothesis 8.3. Taking into account Proposition 8.8, Theorem 8.2 can thus be applied almost surely to the paths of the* d*-dimensional fBm with Hurst parameter* $H > 1/4$.

Acknowledgements S. Tindel is partially supported by the ANR grant ECRU. I. Torrecilla is partially supported by the grant MTM2009-07203 from the Dirección General de Investigación, Ministerio de Ciencia e Innovacin, Spain. I. Torrecilla wishes to thank the IECN (Institut Élie Cartan Nancy) for its warm hospitality during a visit in 2008, which served to settle the basis of the current chapter.

References

1. Alòs, E., Léon, J.L., Nualart, D.: Stratonovich calculus for fractional Brownian motion with Hurst parameter less than *1/2*. Taiwanese J. Math. **5**, 609–632 (2001)
2. Cass, T., Friz, P.: Densities for rough differential equations under Hörmander's condition. Ann. Math. **171**, 2115–2141 (2010)
3. Cass, T., Friz, P., Victoir, N.: Non-degeneracy of Wiener functionals arising from rough differential equations. Trans. Am. Math. Soc. **361**(6), 3359–3371 (2009)
4. Coutin, L., Qian, Z.: Stochastic rough path analysis and fractional Brownian motion. Probab. Theor. Relat. Fields **122**, 108–140 (2002)
5. Decreusefond, L., stnel, A.S.: Stochastic analysis of the fractional Brownian motion. Pot. Anal. **10**, 177–214 (1998)
6. Deya, A., Tindel, S.: Rough Volterra equations 2: Convolutional generalized integrals. Stochastic Processes and Applications **21**(8), 1864–1899 (2011)
7. Deya, A., Gubinelli, M., Tindel, S.: Non-linear Rough Heat Equations. Preprint arXiv: 0911.0618v1 [math.PR] (2009)
8. Ferrante, M., Rovira, C.: Stochastic delay differential equations driven by fractional Brownian motion with Hurst parameter $H > 1/2$. Bernoulli **12**(1), 85–100 (2006)
9. Ferrante, M., Rovira, C.: Convergence of delay differential equations driven by fractional Brownian motion. J. Evol. Equ. **10**(4), 761–783 (2010)
10. Friz, P.: Continuity of the Itô-map for Hölder rough paths with applications to the support theorem in Hölder norm. Probability and Partial Differential Equations in Modern Applied Mathematics, pp. 117–135. IMA Vol. Math. Appl., vol. 140. Springer, New York (2005)
11. Friz, P., Victoir, N.: Differential equations driven by Gaussian signals. Ann. Inst. Henri Poincar Probab. Stat. **46**(2), 369–413 (2010)
12. Friz, P., Victoir, N.: Multidimensional dimensional processes seen as rough paths. Cambridge Studies in Advanced Mathematics, vol. 120. Cambridge University Press, Cambridge (2010)
13. Gubinelli, M.: Controlling rough paths. J. Funct. Anal. **216**, 86–140 (2004)
14. Gubinelli, M.: Ramification of rough paths. J. Differ. Equat. **248**, 693–721 (2010)
15. Gubinelli, M., Tindel, S.: Rough evolution equations. Ann. Probab. **38**(1), 1–75 (2010)
16. Hoff, B.: The Brownian Frame Process as a Rough Path. Preprint (2006)
17. Lyons, T.J., Qian, Z.: System control and rough paths. Oxford Mathematical Monographs. Oxford Science Publications. Oxford University Press, Oxford (2002)
18. Mohammed, S.-E.A.: Stochastic functional differential equations. Research Notes in Mathematics, vol. 99. Pitman Advanced Publishing Program, Boston (1984)
19. Neuenkirch, A., Nourdin, I., Tindel, S.: Delay equations driven by rough paths. Elec. J. Probab. **13**(67), 2031–2068 (2008)
20. Neuenkirch, A., Nourdin, I., Rößler, A., Tindel, S.: Trees and asymptotic developments for fractional differential equations. Ann. Inst. H. Poincaré, Probab. Stat. **45**(1), 157–174 (2009)

21. Neuenkirch, A., Nourdin, I., Rößler, A., Tindel, S., Unterberger, J.: Discretizing the Lévy area. Stoch. Process. Appl. **120**(2), 223–254 (2010)
22. Nualart, D.: The Malliavin calculus and related topics. Probability and Its Applications, 2nd edn. Springer, Berlin (2006)
23. Pipiras, V., Taqqu, M.S.: Integration questions related to fractional Brownian motion. Probab. Theor. Relat. Fields **118**(2), 251–291 (2000)
24. Russo, F., Vallois, P.: Forward, backward and symmetric stochastic integration. Probab. Theor. Relat. Fields **97**, 403–421 (1993)
25. Samko, S.G., Kilbas, A.A., Marichev, O.I.: Fractional Integrals and Derivatives. Gordon and Breach, New York (1993)
26. Samorodnitsky, G., Taqqu, M.S.: Stable Non-Gaussian Random Processes. Chapman and Hall, London (1994)
27. Tindel, S., Torrecilla, I.: Some differential systems driven by a fBm with Hurst parameter greater than 1/4. Preprint arXiv:0901.2010v1 [math.PR] (2009)
28. Zähle, M.: Integration with respect to fractal functions and stochastic calculus I. Probab. Theor. Relat. Fields **111**, 333–374 (1998)

Chapter 9
Transportation Cost Inequalities for Diffusions Under Uniform Distance

Ali Suleyman Üstünel

Abstract We prove the transportation inequality with the uniform norm for the laws of diffusion processes with Lipschitz and/or dissipative coefficients and apply them to some singular stochastic differential equations of interest.

Keywords Dissipative functions • Entropy • (Multi-valued) stochastic differential equations • Transport inequality • Wasserstein distance

9.1 Introduction

Let (W, d) be a separable Fréchet space, for two probability measures P and Q on $(W, \mathscr{B}(W))$, then the Wasserstein distance (cf. [11]) between P and Q, denoted as $d_W(P, Q)$, is defined as

$$d_W^2(P, Q) = \inf\left\{\int_{W\times W} d(x, y)^2 \theta(dx, dy) : \theta \in \Sigma(P, Q)\right\},$$

where $\Sigma(P, Q)$ denotes the set of probability measures on $W \times W$ whose first marginal is P and the second one is Q; note that this is a compact set under the weak topology, hence the infimum is always attained for any d (even lower semi-continuous). It is quite useful to find an upper bound for this distance, if possible dimension independent. There are a lot of works on this subject (cf. [11]), beginning by the contributions of M. Talagrand, cf. [10], where it is shown that the relative entropy is a fully satisfactory upper bound. In [5, 6], it is shown that the relative entropy is again an upper bound when P is the Wiener measure

A.S. Üstünel (✉)
Télécom-Paristech, 46 rue Barrault, 75013 Paris, France
e-mail: ustunel@telecom-paristech.fr

L. Decreusefond and J. Najim (eds.), *Stochastic Analysis and Related Topics*, Springer Proceedings in Mathematics & Statistics 22, DOI 10.1007/978-3-642-29982-7_9,

and d is the singular Cameron–Martin distance using the Girsanov theorem (cf. also [4]). The same method has also been employed in [12] and more recently in [8] to obtain a transportation cost inequality w.r.t. Banach norm for diffusion processes. The former assumes quite strong conditions on the coefficients which govern the diffusion which are superfluous and make difficult the applicability of the inequality, while the latter one treats essentially the one-dimensional case with an extension to the case where the diffusion coefficients are independent and their slight perturbations. Inspired with these works, we have attacked the general case: namely, the case of fully dependent diffusion like processes and their extensions and infinite dimensional diffusion processes governed with a cylindrical Brownian motion. Besides, there is a special class of diffusion processes with singular (dissipative) drifts which are constructed as weak limits of the Lipschitzian case where the approximating diffusions have Lipschitz continuous drifts but the Lipschitz constant explodes at the limit; this last class is particularly interesting because of their applications to physics.

To achieve this program, we need the following result about the stability of the transportation cost inequality under the weak limits of probability measures, which is proved by Djellout et al. in [4]. Since we make an important use of it, we give it with a (slightly different and more general) proof.

Lemma 9.1. *Assume that $(P_k, k \geq 1)$ is a sequence of probability measures on a separable Fréchet space (W, d), converging weakly to a probability P. If*

$$d_W^2(Q, P_k) \leq c_k \int_W \frac{dQ}{dP_k} \log \frac{dQ}{dP_k} dP_k = c_k H(Q|P_k)$$

for any $k \geq 1$, for any probability Q, where $c_k > 0$ are bounded constants, then the transportation inequality holds for P, namely

$$d_W^2(Q, P) \leq cH(Q|P), \tag{9.1}$$

where $c = \sup_k c_k$.

Proof. If $f = dQ/dP$ is a bounded, continuous function, then the inequality (9.1) follows from the lower semi-continuity of the transportation cost w.r.t. the weak convergence and from the hypothesis since $f \log f$ is continuous and bounded. Due to the dominated convergence theorem, to prove the general case, it suffices to prove the case where f is P-essentially bounded and measurable. In this case, there exists a sequence of bounded, upper semi-continuous functions, say $(f_n, n \geq 1)$, increasing to f P-almost surely. By the dominated convergence theorem, the measures $(\tilde{f}_n dP,\ n \geq 1)$ converge weakly to the measure $f dP$, where $\tilde{f}_n = f/P(f_n)$. On the other hand, $H(\tilde{f}_n dP|dP) \to H(f dP|P)$ again by the dominated convergence theorem. Hence, to prove the general case, it is sufficient to prove the inequality with f upper semi-continuous and bounded. Since we are on a Fréchet space, there exists a sequence of (positive) continuous functions decreasing to f which may be chosen uniformly bounded by taking the minimum of each with the upper bound of f, and the inequality (9.1) follows again due to the dominated convergence theorem. □

9.2 Diffusion Type Processes with Lipschitz Coefficients

Let (W, H, μ) be the classical Wiener space, i.e., $W = C_0([0,1], \mathbb{R}^d)$, $H = H^1([0,1], \mathbb{R}^d)$ and μ is the Wiener measure under which the evaluation map at $t \in [0,1]$ is a Brownian motion. Suppose that $X = (X_t, t \in [0,1])$ is the solution of the following SDE (stochastic differential equation)

$$\begin{aligned} dX_t &= \sigma(t, X_t)dW_t + b(t, X)dt \\ X_0 &= z \in \mathbb{R}^d \end{aligned}$$

where $\sigma : [0,1] \times \mathbb{R}^d \to \otimes \mathbb{R}^d$ is uniformly Lipschitz w.r.t. x with a Lipschitz constant being equal to K, $b : [0,1] \times W \to \mathbb{R}^d$ is adapted and such that

$$|b(t,\xi) - b(t,\eta)| \leq K \sup_{s \leq t} |\xi(s) - \eta(s)| = \|\xi - \eta\|_t$$

for any $\xi, \eta \in W$. We denote by d_W the Wasserstein distance on the probability measures on W defined by the uniform norm:

$$d_W^2(\rho, \nu) = \inf\left(\int_{W \times W} \|x - y\|^2 d\gamma(x,y) : \gamma \in \Sigma(\rho, \nu)\right)$$

where $\Sigma(\rho, \nu)$ the set of probabilities on $W \times W$ whose first marginals are ρ and the second ones are ν. We have the following bound for d_W:

Theorem 9.1. *Let P be the law of the solution of the SDE described above; then for any probability Q on $(W, \mathscr{B}(W))$, we have*

$$d_W^2(P, Q) \leq 6\, e^{15K^2} H(Q|P) \tag{9.2}$$

where $H(Q|P)$ is the relative entropy of Q w.r.t. P.

Proof. Due to the rotation invariance of the Wiener measure, we can suppose without loss of generality that σ takes its values in the set of positive matrices. Suppose first that σ is strictly elliptic. From the general results about the SDE (cf. [7,9]), the coordinate process x under the probability P can be written as

$$dx_t = \sigma(t, x_t)d\beta_t + b(t, x)dt$$

with $x_0 = z$ P-a.s., where β is an $\mathbb{R}^d$-valued P-Brownian motion. At this point of the proof we need the following result, which is probably well known (cf. [9] and the references there), though we include its proof for the sake of completeness:

Lemma 9.2. *Any bounded P-martingale can be written as a stochastic integral w.r.t. β of an adapted process $(\alpha_s, s \in [0,1])$, with $E_P \int_0^1 |\alpha_s|^2 ds < \infty$.*

Proof. Let us denote by P^0 the law of the solution of

$$dX_t = \sigma(t, X_t) dW_t,$$

then under P^0, the coordinate process x can be written as

$$dx = \sigma(t, x_t) d\beta_t^0,$$

where β^0 is a P^0-Brownian motion. Let Z be a bounded P-martingale with $Z_0 = 0$, assume that it is orthogonal to the Hilbert space of P-square integrable martingales written as the stochastic integrals w.r.t. β of the adapted processes. Let M be the exponential martingale defined as

$$M_t = \exp\left(-\int_0^t (\sigma^{-1}(s, x_s) b(s, x), d\beta_s) - \frac{1}{2}\int_0^t |\sigma^{-1}(s, x_s) b(s, x)|^2 ds\right)..$$

Then, we know from the uniqueness and the Girsanov theorem that $MdP = dP^0$, since M can be written as a stochastic integral w.r.t. β, our hypothesis implies that ZM is again a P-martingale, hence Z is a P^0-martingale, therefore, from the classical Markov case it can be written as

$$\begin{aligned} Z_t &= \int_0^t H_s . d\beta_s^0 \\ &= \int_0^t H_s . (d\beta_s - \sigma^{-1}(s, x_s) b(s, x) ds) . \end{aligned}$$

This last expression implies that

$$\langle Z, Z \rangle_t = \langle Z, \int_0^\cdot H_s . d\beta_s \rangle_t$$

but Z is orthogonal to the stochastic integrals of the form $\int \alpha_s . d\beta_s$, hence $Z_t = E_P[Z_t] = 0$, which proves the claim. □

Let us complete now the proof of the theorem: If Q is singular w.r.t. P, then there is nothing to prove due to the definition of the entropy. Let L be the Radon–Nikodym derivative dQ/dP, we shall first suppose that $L > 0$ P-a.s. In this case we can write

$$L = \rho(-\delta v),$$

where $v(t,x) = \int_0^t \dot{v}_s(x)ds$, $\dot{v}_s(x)$ is a.s. adapted and $\int_0^1 |\dot{v}_s(x)|^2 ds < \infty$ a.s. and $\delta v = \int_0^1 \dot{v}_s d\beta_s$. From the Girsanov theorem, $z_t = \beta_t + \int_0^t \dot{v}_s ds$ is Q-Brownian motion, hence by the uniqueness of the solution of SDE, if we denote by x^v the solution of the SDE given as

$$dx_t^v = \sigma(t, x_t^v)dz_t + b_t(x^v)dt$$

the image of Q under the solution map x^v is equal to P, consequently $(x^v \times I_W)(Q) \in \Sigma(P,Q)$, hence we have the following domination:

$$d_W^2(P,Q) \le E_Q[\|x^v - x\|^2]$$

where $\|\cdot\|$ denotes the uniform norm on W. Using Doob and Hölder inequalities, we get

$$E_Q[\sup_{r\le t} |x_r^v - x_r|^2] \le (12+3t)K^2 E_Q \int_0^t |x_s^v - x_s|^2 ds$$
$$+3t E_Q \int_0^t |\dot{v}_s|^2 ds\,.$$

It follows from the Gronwall lemma that

$$E_Q[\sup_{r\le t} |x_r^v - x_r|^2] \le 3t\, E_Q \int_0^t |\dot{v}_s|^2 ds\, e^{3K^2(4+t)}$$

since

$$E_Q \int_0^1 |\dot{v}_s|^2 ds = 2H(Q|P)$$

the claim follows in the case $P \sim Q$. For the case where $Q \ll P$ let

$$L_\varepsilon = \frac{L+\varepsilon}{1+\varepsilon}\,,$$

then it is easy to see that $(L_\varepsilon \log L_\varepsilon,\ \varepsilon \le \varepsilon_0)$ is P-uniformly integrable provided $E_P[L\log L] < \infty$. Hence the proof, in the strictly elliptic case, follows by the lower semi-continuity of $Q \to d_W(P,Q)$. The general case follows by replacing σ by $\varepsilon I_{\mathbb{R}^d} + \sigma$, then remarking that the corresponding probabilities $(P_\varepsilon, \varepsilon \le \varepsilon_0)$ converge weakly and that

$$d_W^2(P_\varepsilon, Q) \le 6\, e^{15(\varepsilon+K)^2} H(Q|P_\varepsilon)$$

and hence it follows from Lemma 9.1 that

$$d_W^2(P,Q) \le 6\, e^{15K^2} H(Q|P)\,.$$

□

Since the inequality (9.2) is dimension independent, we can extend it easily to the infinite dimensional case:

Corollary 9.1. *Let M be a separable Hilbert space, suppose that B is a M-cylindrical Wiener process. Assume that $\sigma : [0,1] \times M \to L_2(M, KM = M \otimes_2 M$ (space of Hilbert–Schmidt operators on M) and $b : [0,1] \times M \to M$ are uniformly Lipschitz with Lipschitz constant K. Let P be the law of the following SDE:*

$$dX_t = \sigma(t, X_t)dB_t + b(t, X_t)dt\,, X_0 = x \in M\,.$$

Then the law of P satisfies the transportation cost inequality (9.2).

Proof. Let $(\pi_n, n \geq 1)$ be an sequence of orthogonal projections of M increasing to the identity, define $\sigma_n = \pi_n \sigma \circ \pi_n$, $b_n = \pi_n b \circ \pi_n$, $B^n = \pi_n B$, and $x^n = \pi_n x$. Let then P^n be the law of the SDE

$$dX_t^n = \sigma^n(t, X_t^n)dB_t^n + b^n(t, X_t^n)dt\,, X_0^n = x^n\,.$$

From Theorem 9.1, P^n satisfies the inequality (9.2) with a constant independent of n, since $(P^n, n \geq 1)$ converges weakly to P, the proof follows from Lemma 9.1. □

9.2.1 Transport Inequality with a Singular Cost Function

In the case of Wiener space, we can define a stronger Wasserstein metric using the Cameron–Martin norm as we have already done in [5, 6] as follows:

$$d_H^2(P,Q) = \inf\left\{\int_{W\times W} |x-y|_H^2 \theta(dx,dy) : \theta \in \Sigma(P,Q)\right\}.$$

Note that this distance is strictly stronger than d_W and it is still lower semi-continuous with respect to the weak topology of measures on W. In the above cited references, we have proved the following inequality:

$$d_H^2(\rho,\mu) \leq 2H(\rho|\mu)$$

for any measure ρ, where μ denotes the Wiener measure. This inequality can be extended to the class of diffusions whose diffusion coefficients are constant (it suffices to consider the case where it is equal to the identity matrix):

Theorem 9.2. *Assume that $b : [0,1] \times \mathbb{R}^d \to \mathbb{R}^d$ is a K-Lipschitz map w.r.t. x uniformly in $t \in [0,1]$. Let P be the law of the solution of the following SDE:*

$$dX_t = b(t, X_t)dt + dW_t\,,\; X_0 = x\,.$$

Then the following transport inequality holds:

$$d_H^2(P, Q) \le 2(1 + 2K^2 e^{2K^2}) H(Q|P).$$

Proof. Using the same reasoning as in the proof of Theorem 9.1 and supposing first that dQ/dP is strictly positive a.s., we reduce the problem to calculate (in the canonical space) the expectation of

$$|x - x^v|_{H([0,t])}^2 = \int_0^t |b(s, x_s) - b(s, x_s^v) - \dot{v}_s|^2 ds$$

under the probability Q, the rest of the proof is the same and we get rid of the strict positivity hypothesis again using the lower semi-continuity of the cost function on the space of probabilities. □

9.3 Transport Inequality for the Monotone Case

Assume that the Lipschitz property of the adapted drift coefficient is replaced by the following dissipativity hypothesis

$$(b(t, x) - b(t, y), x_t - y_t) \le 0$$

for any $t \in [0, 1]$ and $x, y \in W$, where, as before $(\cdot, \cdot)$ denotes the scalar product in $\mathbb{R}^d$. The derivative of a proper concave function on $\mathbb{R}^d$ is a typical example of such drift. We shall suppose first that

$$\int_0^1 |b(t, x)|^2 ds < \infty$$

for any $x \in W$.

Proposition 9.1. *Assume that b is of linear growth, i.e., $|b(t, x)| \le N(1 + \|x\|)$ and let P be the law of the solution of the following SDE*

$$dX_t = \sigma(t, X_t) dW_t + b(t, X) dt + m(t, X_t) dt \tag{9.3}$$

with $X_0 = x \in \mathbb{R}^d$ and that σ and $m : [0, 1] \times \mathbb{R}^d \to \mathbb{R}^d$ are uniformly K-Lipschitz w.r.t. the space variable. Then for any $Q \ll P$, we have

$$d_W^2(P, Q) \le \left(c 2^{3/2} \|\sigma\|_\infty^{3/2} e^{\frac{1}{2}(K^2+2K+1)}\right) \sqrt{H(Q|P)} + 2\|\sigma\|_\infty e^{\frac{1}{2}(K^2+2K+1)} \left(1 + K(K+2)) e^{\frac{1}{2}(K^2+2K+1)}\right) H(Q|P), \tag{9.4}$$

where $\|\sigma\|_\infty$ is a uniform bound for σ, K is the Lipschitz constant, and c is the universal constant of Davis' inequality for $p = 1$.

Proof. Recall that under P, the coordinate process satisfies $dx = \sigma(t, x_t)d\beta + (b(t, x) + m(t, x_t))dt$, where β is a P-Brownian motion. Assume that Q is another probability on W such that $Q \ll P$, let L be dQ/dP. Suppose first that $L > 0$ P-almost surely. As explained in the first section, we can write L as an exponential martingale $L = \rho(-\delta v)$, then $x^v(Q) = P$, where x^v is defined as before: $dx^v = \sigma(t, x_t^v)(d\beta_t + \dot{v}_t dt) + b(t, x^v)dt + m(t, x_t^v)dt$. Again by the uniqueness of the solutions, we have $(x^v \times I_W)(Q) \in \Sigma(P, Q)$, hence

$$d_W^2(P, Q) \le E_Q[\|x^v - x\|^2].$$

It follows from the Itô formula, letting $dz = d\beta + \dot{v}dt$, that

$$\begin{aligned} |x_t^v - x_t|^2 &= 2\int_0^t (x_s^v - x_s, dx_s^v - dx_s) + \int_0^t |\sigma(s, x_s^v) - \sigma(s, x_s)|^2 ds \\ &= 2\int_0^t (x_s^v - x_s, b(s, x^v) - b(s, x))ds \\ &\quad + 2\int_0^t (x_s^v - x_s, (\sigma(s, x_s^v) - \sigma(s, x_s))dz_s + (m(s, x_s^v) - m(s, x_s))ds) \\ &\quad + \int_0^t |\sigma(s, x_s^v) - \sigma(s, x_s)|^2 ds - 2\int_0^t (x_s^v - x_s, \sigma(s, x_s^v)\dot{v}_s)ds\,. \end{aligned}$$

By the dissipative character of b, we get

$$\begin{aligned} |x_t^v - x_t|^2 &\le 2\int_0^t (x_s^v - x_s, (\sigma(s, x_s^v) - \sigma(s, x_s))dz_s + (m(s, x_s^v) - m(s, x_s))ds) \\ &\quad + \int_0^t |\sigma(s, x_s^v) - \sigma(s, x_s)|^2 ds - 2\int_0^t (x_s^v - x_s, \sigma(s, x_s^v)\dot{v}_s))ds\,. \end{aligned}$$

Using, the usual stopping techniques, we can suppose that the stochastic integral has zero expectation and taking the Q-expectation of both sides, we obtain

$$\begin{aligned} E_Q[|x_t^v - x_t|^2] &\le (2K + K^2)E\int_0^t |x_s^v - x_s|^2 ds \\ &\quad + 2\|\sigma\|_\infty E\int_0^t |x_s^v - x_s||\dot{v}_s|ds \end{aligned}$$

using the inequality $xy \le \delta(x^2/2) + (y^2/2\delta)$, we get

$$E_Q[|x_t^v - x_t|^2] \le (2K + K^2 + \delta\|\sigma\|_\infty^2)E\int_0^t |x_s^v - x_s|^2 ds + +\frac{2}{\delta}H_t(Q|P)\,,$$

where $\delta > 0$ is arbitrary and $H_t(Q|P) = \int \log \frac{dQ}{dP}|_{\mathcal{F}_t} dQ$ is the entropy for the horizon $[0, t]$, which is an increasing function of t. It follows from the Gronwall lemma that

$$E_Q[|x_t^\nu - x_t|^2] \leq \frac{2}{\delta} H_t(Q|P) \exp\left[t(2K + K^2 + \delta\|\sigma\|_\infty^2)\right]. \tag{9.5}$$

Using now the Davis' inequality, the Lipschitz property, and the boundedness of σ, we get

$$E[\sup_{r\leq t} |x_r^\nu - x_r|^2] \leq (2c\|\sigma\|_\infty + \sqrt{2}H_t(Q|P)^{1/2})E\left[\int_0^t |x_s^\nu - x_s|^2 ds\right]^{1/2}$$
$$+K(K+2)E\int_0^t |x_s^\nu - x_s|^2 ds,$$

where c is the universal constant of Davis' inequality. Note that the right-hand side of the inequality (9.5) is monotone increasing in t, we insert it to the above inequality and minimize it w.r.t. δ for $t = 1$ and the proof is completed. □

In fact we have another version of the inequality (9.5) in the case where σ is not bounded but still K-Lipschitz:

Proposition 9.2. *Assume that all the hypothesis of Proposition 9.1 are satisfied except the boundedness of σ which appears in the SDE (9.3), then we have the following transportation cost inequality:*

$$d_W^2(P, Q) \leq H(Q|P)\frac{2}{(1-acK)^2}\exp\left(\frac{1}{1-acK}\left(\frac{cK}{a} + 1 - acK + 2K + K^2\right)\right) \tag{9.6}$$

where P is the law of the SDE (9.3), Q is any other probability, and $a > 0$ is arbitrary provided that $acK < 1$.

Proof. The proof is somewhat similar to the proof of Proposition 9.1: in fact we control uniformly the stochastic integral term in the Itô development of $|x_t^\nu - x_t|^2$ as follows:

$$E\left[\sup_{r\leq t}\left|\int_0^r (x_s^\nu - x_s, (\sigma(s, x_s^\nu) - \sigma(s, x_s)dz_s)\right|\right]$$
$$\leq cE\left[\left(\int_0^t |x_s^\nu - x_s|^2|\sigma(s, x_s^\nu) - \sigma(s, x_s)|^2 ds\right)^{1/2}\right]$$
$$\leq cKE\left[\left(\int_0^t |x_s^\nu - x_s|^4 ds\right)^{1/2}\right]$$

$$\leq cKE\left[\left(\sup_{s\leq t}|x_s^v - x_s|^2\int_0^t |x_s^v - x_s|^2\right)^{1/2}\right]$$

$$\leq \frac{caK}{2}E\left[\sup_{s\leq t}|x_s^v - x_s|^2\right] + \frac{cK}{2a}E\int_0^t |x_s^v - x_s|^2 ds\,.$$

Hence we get

$$E\left[\sup_{s\leq t}|x_s^v - x_s|^2\right] \leq acKE\left[\sup_{s\leq t}|x_s^v - x_s|^2\right] + \frac{cK}{a}E\int_0^t |x_s^v - x_s|^2 ds$$
$$+(2K + K^2 + \delta)E\int_0^t |x_s^v - x_s|^2 ds + \frac{1}{\delta}E\int_0^t |\dot{v}_s|^2 ds\,,$$

where $a, \delta > 0$ are arbitrary, c is the constant of Davis' inequality. From above, we obtain

$$(1 - acK)E\left[\sup_{s\leq t}|x_s^v - x_s|^2\right] \leq \left(\frac{cK}{a} + 2K + K^2 + \delta\right)E\int_0^t |x_s^v - x_s|^2 ds$$
$$+\frac{2}{\delta}H_t(Q|P)$$

and Gronwall lemma implies that

$$E\left[\sup_{s\leq t}|x_s^v - x_s|^2\right] \leq \frac{2}{\delta(1 - acK)}H_t(Q|P)$$
$$\cdot\exp\left[\frac{t}{1 - acK}\left(\frac{cK}{a} + \delta + 2K + K^2\right)\right].$$

Taking $t = 1$ and minimizing the r.h.s. of the last inequality w.r.t. δ completes the proof. □

It is important to notice that we did not use any regularity property about b except that the integrability of $t \to b(t, x)$ for almost all x in an intermediate step. This observation means that we can deal with very singular drifts provided that they are dissipative. Let us give an application of Proposition 9.1 to multi-valued SDE (cf. [1]) from this point of view

Theorem 9.3. *Let P be the law of the process which is the solution of the following multi-valued stochastic differential equation:*

$$m(X_t)dt + \sigma(t, X_t)dW_t \in dX_t + A(X_t)dt\,,\ X_0 = x \in D(A)\,,$$

where A is a maximal, monotone set-valued function (hence $-A$ is dissipative), such that $Int(D(A)) \neq \emptyset$. Assume that σ and m are uniformly K-Lipschitz and that σ

is bounded. Then P satisfies the transportation cost inequality (9.4). If σ is only Lipschitz, but not necessarily bounded, then P satisfies the inequality (9.6).

Proof. Let b_n be the Yosida approximation of A, i.e., $J_n = (I_{\mathbb{R}^d} + \frac{1}{n}A)^{-1}$ and $-b_n = n(I - J_n)$ then b_n is dissipative and Lipschitz, hence the law of the solution of the SDE

$$dX_t^n = \sigma(t, X_t)dW_t + b_n(X_t^n)dt + m(X_t^n)dt$$

satisfies the inequality (9.4) with the constants independent of n, moreover the law of $(X^n, n \in \mathbb{N})$ converges weakly to P (cf. [1]), hence P satisfies also the inequality (9.4) due to Lemma 9.1. □

As an example of application of this theorem, let us give

Theorem 9.4. *Let P be the law of the solution of the following SDE:*

$$dX_t^i = m(X_t^i)dt + \sigma(X_t^i)dW_t^i + \gamma \sum_{1 \leq j \neq i \leq d} \frac{1}{X_t^i - X_t^j} dt\,,\ i = 1, \ldots, d\,,$$

with σ bounded and Lipschitz, $\gamma > 0$. Then P satisfies the transportation cost inequality (9.4) and if σ is not bounded but only Lipschitz, then P satisfies the inequality (9.6).

Proof. It suffices to remark that the drift term following γ is the subdifferential of the concave function defined by

$$F(x) = \gamma \sum_{i<j} \log(x^j - x^i)$$

if $x^1 < x^2 < \ldots < x^d$ and it is equal to $-\infty$ otherwise. □

Remark 9.1. For details about the equation of Theorem 9.4 cf. [2]. Moreover Theorem 9.3 is applicable to all the models given in [3].

References

1. Cépa, E.: Equations différentielles stochastiques multivoques. Séminaire de Probabilités, tome 29, pp. 86–107. Lecture Notes in Math., vol. 1613 (1995)
2. Cépa, E., Lepingle, D.: Diffusing particles with electrostatic repulsion. Probab. Theor. Relat. Fields **107**, 429–449 (1992)
3. Cépa, E., Lepingle, D.: Brownian particles with electrostatic repulsion on the circle: Dyson's model for unitary random matrices revisited. ESAIM: Probab. Stat. **5**, 203–224 (2001)
4. Djellout, H., Guillin, A., Wu, L.: Transportation cost-information inequalities and applications to random dynamical systems and diffusions. Ann. Probab. **32**, 2702–2732 (2004)
5. Feyel, D., Üstünel, A.S.: Measure transport on Wiener space and the Girsanov Theorem. C.R. Acad. Sci. Paris, Ser. I, 1025–1028 (2002)

6. Feyel, D., Üstünel, A.S.: Monge–Kantorovitch measure transportation and Monge-Ampère equation on Wiener space. Probab. Theor. Relat. Fields **128**(3), 347–385 (2004)
7. Ikeda, N., Watanabe, S.: Stochastic Differential Equations and Diffusion Processes. North Holland, Amsterdam (Kodansha Ltd., Tokyo) (1981)
8. Pal, S.: Concentration for multidimensional diffusions and their boundary local times. arXiv: 1005.2217v2 [math.PR], June 2010
9. Rogers, L.C.G., Williams, D.: Diffusions, Markov Processes, and Martingales, vol. 2, Itô Calculus. Wiley, New York (1987)
10. Talagrand, M.: Transportation cost for Gaussian and other product measures. Geom. Funct. Anal. **6**, 587–600 (1996)
11. Villani, C.: Topics in Optimal Transportation. American Mathematical Society, Providence (2003)
12. Wu, L., Zhang, Z.L.: Talagrand's T_2-transportation inequality w. r. to a uniform metric for diffusions. Acta Math. Appl. Sinica, English Series **20**(3), 357–364 (2004)